ECHOES FROM
THE DEEP

ECHOES FROM THE DEEP

Inventorising shipwrecks at the national scale by the application of marine geophysics and the historical text

INNES MCCARTNEY

© 2022 Innes McCartney

Published by Sidestone Press, Leiden
www.sidestone.com

Lay-out & cover design: Sidestone Press
Photograph cover: U.S. Naval Historical Centre

ISBN 978-94-6426-116-5 (softcover)
ISBN 978-94-6426-117-2 (hardcover)
ISBN 978-94-6426-118-9 (PDF e-book)

Contents

List of Figures

List of Tables

Glossary

The Board of Trade	BOT
Board of Trade Shipping Casualties (various dates)	Shipping Casualties (year)
Digital Terrain Model	DTM
Exclusive Economic Zone	EEZ
Geographic Information System (Esri ArcMap)	GIS
Kriegstagebuche (War Diary)	KTB
Multibeam Echo Sounder	MBES
Lloyd's of London War Losses (various dates)	War Losses (year)
Lloyd's Register Wreck Returns (various dates)	Wreck Returns (year)
Marine Renewable Energy	MRE
The National Archives	TNA
The National Archives and Records Administration	NARA
The UK Hydrographic Office	UKHO
World War One	WW1
World War Two	WW2

Preface

In a process analogous with the impact of aerial photography on landscape archaeology, survey by multibeam (a form of marine geophysics) is pinpointing the remains of thousands of shipwrecks across the seabed of the globe. This volume describes a multidisciplinary research project that set out to establish whether a large tranche of the shipwrecks in a given geographic region could be identified by name through the mutual study of the 3D models of the shipwrecks, alongside the historic text of shipping losses in the same area. The research was carried out at Bournemouth University and funded by the Leverhulme Trust (Grant No. ECF-2017-705).

All of the 273 shipwrecks in a 7,500sqm study area in the Irish Sea were surveyed using multibeam. The detailed scanning of each wreck showed that whereas it was thought 101 (37%) of the wrecks were unidentified, the actual number, once each wreck site had been assessed, was much higher at 161 (59%). The major cause of the difference was a legacy of misattributions dating from the 1970s which had to be corrected first.

The methodologies subsequently developed to identify the 161 wrecks enabled names to be given to 129 (80%) of the unknown ships, verified by their dimensions, their geographic position, and archival descriptions of the sinking of each ship as derived in each case from the historical text. In all 87% of the ships in the study are now identified.

In historic terms, the newly identified wrecks included myriad vessels from trawlers, cargo vessels, submarines, through to the largest ocean liners and tankers. They include rare ship designs, losses of national importance, and naval graves. Several of the wrecks uncovered have potential environmental concerns. The accurate dating of so many wrecks in one area has a major impact on the study of seabed dynamics and site formation processes, creating better models for the placement of windfarms and tidal generators.

This research is important because the seabed of the world is being increasingly mapped in detail, and shipwrecks are being located in large numbers across myriad regions. This research has developed a relatively low-cost means of inventorising shipwreck datasets across entire national zones without costly physical interaction with each site. It should be of key interest to marine scientists, environmental agencies, hydrographers, heritage managers, maritime archaeologists, and historians around the world.

Acknowledgements

I would like to thank the following for their valuable contributions to this research project: Dr Michael Roberts at Bangor University's School of Ocean Sciences, all the crews and survey teams on *Prince Madog*, Anthony Roy at UK Hydrographic Office, Anne Cowne and Max Wilson at Lloyd's Register Foundation Heritage and Education Centre, Matt Skelhorn at MOD Salvage & Marine, Gert Normann Andersen at The Sea War Museum Jutland, Professors Kate Welham and Dave Parham at Bournemouth University, the late Professor Eric Grove, Jan Lettens at wrecksite.eu, Robert Metcalfe, Jon Shaw, Simon Rodger, Patricia McCartney, PK Porthcurno – Museum of Global Communications, The Bartlett Maritime Research Centre & Library, The National Archives and The Leverhulme Trust.

1

Introduction: *Echoes from the Deep*

Introduction: Flying over a shipwreck in an undersea landscape

From the earliest foray by balloon over Stonehenge to the lidar and satellite imaging of today, aerial photography has transformed landscape studies. When the Royal Air Force reconnaissance techniques developed during WW1 were turned to archaeology and publications such as *Wessex from the Air* (Crawford & Keiller 1928) began to appear, a new epoch was born. To this day more archaeological sites are discovered by studying aerial photographs than by any other method (Barber 2011).

At sea in recent years, a technological revolution has been taking place. Using sound in place of light, the submerged landscapes of the world are being similarly revealed from above in ever increasing detail and extent. At the forefront of this revolution is swathe bathymetry, imaging undersea surfaces in three dimensions.

By 2030, the entire global seabed will have been scanned in this way for the first time. As these surveys reveal the ocean's bottom, in the place of earthworks there are shipwrecks; thousands of them. This is the dawn of an epochal transformation in shipwreck archaeology. Interpreting and managing the results on such a vast scale is the untrodden path which lies ahead, to which it is hoped this study contributes.

Echoes from the Deep: A research project

This monograph presents the full results of The Leverhulme Trust funded research project *"Echoes from the Deep: Modern Reflections on our Maritime Past"*. The research aimed to carry out a systematic reinterpretation of a globally important grouping of shipwrecks in one specific geographic region. This was to test whether the wrecks themselves could be inventorised and linked to specific shipping losses represented in the historical text.

This would be done by using the latest geophysical data from a large collection of surveyed shipwrecks. The aim was to develop a novel methodology for the identification of the shipwrecks. In combination with the historical text, this methodology would allow a finer-grained understanding of the extent, distribution, and identities of the lost vessels from which new, archaeology-driven narratives of shipwreck studies in the selected region would emerge.

The interdisciplinary conjunction of modern marine geophysics with detailed studies of the historical text will pioneer an innovative approach aimed at revolutionising research on offshore underwater cultural heritage at the national scale. It was envisaged that the research would signpost the way to how hundreds of shipwrecks in any given region could be identified using the new methodology.

Crucially this approach could then potentially be applied to any area which has been sufficiently surveyed, offering a low-cost approach to assessing large numbers of shipwrecks in any given region. The research is both timely and necessary. Shipwrecks are being located by marine geophysics in increasing numbers around the world. Initiatives such as the Nippon Foundation-GEBCO Seabed 2030 Project are creating increasingly accurate bathymetric maps of the global seabed. This is increasing the need to understand the distribution of the shipwrecks being uncovered, for environmental, legislative, and cultural purposes.

The survey region ultimately selected for the research was the central Irish Sea and the geophysical dataset of surveyed shipwrecks which was being generated by researchers based at Bangor University's School Ocean Sciences, North Wales. A unique collaboration between our two universities has led directly to the results presented here.

Inventorising underwater cultural heritage: what does it mean?

Term employed for this research has been to "inventorise" the shipwrecks surveyed; ergo to make an inventory. In the terms of this research, it means that each shipwreck will have been subject to a consistent standard of survey and assessment, which is the purpose the research. In brief, it encompasses:

1. The Fieldwork: Each site is surveyed with high resolution multibeam,
2. Archival research phase: A detailed positional, textual, and technical history of every shipping loss is compiled,
3. Archaeology research Phase 1: Each shipwreck is assessed against the UK Hydrographic Office (UKHO) shipwreck database to establish the accurate total of the surveyed shipwrecks which were in fact unidentified,
4. Archaeology research Phase 2: Where a wreck site is found to be unidentified, or previously misidentified then it is subject to a prescribed identification process.

The Application of MBES survey to Shipwreck Archaeology

Multibeam Echo Sounder (MBES) is an acoustic based technology commonly used to survey and map the seabed. These systems create the means by which it is now possible to visualise marine environments in very high levels of detail and to analyse shipwrecks in new, exciting, and cost-effective ways (see Figure 1.1). MBES systems emit and collect a fan of acoustic pulses at specific frequencies that form a 'swathe' across the seabed (which is why this technology is sometimes referred to as swathe bathymetry).

By measuring the time taken for each pulse of sound to travel from and to the sonar system after reflecting off the seabed and knowing the speed of sound in the column of water being surveyed, the distance from the sonar to the seabed can be accurately calculated. This process is constantly repeated as the vessel transits along a defined survey track and the results form a collection of points referred to as a pointcloud. The sonar returns are corrected post-survey to compensate for factors such as movement of the survey vessel by wave action, tidal height changes during the survey and the effect of salinity on sound velocity. Each data point is assigned a coordinate value by utilising an accurate position fixing and reference system on the survey vessel. This is the stage in the survey process when the data can be handed over for archaeological study.

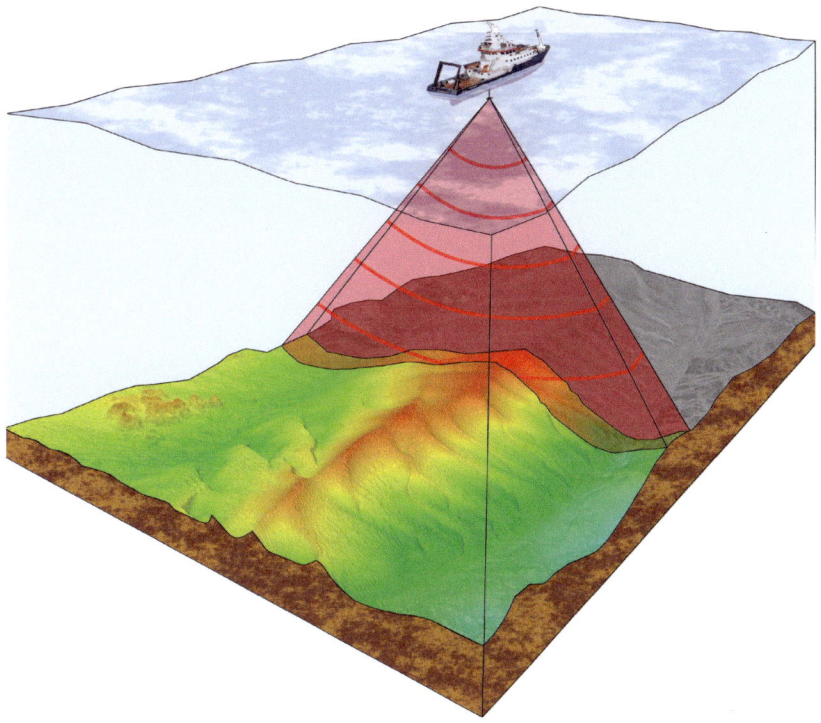

Figure 1.1. Swathe Bathymetry or Multibeam Echo Sounder (MBES) creates a fan of acoustic pulses which create a 3D model of the seabed (Bangor University).

Invariably shipwreck pointclouds require cleaning in order to remove erroneous signals generated by objects in the water column such as fish, debris, and other forms of noise. The cleaned pointclouds can then be used to model 3D images of the shipwreck and also to create an accurate top-down 2D site plan called a Digital Terrain Model (DTM) and the outputs colour-coded based on a height ramp for subsequent interpretation.

The resolution derived is a combination of the number of individual points gathered over the shipwreck and its proximity to the scanner. Good coverage is important, and this can be problematic with respect to obtaining data from vertical or inclined surfaces due to the sonar head being above the object. In order to mitigate this effect and to build up a detailed pointcloud a grid of survey run lines is generally used to pass over the shipwreck. It has been found that targeted oblique passes over the wreck can add point density to areas of interest or otherwise low coverage and this methodology has been adopted when carrying out the surveys (Westley, Plets, Quinn, McGonigle, Sacchetti, Mekayla Dale, McNeary & Clements, 2019).

The proximity of the sonar head to the shipwreck is also important. The closer the head (or the shallower the wreck is) the finer the points are. There is a resolution drop off with increased depth, so when wrecks are being surveyed in 100m or more, an individual point may actually represent a square metre of seabed. All surveys were carried out using a hull-mounted system. Of particular utility for studies of multiple shipwrecks in any given region, DTM outputs can be amalgamated with other sources of information such as GIS charts or satellite images in order to assess the geographic distribution of the shipwrecks in relation to each other and to positional data derived from the historical text.

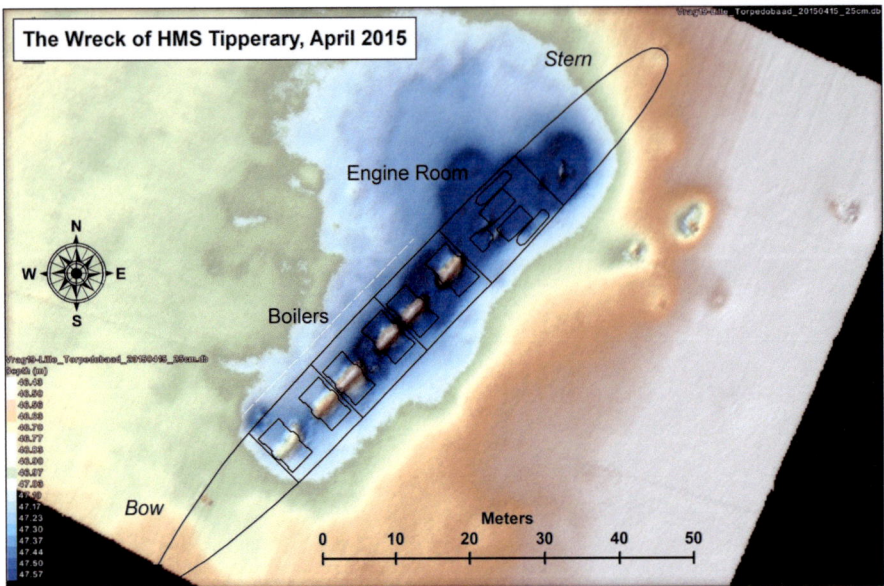

Figure 1.2. The DTM of the wreck of the large destroyer HMS *Tipperary*. The wreck was identified from this site plan. Amongst the destroyers sunk at Jutland, *Tipperary* uniquely had six boilers situated in a single row. These could easily be discerned in the DTM, when an outline of the ships machinery spaces was correctly scaled and overlaid onto it (Innes McCartney).

Shipwreck archaeology at the regional scale

Since 2000, my archaeological research has increasingly been oriented towards regional groupings of shipwrecks. This has involved a number of projects where the objectives have been to identify an assemblage of shipwrecks in a given area, sunk in a series of specific historical events. This began with the U-boat wrecks of *Operation Deadlight* (McCartney 2002). Multi-shipwreck archaeological surveys required a new research approach, where a dataset of physical remains of shipwrecks was reconciled to a larger raft of differing historical texts.

This process was expanded and refined during my doctoral work which focused on 63 U-boat wrecks sunk in action in the English Channel, 1915-1945. Using substantive archival and archaeological investigation, I demonstrated that over a third of these wrecks resided in positions outside of historical knowledge, sunk in events that were overlooked or not recorded at all. The archaeology proved to be the key to understanding how, when, and why inaccuracies appeared in the related text and led to its substantial refinement, ultimately being published as a peer-reviewed book (McCartney 2014).

Both these research projects were based on survey by diving. This method, although productive, was expensive and time-consuming. In order to extend the dataset-driven approach to shipwreck research, a newer technique enabling large-scale coverage was required. Consequently, I developed an innovative typologically-driven shipwreck identification methodology based on analysis of MBES data in the form of DTMs.

I tested this on the wrecks of the Battle of Jutland 1916, the largest naval battle of WW1. This work was carried with The Sea War Museum Jutland, as I am its affiliated

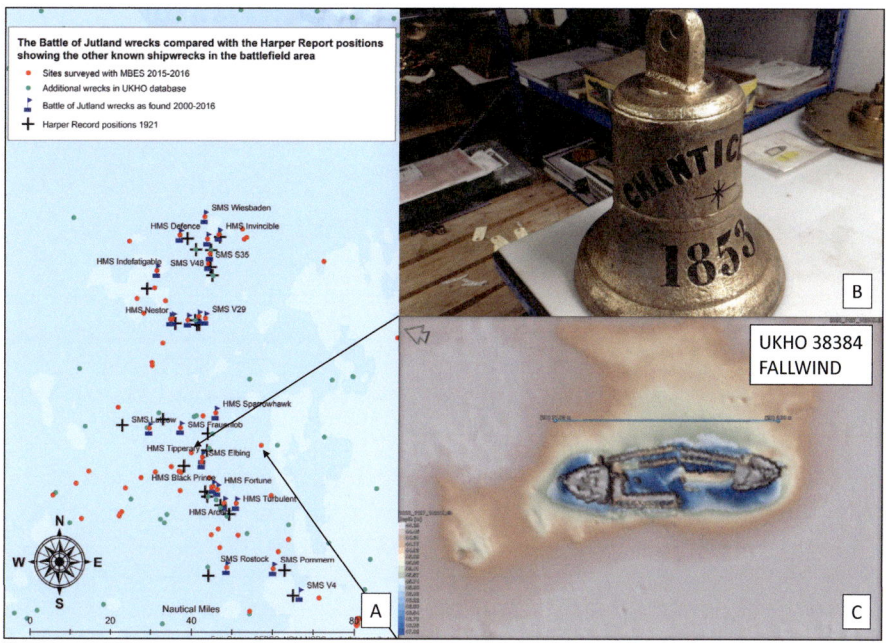

Figure 1.3. The Jutland battlefield with surrounding shipwrecks (A), the bell of the steamer *Chanticleer* which disappeared in 1889 (B) and the DTM of MV *Fallwind*, which sank nearby a century later (A&C Innes McCartney & B Sea War Museum Jutland).

archaeologist. Full identification of the 25 lost ships was achieved, including 13 previously unknown sites that were identified by comparing their geophysical signatures to each vessel's design characteristics (see Figure 1.2). Examination by remotely operated vehicle visually confirmed the results as correct. This proof of concept test led to a more focused version of events during the battle, and once again created a significant revision of the previously known history (McCartney 2016).

It was during the survey cruises at Jutland in 2014-2016 that the idea of a national scale, multi-shipwreck, broad chronology research project took form. The Jutland battlefield does not sit alone on the seabed of the North Sea. It resides aside myriad shipwrecks from the earliest times to the present in the ever-changing palimpsest of a seabed through time. During our hunt for the Jutland wrecks, we surveyed many sites, which were then eliminated as casualties from the battle by the details seen on their individual DTMs. However, the DTMs of each site surveyed began to form a detailed regional catalogue of wreck sites. Some were known losses, others complete surprises.

In Figure 1.3 the location of the Battle of Jutland wrecks is shown in image A. This also shows their proximal relationship to one important historical text, Harper (1927). Its positional data is projected spatially alongside the positions of the wrecks, allowing their interrelationships to be analysed. This is a key component to multi-shipwreck studies and at the heart of this research.

Alongside the wrecks of the battle lie several other shipwrecks. One small wreck was eliminated by its dimensions seen in its DTM. It was later identified by the recovery of its bell (B). The steamer *Chanticleer* built in 1853 disappeared in the North Sea in 1889 while enroute from Blyth to the Baltic (*Shipping Casualties* 1889). Its discovery in the Jutland area, miles off course was unexpected. Another nearby shipwreck is a modern loss, which was recorded at the time. This is the wreck of the MV *Fallwind* which foundered in heavy seas in 1988 when its cargo shifted. The wreck is recorded by the UK Hydrographic Office (UKHO) as UKHO 38384.

So, in this small region of the North Sea we surveyed and identified examples of shipping losses throughout the industrial age, in peace and war, from the earliest steamships to modern motor vessels. The idea which emerged from the work at Jutland was to see if it was possible to extend this type of research to the national scale and test the potential to identify large numbers of shipwrecks from their DTMs in a new unstudied area. Ultimately this idea became the research question which was framed as *Echoes from the Deep*. Aside from academic funding, gratefully received from The Leverhulme Trust, a suitable area for study needed to be identified.

Bangor University's School of Ocean Sciences: SEACAMS2 and the role of archaeology

Bangor University's School of Ocean Sciences is one of the largest marine sciences departments in Europe. Between 2010-2022 it has been running a research programme termed *Sustainable Expansion of the Applied Coastal and Marine Sectors* (SEACAMS and SEACAMS2) and is collecting geophysical data on fixed objects on the seabed of the Irish Sea. The SEACAMS2 project supports emerging economic opportunities in the marine Low Carbon, Energy and Environment sectors through specialisation in the commercial application of research into marine renewable energy (MRE), climate change resilience and resource efficiency in Wales.

One programme is associated with researching seabed processes to identify appropriate sites for MRE infrastructure deployments. This research aims to conduct geophysical surveys at charted shipwreck sites characterised by differing water depths, current flows, and seabed substrates. The research is examining the potential that shipwrecks and other seabed structures have as proxies for understanding interactions between MRE infrastructure and seabed processes within the marine environment. Studying shipwreck sites of varying size, orientation, and duration since sinking is providing new insights into how marine processes respond to the presence of such structures and elucidate the time required for the physical impacts associated with their presence to reach some form of equilibrium with their surrounding environment or crucially, to determine if they ever truly do so.

Shipwreck structures can have anything from extremely limited through to significant long-term impacts on the seabed and its associated physical and biological processes depending on their location, orientation, and localised tidal regime and some structures appear to settle into the seabed within the scour they create. The researchers aim to establish whether this occurs over the short term of weeks and months following submergence, or over more prolonged periods spanning several years, or during isolated periods due to structural collapse, or low frequency storm events etc.

It is well understood that submerged structures such as shipwrecks will result in a change in the flow of water, quite often resulting in the creation of seabed scouring and also potentially modifying the distribution of sediments around them, see Figure 1.4. These impacts are all highly dependent on the structure itself, the strength and direction of the tidal flow regime, the type of sediment available, and the period of time that has elapsed since the structure became submerged.

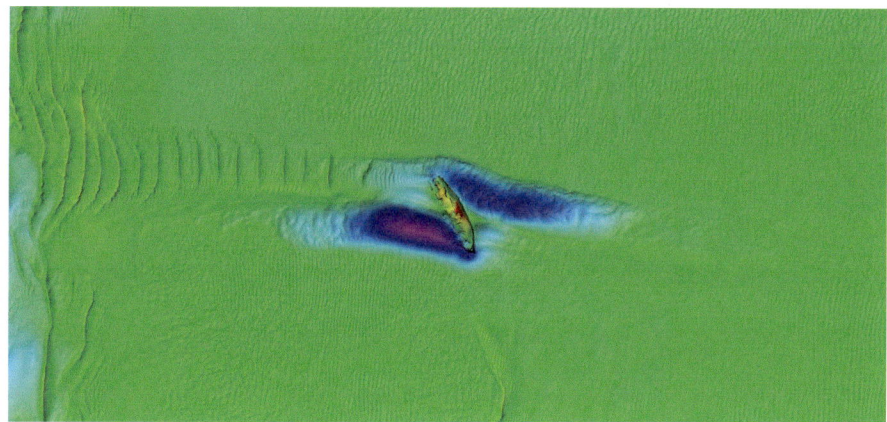

Figure 1.4. The scour pits and seabed ripples created by the ebb and flood tides circulating around the wreck of the 7,832 ton cargo-liner SS *Apapa* (UKHO 7332) which sank in 1917 off Anglesey. These have accumulated over time. But how long did it take to occur? Has this effect stabilised? The questions are being addressed by SEACAMS2. Dating the wrecks is crucial to understanding the temporal and spatial scale of the effect observed (Bangor University).

Crucially, knowing how long a shipwreck has been submerged is a key component to understanding its temporal effect on the marine environment. This is not always possible because so many are either completely unidentified or have been previously misidentified and therefore are undated. Bangor University's School of Ocean Sciences knew that a large, but unquantified number of the shipwrecks it was studying were not reliably identified. Consequently, this created a requirement which could only be fulfilled by the archaeological study of each of the shipwrecks.

This requirement to establish a date for when a shipwreck was created, coupled with my requirement to access a large regional MBES dataset led to the unique interdisciplinary collaboration which formed the basis of this research. The results clearly demonstrate that the dating of shipwrecks by archaeological research has an important role to play in the development of the MRE sector going forward into the future.

The selected shipwreck dataset for the research

Bangor University's School of Ocean Sciences operates a specialist survey vessel, *Prince Madog* which is equipped with an MBES system. This has been used to scan over 400 shipwrecks around Wales and in the Irish Sea, which made for an ideal research dataset in size and scope.

"Offshore" shipwrecks

For obvious operational reasons *Prince Madog* cannot survey in proximity to the shore, nor close to submerged objects or other navigational hazards. As a general rule, this has meant that the shipwrecks surveyed are in water depths greater than 20m. Therefore, shipwrecks caused by being beached, or stranded on rocks etc. before sinking are not surveyed. The shipwrecks are in effect, offshore and this has become the term to describe the dataset used in the project, the offshore shipwrecks of the Irish Sea.

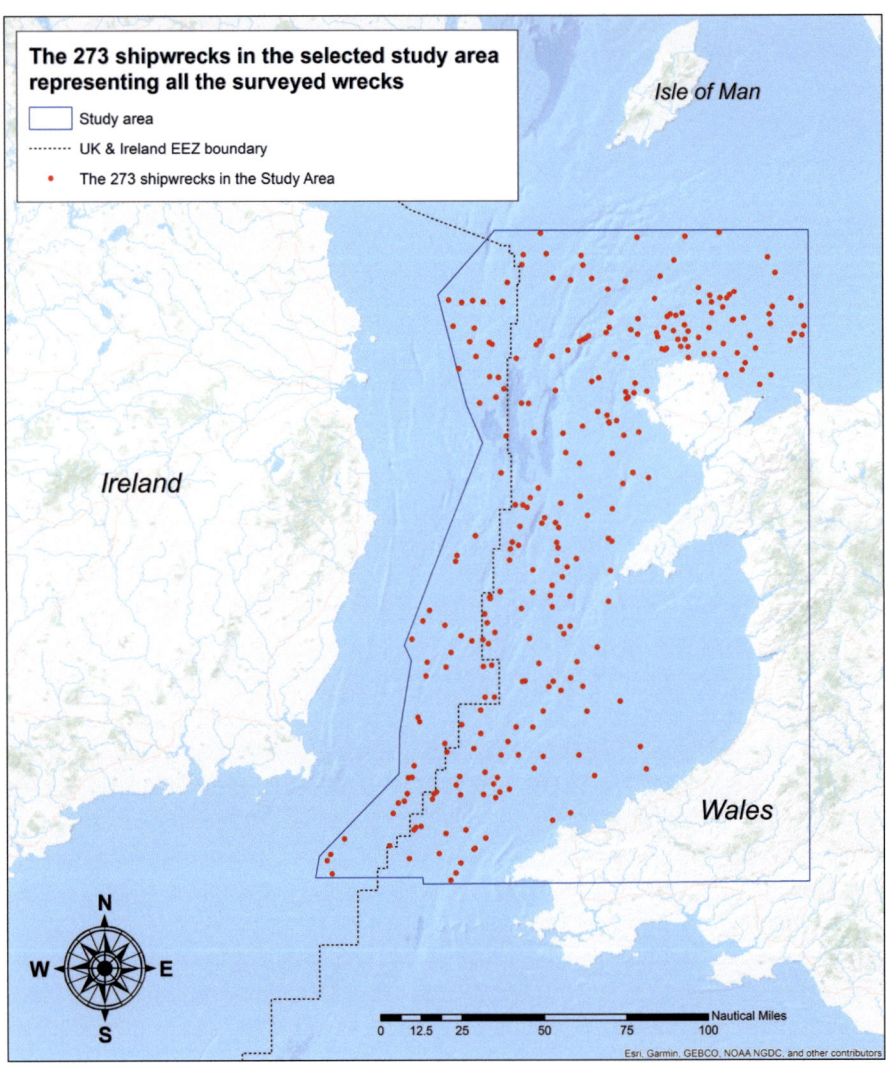

Figure 1.5. The distribution of the 273 accessible charted shipwrecks in the study area, forming the archaeological basis of the research undertaken. This represents full coverage of the offshore shipwrecks in the area (Innes McCartney).

The need for 100% coverage

It was an essential requirement of the archaeological research that the area to be studied must have 100% coverage of all of the charted offshore wrecks within it. This was because the wrecks could only be satisfactorily identified if all the potential candidate sites for any sinking event had been surveyed, measured, and catalogued. Only then could other potential contenders for any specific event be safely eliminated, based on knowing the details of each case. In many instances it is likely to be pointless to try to establish the identity of any shipwreck without first knowing what all the other nearby shipwrecks look like and what is also known of them as well.

Figure 1.5 shows the extent of the area of Bangor University's School of Ocean Sciences surveys in which 100% of the wrecks have been surveyed. This became the study area of the research. However, it must be noted that the western border does not extend to the Irish coast, this is because Irish regulations forbid marine survey inside its 12 mile limit.

Historical importance of the Study Area

It was very desirable that the sea area to be selected for study was of notable historical importance. The study area fulfilled this requirement because of its geographic location, as it has been the key transit route for shipping heading northwards to the cities of Dublin, Liverpool, and Glasgow. Shipping heading south would be bound for global destinations and to the cities of Cardiff and Bristol. The strategic importance of these shipping lanes can be traced back to the iron age and beyond, the Roman period port at Chester being an example.

Shipwrecks of iron & steel

The historical time period to be studied was defined by the capability of the MBES technology. Experience of surveying with MBES and working with differing MBES datasets in differing depths and on varying seabed substrates has shown that it lends itself very well to the study of ships built of metal, as iron and steel hulls reflect sound well. The same result is not consistently seen with shipwrecks made of timber. Unless heavier objects such as cannon are present, it seems that most offshore timber shipwrecks have been flattened by time and the effects of trawling. If not already buried, a spread of ballast stones may be all that survives and therefore, in most cases they are practically impossible to survey by using MBES, unless the sonar head is in close proximity to the wreck.

Prince Madog's MBES system is hull mounted, therefore the minimum distance to the wreck under study was never less than 21m and for these reasons, timber-built shipwrecks are not included in this study. It is solely aimed at vessels of the industrial age, beginning with the iron-hulled ships which began to appear and subsequently sink from the 1830s onwards. Timber shipwrecks cannot be satisfactorily identified using the methodologies developed during the research. In fact, of all the 273 shipwrecks surveyed, none of obvious timber construction were seen.

Framing a research methodology

The research process as outlined in the next three chapters was complementary to the fieldwork. Prior to the COVID pandemic, I was able to work with Bangor University's School of Ocean Sciences in the field, getting a feel for the key navigational features used by the mariners of the past and the sheer geographic scale of the project we had taken on. Once the survey results had been processed to produce the final pointclouds, that is when the work of the research began. The main stages formed a workflow:

1. The wreck site modelling. As a minimum each wreck pointcloud was cleaned and a DTM produced, and the shipwreck dimensionally measured for length and width. It was then inserted into the GIS map of the study area,
2. The historical text (Chapter 2). The research sought out as many sources relating to ships lost in the study area as needed to understand how each sinking event occurred.

Special importance was placed on sources which contained spatial and positional details of ships sunk and textual accounts of how the ships sank. The positional data was digitised and converted into shapefiles, so that they could be projected spatially on the GIS research map, alongside the DTMs and the UKHO charted shipwreck data,

3. The archaeology research Phase 1 (Chapter 3). It was essential to accurately establish the number of surveyed sites which had already been identified with 100% certainty. All the wrecks exist in the UKHO shipwreck database. However, legacy issues pertaining to the misattribution of shipwrecks within it necessitated that each site be individually assessed against it and amendments made,

4. The archaeology research Phase 2 (Chapter 4). This stage of the research was conducted in conjunction with the historical text in order to place identities on the wrecks which Phase 1 had revealed as previously unidentified. Each unidentified shipwreck was subject to detailed spatial and dimensional analysis. Evidence was sought in the historical text to find dimensional, spatial, and archival matches for each wreck. A consistent, prescribed methodology was developed, against which each shipwreck was assessed,

5. Collating the results. The research from each shipwreck was compiled into the final inventory. This was used to provide an overview of the entire project across the study area and to provide conclusions and an assessment of the project's impact and significance once completed. This led directly to publication.

References

Barber, M., 2011. *A History of Aerial Photography and Archaeology*. Swindon: English Heritage.

Board of Trade, 1889. *Wreck Report No 3925 CHANTICLEER (SS)*. Board of Trade.

Crawford, O.G.S. and Keiller, A., 1928. *Wessex from the Air*. Oxford: Clarendon Press.

Harper, J. E. T., 1927. *Reproduction of the Record of the Battle of Jutland*. London: HMSO.

Lloyd's Register of Shipping (various dates). *Lloyd's register of Shipping*. London: Lloyd's Register of Shipping.

McCartney, I., 2002. "Deadlight U-boat Investigation" *After the Battle* No. 116.

McCartney, I., 2014. *The Maritime Archaeology of a Modern Conflict: Comparing the Archaeology of German Submarine Wrecks to the Historical Text*. New York: Routledge.

McCartney, I., 2016. *Jutland 1916: The Archaeology of a Naval Battlefield*: London: Bloomsbury.

UK Hydrographic Office (various dates). *Wreck Record No. 38384 MV FALLWIND*. Taunton: UKHO.

Westley, K., Plets, R., Quinn, R., McGonigle, C., Sacchetti, F., Mekayla, D., McNeary, R., and Clements, A "Optimising protocols for high-definition imaging of historic shipwrecks using multibeam echosounder." *Archaeological and Anthropological Sciences* 11 (2019): 3629-3645.

2

Researching and databasing the historical text

Introduction

Prior to carrying out any analysis on the DTMs in the MBES dataset, it was essential to undertake historical research into the spatial, technical, and textual details of the shipping losses in the study area. The DTMs were of ships constructed of iron or steel. Wooden wrecks were not seen in the survey dataset and therefore were not a feature of the research. With this in mind, the archival research focused specifically on the records of iron and steel ships lost in the study area. Research reached back as far as 1830, when iron was introduced into ship construction, to ensure all possible cases were captured as accurately as contemporary records permit.

In order to do this, the most informative sources on shipping losses were consulted and the positional data of each was entered into a GIS shapefile. The sources selected, how they were interpreted and how the data was plotted spatially is described in this chapter. The circumstances of the loss of each vessel were important to note, and it was also crucial to establish with as much accuracy as the historic text permits, the position of where each ship sank. This is because it was anticipated that the historic loss positions when plotted spatially could give some of the most important clues as to the identities of the unknown shipwrecks in the MBES dataset.

In reality work of this type cannot ever be 100% definitive. The ad hoc means by which shipping losses have been recorded in the past and the fact that many ships simply vanished without trace, such as SS *Chanticleer*, described in Chapter 1 handicap the chances of identifying every wreck in any given area. Nonetheless it offers a way of looking at shipping losses which provides a spatial means of searching the various sources which would simply be too difficult to undertake manually, by rifling through the original published indices, attempting to find nearby shipping loss positions to each DTM as they were examined.

The *Shipwreck Index of the British Isles* (Larn & Larn 2000 & 2002) provided an excellent starting point from which to spatially project the number and extent of the offshore shipping losses in the study area and to understand their chronological development. Once assessed, further sources were brought into the GIS, targeting the specific time periods which accounted for greater portions of the losses. For example, due to the high proportion (46%) of the losses being sunk by U-boats in WW1, it was decided to consult

Allied and German navy sources to enhance the loss data, giving positional and textual evidence from both sides of each sinking event.

Alongside the development of the spatial databases, technical details of the ships lost and textual accounts of the circumstances of each loss were also sought. It was important to capture as much evidence as possible from the historical text which could be of utility when identifying the shipwrecks.

Creating archival shipwreck databases and shapefiles

The archival study area

Due to the relative inaccuracy of navigation prior to 1944, when electronic means began to be used, it was necessary capture positional data which plotted close to as well as inside the study area boundary. In some cases, such as Larn & Larn (2000 & 2002) it meant capturing the entire chapters which only partially intersected the study area. Whereas with other published indices it meant plotting the results which in lat/long were within a few minutes of the study area boundary or given in range and bearing to plot in the periphery of the study area. See Figure 2.1 for the extent of the archival study area.

Historic navigational points

The range and bearings given in British and German records were given from prominent navigational marks, such as a headlands or lighthouses. Whereas some, such as the Tuskar Rock can easily be discerned on modern charts, the names of some of these places have changed or fallen out of use. A shapefile was created to capture all of the historic navigational points referenced in the historic text.

A good source in assisting in locating the most common was found in Tennent (2006, p253-257). Interestingly it was also discovered that the German navy charts noted the positions of all of the active lightships (LVs) of the period. This proved very useful as several of them, especially along the coast of Ireland are no longer on station.

It was also noted that there were in fact two "Skerries" in the study area; one off Dublin and the other off Holyhead. This had to be taken into consideration during the plotting phase. Moreover, there are in fact several "Skerries", "Bishop Rocks" etc. around the UK and elsewhere, so during the database building phase it was necessary to assess which related to navigational points in the study area.

Timing and measurements

All sea distances and measurements were made in nautical miles using the geodesic plane. Distances of depth are given in metres. All dimensional measurements of ships and shipwrecks are given in metres. Where necessary they have been converted from the original imperial. All timings are given from the UK system, being either Greenwich Mean Time (G.M.T.) or British Summer Time (B.S.T.) as appropriate. Where timings have come from German military sources given in Central European Time (C.E.T), they have been converted to G.M.T or B.S.T when necessary to do so.

Range and bearing conventions and plotting

Plotting the lat/long positions was straightforward. The positions were converted to the more modern degrees and decimal minutes and incorporated into the GIS so that they could be presented spatially on a map of the study area. However, alongside the lat/long positions were many recorded as a range and bearing from a navigational point. These needed to be considered more carefully. Positions given from a navigational point would have used a bearing recorded from a compass. Two types of compass were mainly in use during the period under study, the magnetic compass, and the gyrocompass.

The magnetic compass will find magnetic north, which can be corrected to true north by reading the amount of correction of "variation" required off the chart the vessel is currently using for navigation. However, iron and steel ships also affect magnetic compasses due to the distribution of ferrous metals around the compass itself; this is "deviation". Deviation was usually corrected by using magnets, iron spheres or a "Flinders Bar" on or inside the compass binnacle.

Magnetic compass readings could potentially be in three different conventions: a) true course, with the bearing being corrected for both deviation and variation; b) magnetic course, where deviation has been corrected but the reading is not corrected for variation, and c) compass course, where neither variation nor deviation had been corrected (Admiralty 1904, 143).

In relation to magnetic variation in the historic text, it was noted that during the database building there were very few range and bearing readings given which stated the convention being used. However, a very occasional reading given with a "(mag)" suffix was observed, leading to the conclusion that true course was the default convention being used.

In relation to magnetic deviation, there is now, of course no way of being able to reverse correct a compass course because each ship's unique deviation cannot now be known. In reality no cases were seen recorded in this way, although the possibility remains, especially among smaller vessels that uncorrected bearings may have been in use.

The gyrocompass, which uses a gyroscope to detect the rotation of the earth and is magnetism-free giving a true bearing, had been in development in the nineteenth century. The Anschütz Company of Germany and Sperry of the USA produced the first examples adopted by the world's navies. Both systems were also commercially available to the non-military market from 1913. Germany adopted the Anschütz in 1908 and the US Navy was widely using Sperry by the end of WW1. The Royal Navy trialled both systems prior to WW1 and opted for Sperry as well (Admiralty 1931).

In order to be consistent in plotting the data, it was decided to adopt true north as the default plotting convention for all range and bearing readings in the historic text. It is simply impossible to know for certain in most cases if corrections for variation had been made, or if in the twentieth century a gyrocompass was installed. Therefore, to have made variation corrections to all would have served no purpose because it would have made the originally corrected cases incorrect. Moreover, the confined nature of the Irish Sea means that the distances involved would not make a significant difference. The range component could also add its own distance errors to any given position anyhow.

Understanding the set rules by which bearings were reported is important. The basic Admiralty manuals on seamanship stressed the conventions to be used. For example, never read from a point which begins and ends with the same letter, such as WNW and always start with N or S and then E or W (Admiralty 1904, 145-146). Caution needs to be taken when addressing the compass bearing conventions that may have been reported. For example, "S by W" means that the bearing is actually one point of the compass (11deg 15min) west of due south, or 191deg 15min. The addition of half and quarter points is not uncommon in the Lloyd's records. For example, "S by W, half W" would mean a point and a half west of south, i.e., closer to 197deg.

The distance component could be made up of either a calculation of distance based on a visual means or as the last point of land fixed. Visual fixing could be done quite accurately if the navigation point had a known height. A vertical sextant could be used to calculate the angle from the horizon to the top of the object being measured. A manual calculation (based on Pythagoras theorem) could be made, or a reference system such as Lecky's Tables could be consulted to ascertain an accurate distance (Admiralty 1964).

Some cases in the study area referred to navigational points that would have been out of sight. In these cases, it seems most likely that the course of the ship would not have deviated from the time at which the navigation point disappeared from view. Clearly speed would have been monitored by use of a chip or taffrail log to ascertain the range from the navigation point.

The historic text and the extent of shipping losses in the study area

In the UK, there is no national register of the nation's shipping losses. Only a few countries, such as Denmark have done this. As English Heritage has acknowledged, the compilation of a complete database of shipping losses is impossible (Cant 2011, vii). The reasons are multifarious and in general, the further back in history one travels, the weaker the historic text becomes. Even when published lists of shipping losses began to emerge in the 18th century, there were significant omissions in both official and unofficial sources. It is therefore important to recognise that for the period up to at least 1900, the record is not 100% reliable.

Two of the most important sources on shipping and shipping losses are Lloyd's of London (publishers of the *Lloyd's List* since around 1740) and Lloyd's Register of Shipping, (publishers of the *Lloyd's Register of Shipping* since around 1764) giving details of ship design and construction and from 1890 publishing the *Lloyd's Register Wreck Returns* (*Wreck Returns*), an annual digest of shipping losses, as they pertained to ships struck from the *Lloyd's Register of Shipping*. Financially and commercially independent, the two companies trace their origins to a 17th century coffee house owned by a Mr Edward Lloyd. The *Lloyd's List* (and its various weekly summaries) reported shipwreck incidents, although up to around 1800 it is thought that only about half of the actual losses were being listed.

The large number of losses of the UK's ships at sea in the early 19th century led to the election of the *Select Committee into the Causes of Shipwreck*. It reported its findings in 1836 and revealed the true extent of shipping casualties. For example, from 1833-35 more than one UK registered ship a day was being sunk, in many cases because the vessels themselves were defective. It led directly to major reforms to the Merchant Shipping Act 1786, which continued throughout the 19th century, introducing such innovations as the Plimsoll Mark in 1876, and importantly for shipwreck research the first annual shipwreck statistics being returned to parliament in 1851 by the Board of Trade (BOT), as what has become known as the *Board of trade annual returns* (or *Shipping Casualties*). The Admiralty took over from 1852 to 1855 and then in 1856 the Board of Trade resumed the task, presenting annual statistics through to the end of WW1.

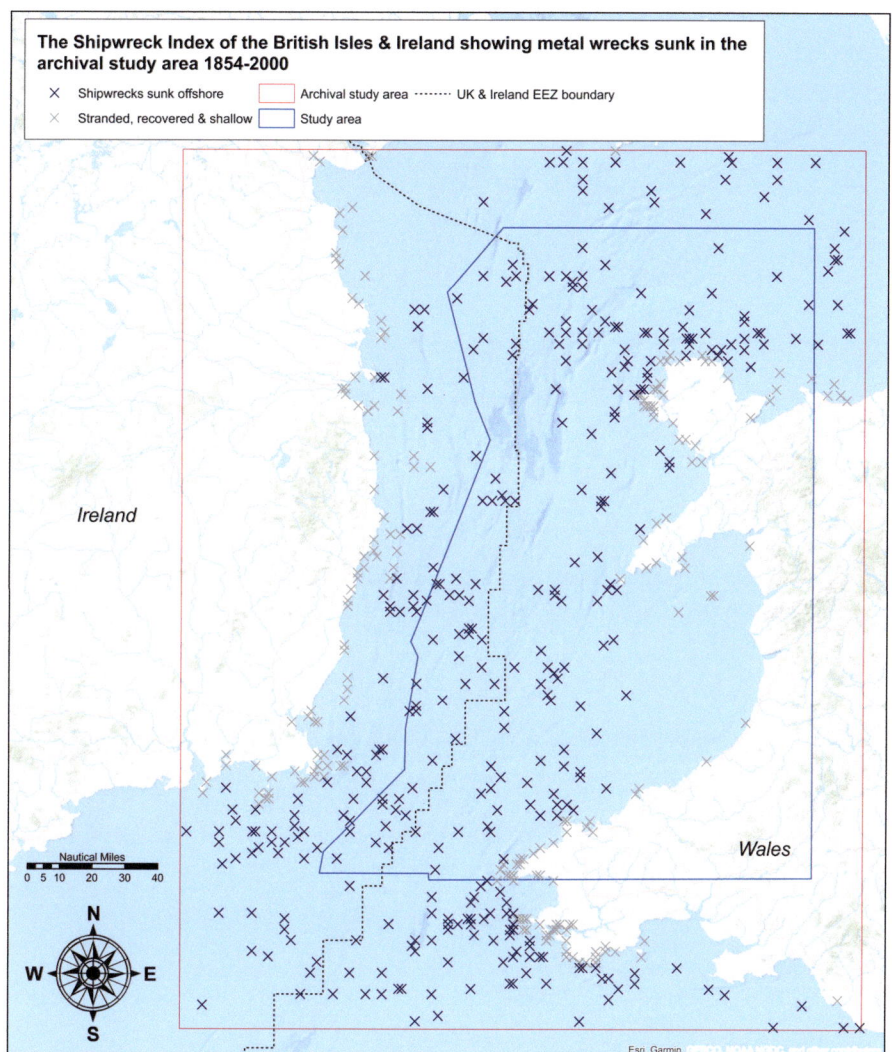

Figure 2.1. The positions of the 724 shipwrecks listed in the Shipwreck Index of British Isles, which plot in the archival study area, separated into the 298 inshore and 426 offshore cases (Innes McCartney).

Despite the *Shipping Casualties*, and the *Lloyd's List* there are, according to Larn & Larn (2002 xiii) omissions in the sources up to least 1900. Quite often shipwrecks appear in one source and not the other and sometimes even important losses appear absent from both. When studying shipping losses holistically, as opposed to specifically, Larn & Larn (2000, xii) have stated "*Shipwreck research is always a compromise between one's resources, time, patience and physical stamina. Searching in depth for every detail of just one ship-wreck is simply a matter of time and persistence, but when searching for details of many thousands of vessels en masse, the task is impossible...*" Therefore, to have attempted to build a spatial database prior to at least 1900 from primary sources specifically tailored for this research was simply not practicable in the time allowed and secondary published sources needed to be relied upon, primarily the *Shipwreck Index of the British Isles*.

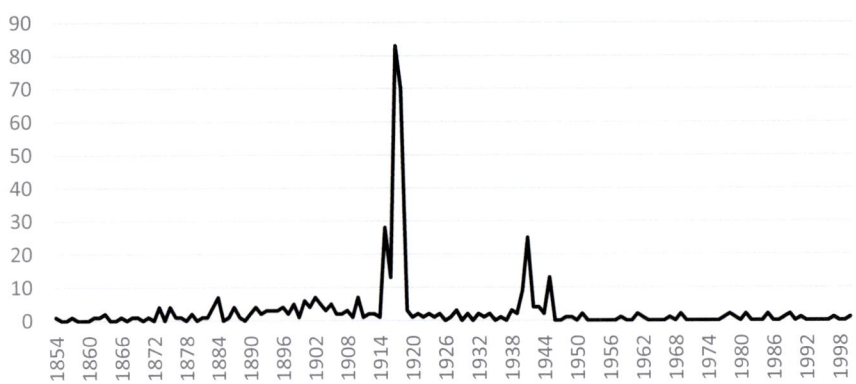

The 426 offshore shipping losses in the archival study area 1854-2000 shown by year

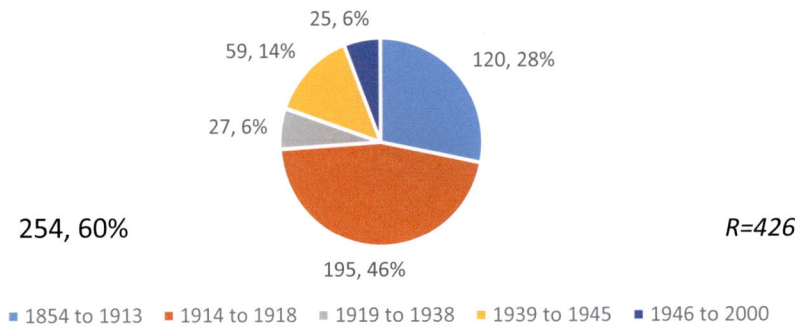

The 426 shipping losses in the archival study area 1854-2000 segmented by historical period

25, 6%

59, 14%

120, 28%

27, 6%

254, 60%

195, 46%

R=426

■ 1854 to 1913 ■ 1914 to 1918 ■ 1919 to 1938 ■ 1939 to 1945 ■ 1946 to 2000

Figure 2.2. The 426 offshore shipping losses in the archival study area analysed by year and historical period of loss. The two world wars are the dominant cause of these losses (Innes McCartney).

The Shipwreck Index of the British Isles

The National Maritime Museum's *Research Guide C8* (2021) recommends a number of published indices of shipping losses, such as Hocking (1969). However, it was found that the most comprehensive is the *Shipwreck Index of the British Isles*, specifically volumes 5 and 6, which cover the archival study area (Larn & Larn 2000 & 2002). In these volumes the authors have searched a wide array of sources, crucially including the *Lloyd's List*, *Wreck Returns* and *Shipping Casualties*, from which it appears that most cases (omissions, oversights, and double entries notwithstanding) have been incorporated into the text. Importantly for this study the authors have given, as best as it is possible to do, a position for most of the cases recorded in its pages, even if they are merely indicative of an area. This made it the most practical source to develop into a spatial database covering the historical timeline of this research. Indicative positions notwithstanding, it was anticipated that it would give a starting point to view the mix and extent of the shipwrecks likely to be encountered during the fieldwork, from which further research could be extended, where necessary and practicable.

The volumes are divided up into sections by region. The data entered into the GIS came from seven sections (3, 4, 5, 6, 7, 9, and 11) of volume 5, which in total comprise entries on the loss of 3,660 ships, and two sections (2 and 3) of volume 6 which comprise a further 2,073 entries. Each entry contains details of the vessel in question and the circumstances of its loss. All of these 5,733 were searched for cases involving composite, iron, and steel-hulled vessels and these were entered into the database recording the name, date, type of ship, nationality, length, width, cause of loss, additional notes, casualties, and position. In total, by using this methodology 724 shipwrecks are recorded as lost in the archival study area, spanning from 1854 to 2000.

Further analysis of these shipwrecks shows that 298 of the losses are recorded as strandings or sank in water too shallow to survey. As such, they were not likely to be detected within the MBES dataset, because, as described in Chapter 1, *Prince Madog* did not, as a rule operate inside the 20m line. The extent and distribution of the 724 cases is shown in Figure 2.1. It shows that Larn & Larn (2000 & 2002) list a total 426 shipwrecks sunk offshore in the archive study area. Up to 273 of these cases would be expected to be among the shipwrecks surveyed during the fieldwork for this research. These would be cases within the study area and potentially some of the outliers in the archival study area.

Focusing on the 426 offshore shipping losses, the data can be broken down by year to give an insight into the numbers of ships sunk throughout the period since the first recorded loss in 1854. The results are shown in Figure 2.2. It becomes immediately apparent that the dataset is dominated by ships sunk during WW1. This accounts for 195 (46%) of the wrecks. The losses in WW2 account for a further 59 (14%) of the wrecks. In total 254 ships sunk offshore were sunk in wartime. This amounts to 60% of the total.

Where possible, further positional sources of shipping losses were sought out in order to develop complimentary shapefiles to use in the research alongside the ones made from the *Shipwreck Index of the British Isles*. The reality is that for the peacetime losses up to 1946, the only consistent publications are the *Lloyd's List*, *Wreck Returns* and *Shipping Casualties* which are incorporated into the *Shipwreck Index of the British Isles* and there was little to be gained in attempting to replicate it.

The greatly reduced numbers of peacetime shipping losses after 1945 begin to also show up as charted losses in the UKHO wreck cards, as described in Chapter 3. So, for the losses in peacetime, the *Shipwreck Index of the British isles* was the main source used to plot spatially an indication of where these losses occurred, with each unidentified case being checked against the original entries in the *Wreck Returns* and *Shipping Casualties* during Phase 2, as described in Chapter 4.

However, due to the dominance of the wartime losses among the shipwrecks in the archival study area, additional positional sources were sought to provide greater positional detail during the war years. Fortunately, good sources of positional data are available. The key sources selected to develop three further shapefiles were *Lloyd's War Losses* (*War Losses*) and the actual U-boat patrol logs from WW1.

Lloyd's War Losses 1914-1918 and 1939-1945

Lloyd's of London, as underwriters of ships and publishers of the *Lloyd's List* naturally recorded the losses of shipping during wartime. Its role, while that of an independent company was closely linked to the Admiralty. For example, in WW1 the *Lloyd's List* was initially published in the peacetime format by posting daily reports on movements and losses. But in preparation for the 1917 attack on shipping by U-boats this was made secret under the title of "Oversea Shipping Intelligence" (Lloyd's of London 1990, v-vii).

Nevertheless, Lloyd's retained detailed statistics on the losses of ships during WW1. In the immediate post-war years, it compiled them into a single reference source, *Casualties to Shipping Through Enemy Causes 1914-1918*, later published as *War Losses* (1990). This is regarded as the most comprehensive source on UK, Allied, and neutral ships sunk in WW1, with the least omissions. It also benefits from providing details of routes and cargoes and post war information on salvage and occasionally listing the U-boat involved. Crucially, for this study, Lloyd's also recorded a position for each ship sunk. By contrast the Admiralty lists of ships sunk, published as *British Vessels Lost at Sea* (Admiralty 1988) is not so useful, especially for neutral shipping which quite often only gives the ocean in which a ship was noted as lost (*War Losses* 1990, v-vii).

When compared to Admiralty sources, the accuracy of the Lloyd's records is even more marked when examining WW2. As is noted in the introduction to Volume 1, in the decades since its compilation: *"nothing has appeared in print which approaches so comprehensive a record"* (*War Losses* 1989, v). The key reason for its authoritativeness is that from the war's outset, Lloyd's Intelligence Department had been delegated to collect shipping information on behalf of the Ministry of War Transport but also retained its own business and intelligence sources. This allowed it to build data both on merchant and naval vessels of all Allied and neutral countries during WW2. It complied *War Losses: The Second World War* during the 1950s and it was published 40 years later in two volumes (1989 & 1991).

Because of the comprehensive nature of these two wartime volumes, they were selected to be converted into two further spatial databases inserted into the research GIS, which would list all of the wartime losses of merchant ships in the archival study area. These shapefiles record the name, date, tonnage, cause of loss, heading, cargo, notes, and position of each ship. When all the relevant Lloyd's records were entered into the GIS, there were 184 records for WW1 and 55 for WW2, both being slightly lower than the *Shipwreck Index of the British Isles*, primarily because only losses by war action are listed in *War Losses*, so sinkings by collision etc., such SS *Burutu* in 1918 (see Chapter 4) are not present.

Both volumes also contain lists of missing vessels. It was noted that 17 in WW1 and two in WW2 went missing on voyages which took them through the study area. This does not mean they would be found in the study area, but theoretically they could present themselves during the research. Obviously, with no positional or textual details for these losses, nothing could be done except to note that they may be present, but that a lack of historical detail would make them impossible to identity by the methodologies used during this research.

The German Naval Records of WW1

When studying allied shipping losses in WW1 from the German perspective, there is no published source that provides the crucial positional data describing where each ship was attacked by a U-boat. Published sources have traditionally used British records. Therefore, primary research using the original German navy records had to be undertaken. In this case it entailed accessing the patrol log, or Kriegstagebuch (KTB) of each U-boat that patrolled the waters of the study area.

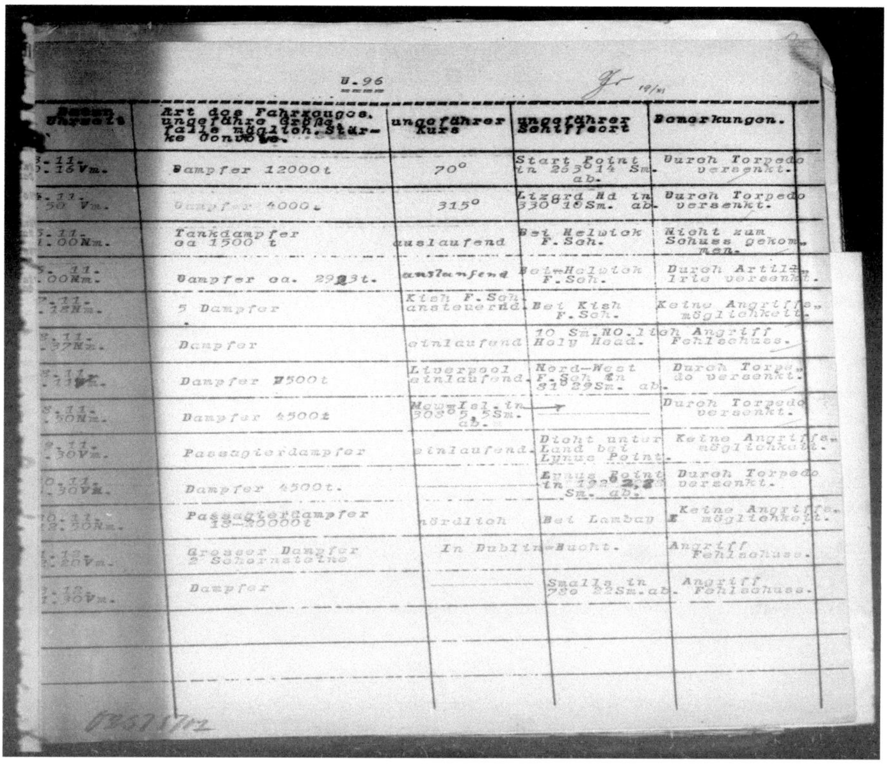

U.96

Datum Uhrzeit	Art des Fahrzeuges, ungefähre Größe, falls möglich, Stärke Convoys.	ungefährer Kurs	ungefährer Schiffsort	Bemerkungen.
3.11. .16Vm.	Dampfer 12000t	70°	Start Point in 253° 14 Sm. ab.	Durch Torpedo versenkt.
.11. .50 Vm.	Dampfer 4000.	315°	Lizard Hd in 330° 10Sm. ab.	Durch Torpedo versenkt.
5.11. 1.00Nm.	Tankdampfer ca 1500 t	auslaufend	Bei Helwick F.Sch.	Nicht zum Schuss gekommen.
5.11. .00Nm.	Dampfer ca. 2923t.	auslaufend	Bei Helwick F.Sch.	Durch Artillerie versenkt.
7.11. .18Nm.	5 Dampfer	Kish F.Sch. ansteuernd.	Bei Kish F.Sch.	Keine Angriffsmöglichkeit.
8.11. .37Nm.	Dampfer	einlaufend	10 Sm. NO. lich Holy Head.	Angriff Fehlschuss.
.11. .11Nm.	Dampfer 7500t	Liverpool einlaufend.	Nord-West F.Sch in 81° 29Sm. ab.	Durch Torpedo versenkt.
8.11. .50Nm.	Dampfer 4500t	Mew Isl. in 303° 5,5Sm. ab. →		Durch Torpedo versenkt.
9.11. .30Vm.	Passagierdampfer	einlaufend.	Dicht unter Land bei Lynus Point.	Keine Angriffsmöglichkeit.
0.11. .30Vm.	Dampfer 4500t.		Lynus Point in 192° 2,2 Sm. ab.	Durch Torpedo versenkt.
0.11. 12.50Nm.	Passagierdampfer 18-20000t	nördlich	Bei Lambay I	Keine Angriffsmöglichkeit.
1.12. .20Vm.	Grosser Dampfer 2 Schornsteine	In Dublin-Bucht.		Angriff Fehlschuss.
1.12. 1.30Vm.	Dampfer	————	Smalls in 78° 22Sm. ab.	Angriff Fehlschuss.

Figure 2.3. The summary sheet of *U96's* patrol of November 1917. The "Dampfer 7500t" is SS *Apapa* with the position given as a range and bearing (NARA).

In practical terms this work would have been prohibitively time consuming if the process of reconciling each ship sunk to a reported attack by a U-boat had not been previously undertaken. It was carried out during the compilation of the German Official History of the first U-boat war, *Der Handelskrieg mit U-Booten* (Spindler 1932, 1933, 1934, 1941, 1966). In compiling this five volume history of the first U-boat war, Spindler combed through every U-boat KTB in the German naval archives and matched its movements and reports of ships sunk to published British records, and via correspondence with Allied record keepers, such as the Royal Navy Naval Historical Branch.

In order to associate a specific U-boat to each ship sunk in the Irish Sea during WW1, it was first necessary to know from British sources which ships were lost in the area. This meant that the databases of shipping losses using British sources had to be compiled first. From there Spindler's association to a specific U-boat could be traced and the original KTB consulted to find where the U-boat reported the sinking incident taking place. Spindler had worked through these same KTBs. They bear his handwriting and analysis on many pages.

Due to the capture of the entire German navy archive at Castle Tambach in 1945, nearly every KTB can be accessed on digitally reproduced microfilm rolls because the entire U-boat records were photographed in 1945-1947. Most originals have been repatriated to the Bundesarchiv, Freiburg, but the entire photographed collection is available to order on over 4,000 microfilm rolls from the National Archives & Records Administration (NARA), Washington DC (NARA 2008, vi-viii). In all there are 104 microfilm rolls of U-boat KTBs for WW1 (NARA 1984). These were used to compile the shapefile of positions and textual descriptions of ships reported sunk by U-boats in the Irish Sea.

Working with the microfilmed U-boat KTBs

Extracting positional data from the KTBs themselves is not a straightforward process. This is because the KTBs do not always follow a prescribed format and the amount of detail given is quite often dependent on the individual commanders. Consequently, there is a varying range of detail given on the attacks on shipping from U-boat to U-boat. As a general rule, there is a great deal more information to be found in the KTBs of the U-boats which operated as part of the German fleet, as opposed to those of the smaller flotilla based in Flanders and the details tend become more user friendly in the latter months of the war, due to standardisation of reporting and the wider provision of typewriters. Positions were recorded by lat/long, range and bearing and by grid chart (Quadratkarte).

There were four key means by which positional data was extracted from KTBs.

1. Some commanders recorded the position of a sinking as a matter of course in the text,
2. They can also usually be extracted from the summary sheets of ships encountered, which U-boat commanders submitted with each KTB,
3. All U-boats submitted a chart tracing, summarising the track of the U-boat patrol, and marked on it are the locations of the ships sunk,
4. Where the U-boat laid mines, a chart of the position of the minefield was submitted with the KTB.

An example of a summary sheet can be seen in Figure 2.3, where U96 reported its encounters with shipping, with the positions given by range and bearing. The *"Dampfer (steamship) 7500t"* is the liner SS *Apapa* which was sunk in November 1917 off the north coast of Anglesey. The position is given as 81deg 29sm (nautical miles) NW lightship, which was at the western end of the Liverpool roadstead.

Results of plotting positional details from KTBs in the archive study area

After extracting and plotting the reported positions into the GIS and cropping the file to the boundaries of the archival study area the following results were observed.

1. The position of the minefields laid was traced into the GIS from the minefield charts given in the KTBs,
2. The attack positions were recorded for 178 shipping losses. These could be attributed to 45 different U-boats,
3. The positions of several ships could not be plotted spatially because the U-boat did not return home, or no positional data could be extracted from the KTB. For example, *U38's* KTB from August 1915 in which it sunk 10 ships gives only very basic details.

It should be noted that the positions extracted from the georeferenced individual patrol KTB summary maps are more likely to be indicative of an area, being merely descriptive, and therefore not necessarily as accurate as positions recorded in the KTBs by lat/long, range and bearing or Quadratkarte.

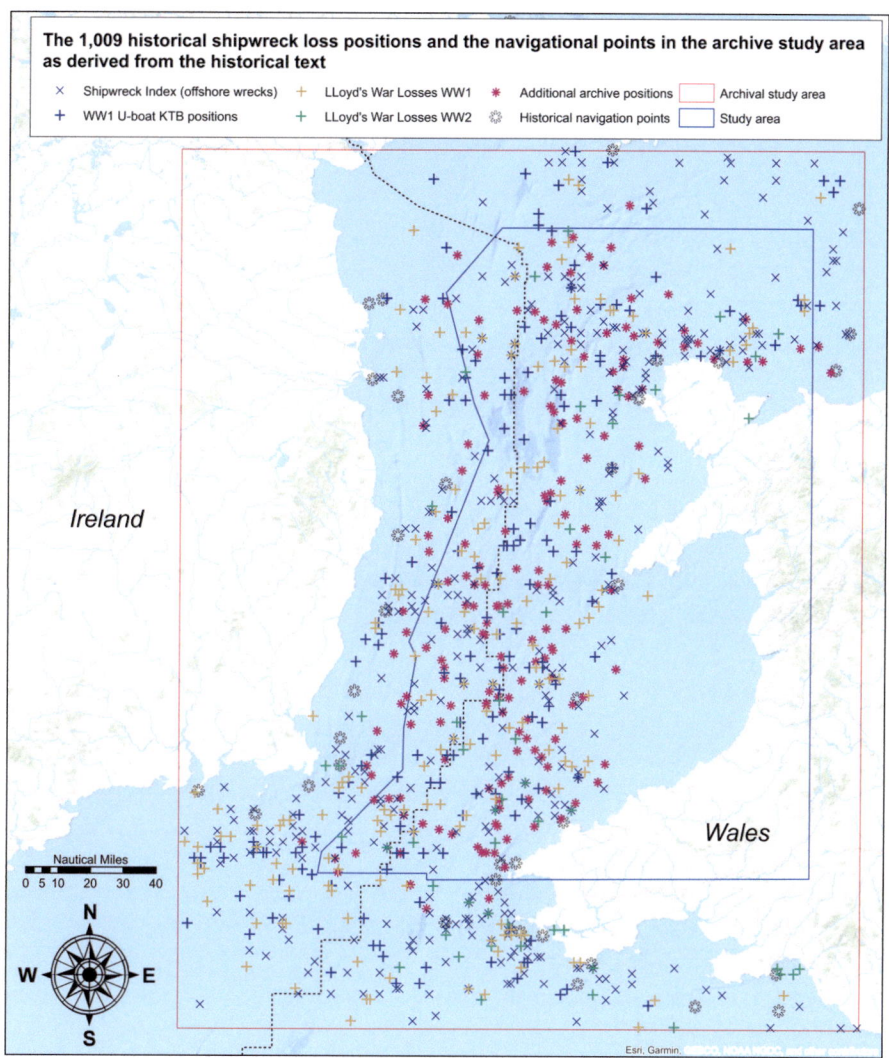

The 1,009 historical shipwreck loss positions and the navigational points in the archive study area as derived from the historical text

| × | Shipwreck Index (offshore wrecks) | + | LLoyd's War Losses WW1 | ✳ | Additional archive positions | | Archival study area |
| + | WW1 U-boat KTB positions | + | LLoyd's War Losses WW2 | ✿ | Historical navigation points | | Study area |

Figure 2.4. The final extent of datapoints for shipping losses which were plotted in the GIS. Also included are the historical navigational points from which range and bearing positions were plotted. The final total was 1,009 historical positions plotted in GIS (Innes McCartney).

Additional positional sources

In order to provide additional historical data on shipping losses, a range of alternative sources was consulted. While some proved to contain interesting textual accounts, most such as Hocking (1969), Huntress (1979) and Hooke (1997) do not generally provide positional information, so are only useful for background details and not for populating spatial databases.

The HMSO publication of Admiralty records, *British Vessels Lost at Sea* (Admiralty 1988) contains positional data, but as previously noted it is not considered as authoritative as *War Losses*, although it provides a valuable alternative official source. A well-researched volume, based on fully referenced primary sources on Royal Navy losses in the steam age (Hepper 2006) occasionally proved useful.

The National Archives holds a very wide collection of files on wartime shipping losses, for example the Admiralty records from both world wars (TNA ADM137/2959-2964 for WW1 and ADM 199/2130-2148 for WW2). These files comprise reports from the senior survivors of ships sunk and provided further sources of positional and textual data, which were consulted as needed.

In reality there were a wide range of other positional sources consulted as and when required. This necessitated the development of a further shapefile into which additional positions from Admiralty files, contemporary newspapers, official inquiries etc. could be added to provide evidence during Phase 2 of the research. The sources used are cited in the relevant shipwreck datasheets given in Appendix 2. This final shapefile of "Additional archive positions" contained a further 166 datapoints once Phase 2 had been completed. The final, full extent of the plotted datapoints collated during the research is shown in Figure 2.4.

For the wartime losses there were usually a minimum of two plotted historical positions to consult. One benefit of having more than one sinking position plotted for each shipping loss was the effect it had on mitigating any historical errors or data entry errors. It therefore was desirable whenever practicable, to find additional sinking positions for the ships lost in peacetime as well. Those positions were also added to the "Additional archive positions" shapefile.

The spatial element of the historical text in practice

In practice, the positional data projected spatially by the shapefiles formed only a facet of the historical research applied to identifying the shipwrecks. The primary reason for developing the shapefiles was to ensure that any shipwreck being identified was situated in an area where it might be expected to be located.

Dead reckoning navigation in the steam age could be very accurate. However, positions may not necessarily have been recorded with accuracy during the stressful occasion of either being sunk or sinking a ship. The question was how many recorded sinking positions would be merely indicative of place? Clearly, historical positions do not necessary plot exactly where the wrecks themselves lie. However, it was not known what ranges of discrepancy between the actual wrecks and the historical positions might be encountered. The expectation was that the distances would fall within a few miles, as had proven the case at the Jutland battlefield (McCartney 2016, p245). But this could not be tested until the research was completed (see Chapter 4).

An example of how the historical positions actually plot in relation to a known shipwreck is shown with the case of SS *Apapa* in Figure 2.5. In this instance this well-known wreck (as seen in in its wider seabed context in Figure 1.4) lies to the west of all three of plotted positions in the GIS. The distances are shown and the nearest is the position derived from the U96 KTB, as shown in Figure 2.3. The source of the *Shipwreck Index* position is not known and the *War Losses* position plots 6.6nm from the wreck.

Therefore, as an initial guide to where the wrecks might be found, the distance from an historical position had to be considered to be several miles. For example, a 10 mile circle drawn around any shipwreck amounts to 314.2sqnm. In the study area, which has a high density of shipwrecks, this leads to numerous possible shipping losses. Therefore, the approach of using historical positions to pinpoint a wreck was on its own, not going to be enough to resolve an identity in cases in the study area where there are particularly high densities of shipwrecks.

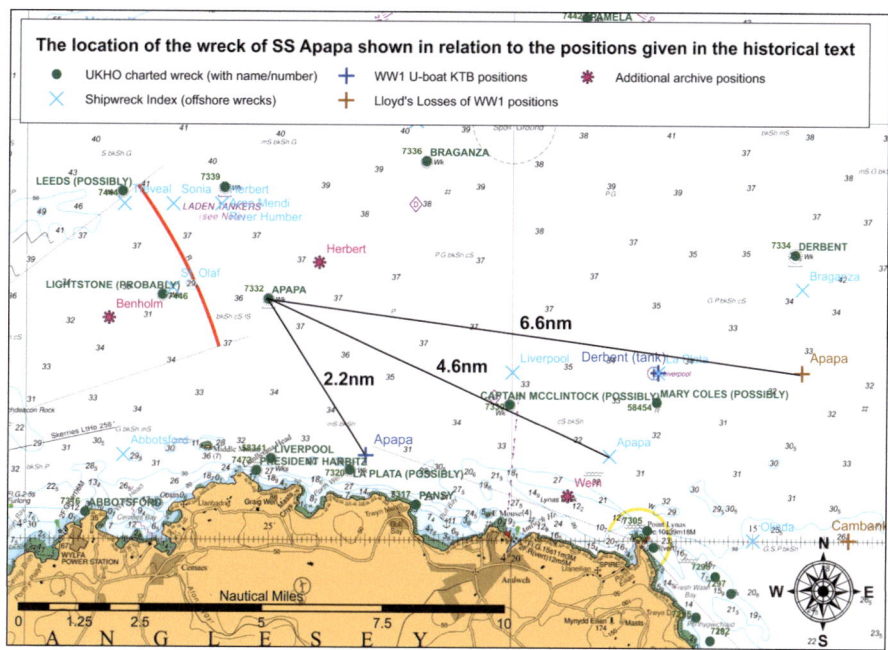

Figure 2.5. The wreck of SS *Apapa* shown in relation to the historical positions as plotted in the shapefiles. (Innes McCartney).

Technical and textual data from the historical text

More will be said of how the historical text was employed in Phase 1 and Phase 2 of the research in their relevant chapters. However, it should be noted here, that the use of the historical text during the research was not simply limited to the development of spatial databases. In reality it was necessary to consult a wide range of technical, textual, and descriptive sources of information on the losses themselves. These came mainly from primary sources, newspapers, and published shipping indices.

The wrecks surveyed during the fieldwork were processed to provide a "top down" 2D site plan of each wreck, the DTM. From this, its dimensions could be measured. This provided a key means of comparing each shipwreck to the historical text in a means which was entirely distinct from the spatial approach. Nearly every non-military vessel sunk in the study area over 100 tons displacement was found in *Lloyd's Register of Shipping*. It provides an irreplaceable resource for researching the technical aspects of shipwrecks.

Every shipwreck was searched for in the *Lloyd's Register of Shipping* for the year closest to when it sank. The dimensions of the vessel were recorded and converted into metric for easy comparison with the dimensions measured in the DTMs. Each relevant entry from the register was copied and placed into a digital file created for each shipwreck, which was also populated with other sources consulted.

The *Lloyd's Register of Shipping* contains a range of technical details about each ship, which aside from its basic dimensions can be useful when examining the DTMs. For example, the measurements which are often given on the lengths of various structures, such as the poop deck can be useful to compare with the DTM of a wreck if it is in a suitable condition to make a comparison. Sometimes even when a wreck is collapsed, knowing where the engine was situated was crucial in some cases.

In 2019 when the research began, The Lloyd's Register Foundation Heritage & Education Centre was in the process of digitising its vast archive of material on each ship it has registered as far back as its records go. At the time of writing, this resource was being steadily uploaded to its web portal and is rapidly becoming an indispensable resource for the study of historic ships (Lloyd's Register Foundation | online catalogue | Archive & Library, 2021). I am extremely grateful to the Lloyd's Register Foundation Heritage & Education Centre for allowing me to search its archive for several days seeking out ship plans while the digitisation project was going on around me. The ship plans have proven to be very useful in examining several shipwrecks, as described in Chapter 4 and in the datasheets in Appendix 2.

Finally, aside from positional and technical details of each shipwreck, it was also important to capture as much textual history as could be found which pertained specifically to the circumstances in which each vessel sank. This was because quite often the DTMs could provide information about how a vessel sank, which could be compared to the historical text. For example, the type of damage sustained during the sinking and its location could quite often be seen in the DTMs. A large hole in a certain hold on the port or starboard side of the ship could have been caused by a torpedo and could therefore be potentially matched to the historical text. So, it is very useful to capture sinking accounts where survivors describe what happened when a ship was attacked. Equally important was knowing the voyage a vessel was taking, because the heading of the ship could in many circumstances be easily discerned from the DTMs. If the ship rapidly sank, there is the possibility that the wreck is pointing in on the heading it was taking when sunk.

Textual accounts came from numerous sources, these included the Admiralty, *Shipping Casualties*, newspapers, and KTBs. Each was photographed and placed in the shipwreck's relevant digital folder, so that by the time the process of researching each DTM through both Phase 1 and Phase 2 began there was on hand detailed positional, technical, and textual information on each loss. In reality some cases needed further research, especially during Phase 2, but the foundations of the historical evidence needed to resolve the majority of cases was compiled prior to embarking on the research into each DTM.

Conclusions

As described in Chapter 1, the shipwrecks under study are of iron and steel construction which show up well on MBES scans. They are termed "offshore", being in depths of over 20m, which can be safely surveyed by Bangor University's School of Ocean Sciences survey ship. The research set out to develop a spatial means in GIS of databasing all of the recorded shipping losses in the archival study area which are of iron and steel and are offshore.

Due to the fact that the UK has no national database of historic shipping losses, it had to be built from scratch. This involved extracting data of a wide range of British-based sources such as (Larn & Larn 2000 & 2002), *War Losses*, *Wreck Returns*, *Shipping Casualties* and Admiralty records. It was established that 426 shipwrecks fell into the iron & steel offshore category. It was noted that ships sunk during both world wars make up a dominant 60% of these losses, mainly being from WW1.

In order to provide as much positional evidence as could be found for each recorded loss, the German navy records from WW1 were also incorporated into the GIS. Further ad hoc sources were introduced when they were found during the archival research. By the time the databasing phase was complete 1,009 datapoints had been mapped.

The research also aimed to capture as much textual data in the form of survivor reports, ship plans etc. which would form part of the identification of each wreck. Each recorded shipping loss had a computer file opened for it and into each of these were placed the records gathered in each case. These were aimed to provide a ready means of accessing them once the identification (Phase 2) began.

There is no doubt that the UK is well served archivally for this type of research. It was anticipated that most of the shipwrecks we had surveyed would show up in the historical text and that in most of those cases the historical text would yield enough detail in positional, dimensional, and textual terms to be able to match the shipwreck DTMs to these recorded sinking events, but exactly how many was not known. What was needed next was to establish the extent of the challenge by knowing exactly how many of the surveyed ships were of unknown name.

References

Admiralty, 1904. *Manual of Seamanship for boys and seamen of the Royal Navy 1904*. London: HMSO.

Admiralty, 1931. *Admiralty Manual of the Sperry Gyro Compass, 1931*. London: HMSO.

Admiralty, 1964. *Admiralty Manual of Navigation, Vol. 1*. London: HMSO.

Admiralty, 1988. *British Vessels Lost at Sea 1914-18 and 1939-45*. London: Patrick Stephens.

Board of Trade, Marine Department (various dates). *Return of Shipping Casualties and Loss of Life*. London: HMSO.

Cant, S., 2011. *England's Shipwreck Heritage: From Logboats to U-boats*. London: English Heritage.

Hepper, D., 2006. *British Warship Losses in the Ironclad Era, 1860-1919*. London: Chatham.

Hocking, C., 1969. *Dictionary of Disasters at Sea During the Age of Steam: Including sailing ships and ships of war lost in action 1824-1962. Vol I, A to L*. London: Lloyd's Register of Shipping.

Hocking, C., 1969. *Dictionary of Disasters at Sea During the Age of Steam: Including sailing ships and ships of war lost in action 1824-1962. Vol II, M to Z*. London: Lloyd's Register of Shipping.

Hooke, N., 1997. *Maritime Casualties 1963-1996 second edition*. London: LLP.

Huntress, K. A., 1979. *A Checklist of Shipwrecks and Disasters at Sea to 1860 with Summaries, Notes and Comments*. Ames: Iowa State University Press.

Larn, B. & Larn, R., 2000. *Shipwreck Index of the British Isles, Volume 5: West Coast & Wales*. Redhill: Lloyd's Register.

Larn, B. & Larn, R., 2002. *Shipwreck Index of Ireland: Volume 6 – All Ireland. Part of the Shipwreck Index of the British Isles series*. Redhill: Lloyd's Register.

Lloyd's of London, 1989. *Lloyd's War Losses: The Second World War Volume 1: British, Allied and Neutral Merchant Vessels Destroyed by War Causes*. London: Lloyd's of London Press.

Lloyd's of London, 1990. *Lloyd's War Losses: The First World War: Casualties to Shipping Through Enemy Causes 1914-1918*. London: Lloyd's of London Press.

Lloyd's of London, 1991. *Lloyd's War Losses: The Second World War Volume 2. Statistics Showing Monthly Losses of British, Allied and Neutral Merchant Vessels*. London: Lloyd's of London Press.

Lloyd's of London (Various dates). *Lloyd's List*. London: Lloyd's of London.

Lloyd's Register Foundation., 2021. Online Catalogue | Archive & Library | Heritage & Education Centre. [online] Available at: <https://hec.lrfoundation.org.uk/archive-library> [Accessed 22 November 2021].

Lloyd's Register of Shipping (various dates). *Lloyd's Register Wreck Returns*. London: Lloyd's Register of Shipping.

Lloyd's Register of Shipping (various dates). *Lloyd's Register of Shipping*. London: Lloyd's Register of Shipping.

McCartney, I., 2016. *Jutland 1916: The Archaeology of a Naval Battlefield*: London: Bloomsbury.

National Archives and Records Administration, 1984. *Guides to the Microfilmed Records of the German Navy, 1850-1945: No. 1. U-Boats and T-Boats 1914-1918*. Washington DC: NARA.

National Archives and Records Administration, 2008. *Guides to the Microfilmed Records of the German Navy, 1850-1945: No. 6. Selected Records of the Imperial German Navy Relating to World War I*. Washington DC: NARA.

Royal Museums Greenwich, 2021. *Research guide C8: The Merchant navy: WRECKS, losses, and casualties*. Royal Museums Greenwich: Available at: https://www.rmg.co.uk/collections/research-guides/research-guide-c8-merchant-navy-wrecks-losses-casualties [Accessed September 27, 2021].

Spindler, A., 1932. *Der Handelskrieg mit U-Booten*, Bd.1, *Borgeschichte*. Berlin: Verlag von Mittler & Sohn.

Spindler, A., 1933. *Der Handelskrieg mit U-Booten*, Bd.2, *Januar bis September 1915*. Berlin: Verlag von Mittler & Sohn.

Spindler, A., 1934. *Der Handelskrieg mit U-Booten*, Bd.3, *Oktober 1915 bis Januar 1917*. Berlin: Verlag von Mittler & Sohn.

Spindler, A., 1941. *Der Handelskrieg mit U-Booten*, Bd.4, *Januar bis Dezember 1917*. Berlin: Verlag von Mittler & Sohn.

Spindler, A., 1966. *Der Handelskrieg mit U-Booten*, Bd.5, *Januar bis November 1918*. Berlin: Verlag von Mittler & Sohn.

Tennent, A, J., 2006a. *British Merchant Ships sunk by U-Boats in the 1914-1918 War*. Penzance: Periscope.

The National Archives (various dates). *British Merchant Vessels sunk and captured by the enemy, August 1914 to February 1918. ADM 137/2959 to ADM 137/2964*. London.

The National Archives (various dates). *Survivors Reports Merchant Vessels*, Sept *1939 to May 1945. ADM 199/2130* to 199/2148. London.

3

The Research Phase 1: Establishing the extent of the unknown shipwrecks

Introduction

Every one of the 273 shipwrecks surveyed during the research is a known object. No prospection was undertaken, and no new wreck sites were found during the fieldwork. Each shipwreck we surveyed has a record which is kept and maintained by The UK Hydrographic Office (UKHO). The existence of the UKHO database made this project possible. It became an indispensable component of the research and as such it became imperative to understand what the database actually represents in practice and to compare our surveyed wrecks against it. After this had been done, 40% of the wrecks in the UKHO database were found to be legacy misattributions, inconsistent with the survey results and were reverted to UNKNOWN status, later to be subject to the identification process developed in Phase 2. This section describes the workings of the UKHO database and the process by which the survey results were compared to it and why the differences came to exist.

The UK Hydrographic Office shipwreck database – its history and purpose

Background

Based in Taunton, UKHO maintains a globally centralised database of shipwrecks and shipwreck data which is managed by its dedicated wrecks officer. The database informs the wrecks "layer" on various charting products available from UKHO. The UKHO has been a major source of seabed survey and charting for the world since 1795.

The central file of known shipwrecks came into existence in 1913 to produce data on ships sunk that may have become navigational hazards. This information was distributed via the well-established process of Fleet Notices, Route Instructions and Notices to Mariners already in place to promulgate navigational warnings and chart updates (Roy 2018a). The database, as it exists today still contains the traces of its records from this period, but now the means of distribution is mainly electronic.

During WW2 it was recognised that the heavy number of losses of ships required a wrecks officer to be employed solely on maintaining the records and promulgating details of shipwrecks. Initially wrecks were only charted when they presented a hazard

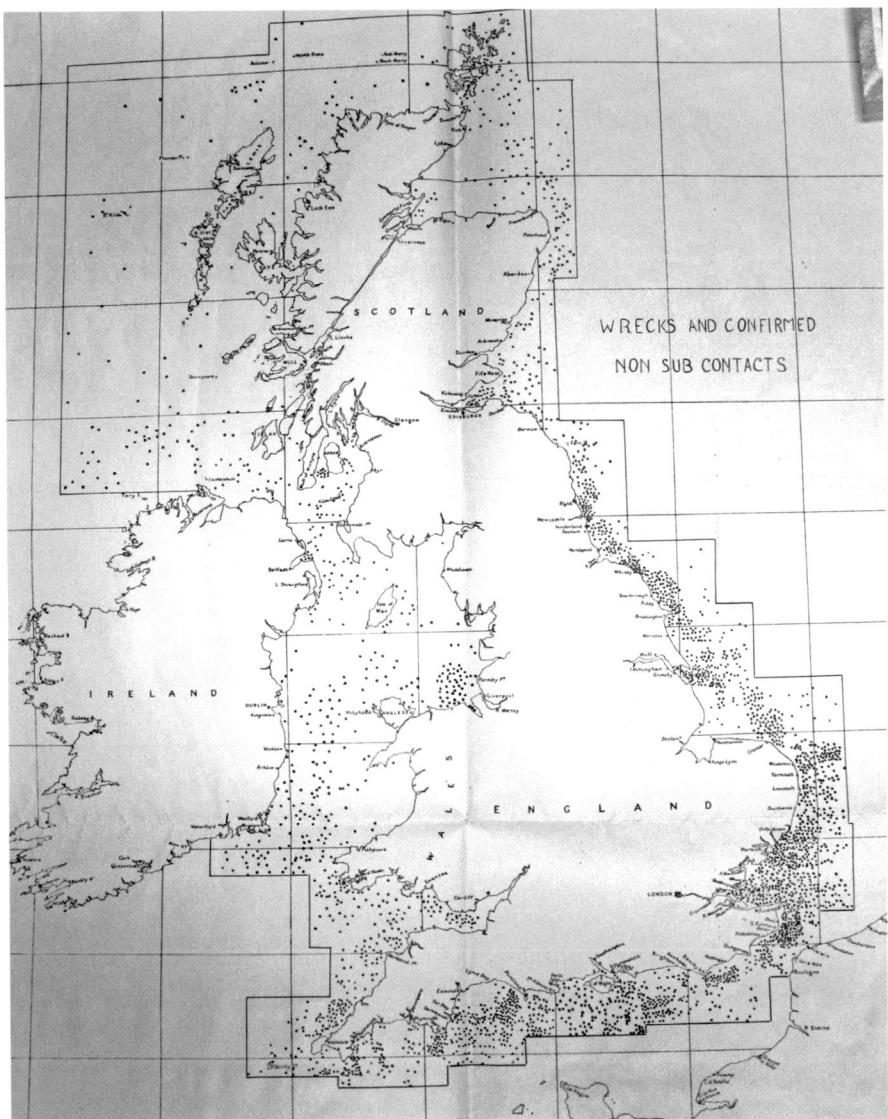

Figure 3.1. A summary chart from December 1944 showing the extent of the 3,200 known "non-sub" contacts in UK waters at that time (Admiralty).

to navigation. However, the Fleet requested data on non-dangerous wrecks because their locations were useful to know when hunting enemy submarines. This led to a series of 20 home waters charts being produced showing all known wreck and "non sub" sonar echoes in over three fathoms of water (Morris 1995, 113-114). The "non-sub" title meant any sonar contact determined to be fixed permanently to the seabed and therefore not an enemy submarine. In subsequent years, non-dangerous wrecks became a common feature on Admiralty charts and bathymetric features have largely been distinguished from wreck sites so that "non sub" is no longer a term in general use.

The dawn of the modern shipwreck database

From June 1944, the use of a radio-based navigational system, "QH" gave accurate and repeatable positional data to antisubmarine vessels at sea. With the combined use of "QH" and antisubmarine sonar (known as ASDIC) wrecks could be found and then relocated later. This ultimately led to the collation of a positionally accurate list (referred on UKHO wreck cards as the "Plymouth List") of all known "non-sub" contacts in UK waters and formed one of the key building blocks of the modern shipwreck database.

Research visits to the UKHO archive at Taunton have been unsuccessful in finding the original Plymouth List and despite wider searches, it is still unconfirmed whether a copy survives in the UK now. However, the extent to which the Plymouth List was developed is evidenced in a report issued by The Directorate of Naval Operational Studies in December 1944. It stated that there were 3,200 "non-sub" contacts in UK waters and around 2,800 of those were known to be shipwrecks, see Figure 3.1 (TNA ADM 219/166).

The process of maintaining the shipwreck files was carried out manually until 1976 when it was computerised using the Navy's Ingress system, from which the earliest fanfold hydrographic "printouts" became available to researchers. Although extensively modified over the years, Ingress remained the software platform for the database until 2014 when it was migrated to the Caris system (Roy 2018a).

The function and purpose of the shipwreck database

Since the 18th century, UKHO, has continued to survey and resurvey the waters of the UK Exclusive Economic Zone (EEZ) and elsewhere. Equally, the technologies employed in these surveys has continued to evolve over this period, from lead line to MBES so that the data available on the distribution and extent of the shipwrecks in UK waters, is constantly improving. Periodic issues of the chart *Q6090 United Kingdom British Isles hydrographic survey status* show the current state of survey in UK waters.

The wrecks officer takes in shipwreck information from both formal sources, such as its hydrographic surveys, and informal ones, such as telephone conversations with divers. Each shipwreck is given a unique number (the UKHO number), issued sequentially and it is the means by which each wreck can be clearly differentiated. Each record has its own "wreck card" which contains the history of data compiled on that record and the current state of survey. On 21 November 2018, the number of records in the entire database had reached 90277 (Roy 2018a). All of the UKHO LIVE, DEAD and LIFTED wreck cards are publicly available on the internet at www.wrecksite.eu (2021).

It is the duty of the wrecks officer to determine which wrecks are displayed on Admiralty charts. Importantly, all shipwrecks verified by survey are displayed. Moreover, reported wrecks, which are considered navigationally significant, would also be charted. All wrecks are flagged as "LIVE" in the database until they are disproved by survey or salvaged. At this point, they are declared either "DEAD" if disproved or "LIFTED" if they have been salvaged (Roy 2018b). The wreck surveys we carried out were on "LIVE" proven wreck sites only. Please note the capitalisation of the terms used by UKHO, such as "UNKNOWN" has been retained in the text from this chapter onwards, to make it easier for the reader to spot their relevance and context.

Today the UKHO database for the entire UK Exclusive Economic Zone (EEZ) records 9,845 shipwrecks, as depicted in Figure 3.2. This is a far greater number than were recorded during WW2 (the 3,200 "non sub" sites on the Plymouth List shown in Figure 3.1)

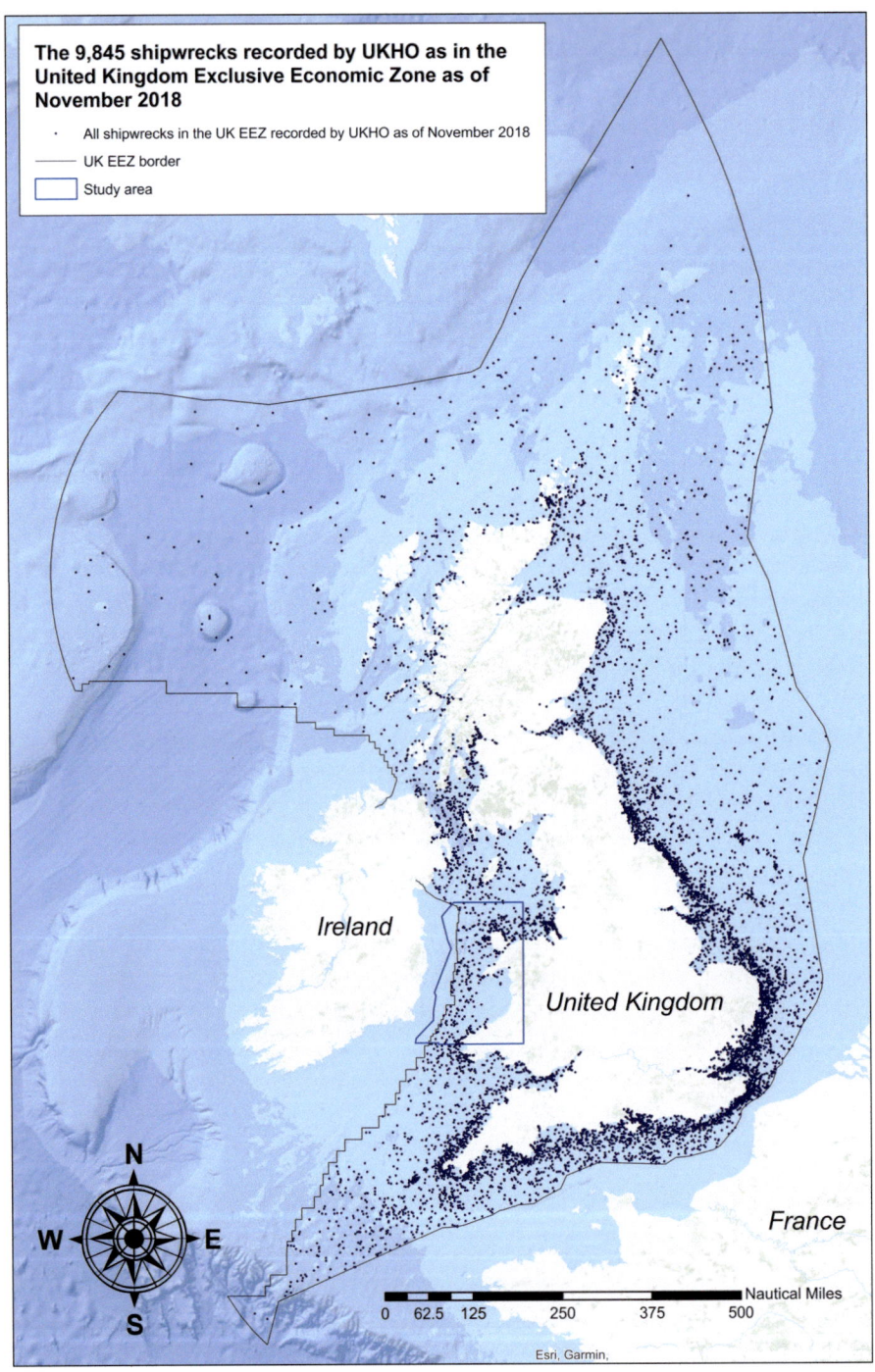

The 9,845 shipwrecks recorded by UKHO as in the
United Kingdom Exclusive Economic Zone as of
November 2018

·   All shipwrecks in the UK EEZ recorded by UKHO as of November 2018

—— UK EEZ border

☐ Study area

Figure 3.2. The distribution of the 9,845 shipwrecks in the UKHO database in the UK
EEZ as of November 2018, showing the research study area (Innes McCartney. Contains
public sector information, licenced under Open Government Licence v3.0 from UKHO).

and shows the extent to which the database has developed. The reason for the notable increase in the numbers of extant wrecks found is mainly due to the employment of improving survey technologies since around 1980, when side-scan sonar came into use on Royal Navy survey ships. The decreasing annual numbers of shipping losses since WW2 have made little contribution to this increase. The reality is that the wrecks found since WW2 are mainly the remains of iron and steel hulled ships, sunk after 1830.

The value of the UKHO database to archaeology

The value of the UKHO wrecks database to archaeological studies cannot be overstated. It is the most comprehensive resource available on UK shipwreck sites and what is physically known of them. Crucially the added benefit beyond the surveying details amassed on each site through time, is the additional information on each wreck site that has come from other sea users, mainly divers and salvage companies.

In my own working experience with UKHO since the late 1980s, sharing my diving information from sites previously unseen in return for details of new wreck sites to go and dive was a transaction which benefitted both parties. I could explore wrecks being better prepared for what I would find, and UKHO would benefit from my findings. This was invariably the same relationship UKHO had with many other divers and sea users, benefitting all concerned. Over more than three decades, it became evident to me and to many other divers and dive boat skippers that quite often shipwrecks with given names had been misidentified in the past. This only came to light when divers had examined and identified them. Therefore, the reporting by divers of their discoveries enhanced the UKHO database over time and eliminated previous misattributions.

Therefore, it was a surprise when in 2014, UKHO stopped taking in diver information. Although this decision was reversed sometime later, the motivation for divers to report findings to UKHO may not be what it was, because the wreck cards are all now publicly accessible online at wrecksite.eu.

UKHO categorisation of shipwreck identities

It is important to remember that the actual identity and type of vessel any shipwreck may actually be is not the database's core purpose. This is to record the properties of an obstruction on the seabed at a known position for charting and safe navigation. The default name for any shipwreck is "UNKNOWN", unless there is evidence which will give the wreck a name. Today the wrecks officer does not attempt to reconcile the historical text on shipping losses to known shipwrecks, but a legacy of this activity in past remains and it is problematic to archaeology.

UKHO legacy misattributions

When one studies the wreck cards, there is evidence that some attempt to reconcile history and text was attempted, albeit many years ago. For example, at the time the UKHO shipwreck database was incorporated into the Ingress system, a significant number of wreck positions were plotted from data based solely on the historical text (they appear to have been mainly derived from Admiralty sources), but these have now largely been declared "DEAD" in favour of better "LIVE" data in the form of known extant shipwrecks located by geophysical survey (Roy 2018b).

Importantly though, sometimes when this occurred, the newly found wreck, instead of being entered as an "UNKNOWN" would take on the name of a nearby historical wreck position which was originally plotted based on the historical text. This name would remain unless overturned by better evidence later. This process although abandoned many years ago, has left the problematic legacy issue for archaeology, because many of the attributions are incorrect. They have remained so in every case where recreational divers have not subsequently corrected them. In reality the legacy misattributions obscure the real identities of many shipwrecks in the UKHO database and have given a false impression of just how many of the UK's shipwrecks have actually been identified.

UKHO grading criteria

Another unique feature of the UKHO shipwreck database is the grading of the validity of the actual name of the shipwrecks. Some ship's names in the database are listed either with a "(PROBABLY)" or "(POSSIBLY)" suffix after them, showing that some grading of the accuracy of wreck identities routinely takes place when a wreck is reported. In my own experience, I would occasionally be asked about the accuracy of wreck identities I discussed with the wrecks officer and whether they should be noted just by name (a CONFIRMED wreck) or with the "(POSSIBLY)" or "(PROBABLY)" suffix. This was particularly important with some WW2-era U-boat wrecks which will always prove nigh on impossible to accurately identify.

Therefore, as a general rule if a wreck is seen in the UKHO database without a suffix, it is considered to be an accurate CONFIRMED attribution, not requiring the qualification of a "(POSSIBLY)" or "(PROBABLY)" suffix. However, it is not known whether this was consistently applied across the entire database, or even when the suffix nomenclature was first introduced. The suffixes (or lack of them) are helpful in understanding what UKHO thought of the accuracy of the shipwreck's name in some cases, but they should be treated carefully by archaeologists because it seems they have been applied inconsistently in the past. In some cases, the wreck card gives an indication as to why the grading was made, but this is not always the case. Many cases will relate to the issue of legacy misattributions.

Figure 3.3 shows the 273 surveyed wrecks which feature in this research. They are all listed in the UKHO database specifically as shipwrecks. There are 101 UNKNOWN, 99 (POSSIBLY) and (PROBABLY) and 73 wrecks with CONFIRMED identities. There is a high concentration of CONFIRMED cases to the east of the 50m depth contour line, especially in the north. This undoubtedly due the activities of recreational divers and their reporting of discoveries to UKHO in the past. This has made a significant difference to the accurate identification of wreck sites in shallower waters. The CONFIRMED wrecks in deeper waters are mainly made up of sites commercially salvaged, sites discovered by salvage companies, recent shipwrecks and ships sunk when the identity and location of the wreck was accurately fixed at the time.

The UKHO wreck cards

Table 3.1 shows an example of a UKHO wreck card. This wreck is UKHO 7347, which lies within the study area, northwest of Anglesey. It serves to illustrate the types of data captured by UKHO and some of the considerations archaeologists should use when reading the cards. The first sections of the card are made up of the basic information about the wreck site, giving the number, name, position, depth, dimensions, and sinking circumstances etc. In this case, the identity of the wreck is given as MANCHESTER without a "(POSSIBLY)" or "(PROBABLY)" suffix, meaning the wreck is considered CONFIRMED. For the purposes of archaeological evaluation and historical research, it is the information

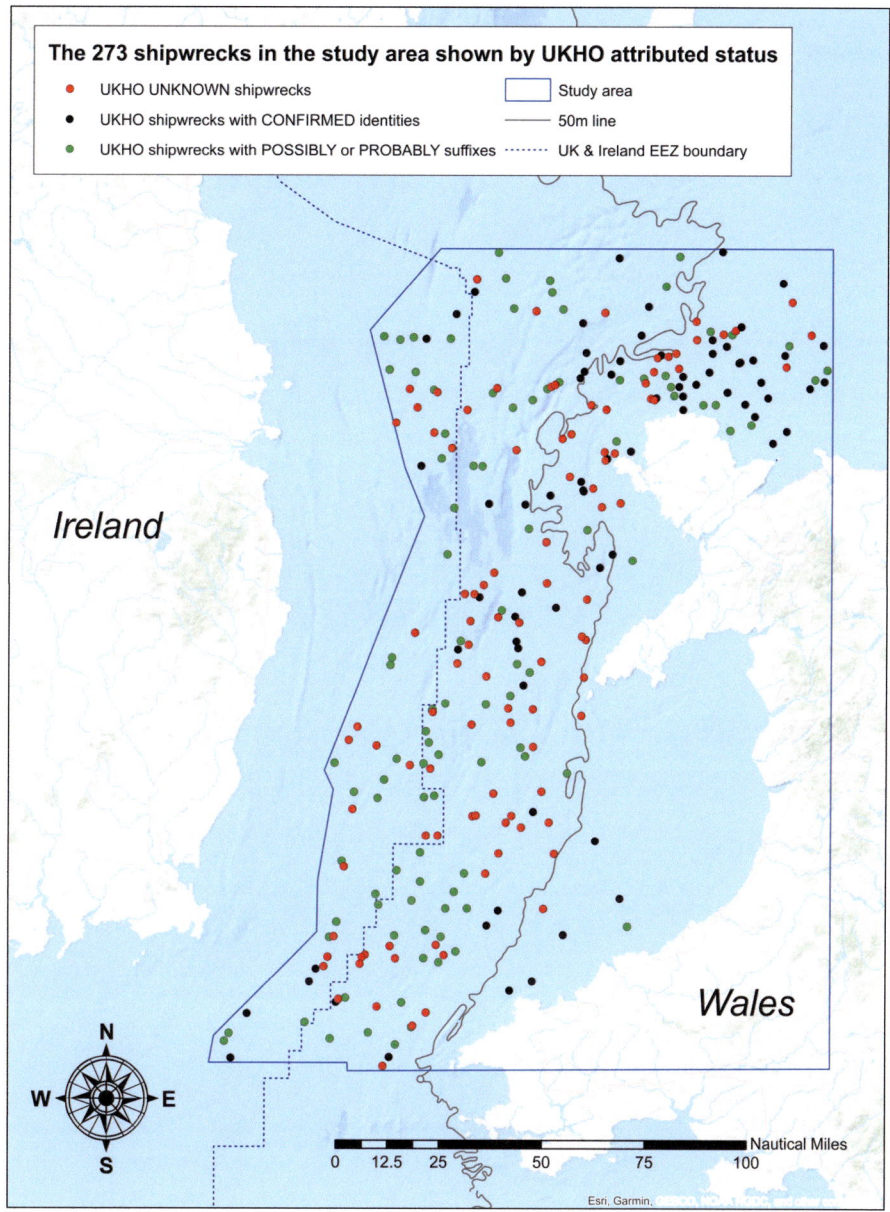

Figure 3.3. The 273 shipwrecks listed by UKHO in the study area shown by the attributed status of the wrecks (Innes McCartney. Contains public sector information, licenced under Open Government Licence v3.0 from UKHO).

shown from the name of the vessel downwards, and in particular, the "Surveying Details" section, which can very often be of unique value.

The first entry in "Surveying Details" shows that the wreck was originally located in 1976, during which time it was surveyed by echo sounder (shown as "E/S") and measured as being 93m long and 11m high. This is when the wreck was inserted on Admiralty charts as "WK 56m". The next report came via a letter in November 2004 which gives some

Wreck Number:	7347		State:	LIVE	Classification:	Unclassified
Chart Symbol:	WK 54.0				Status:	--
Existence Doubtful:	--				Reported Year:	--
Date Last Amended:	3/11/2014				Charting Comments:	--
Wreck Category:	Non-dangerous wreck					
WGS84 Position:	Latitude:		53°30,388'N		Longitude:	04°50,027'W
WGS84 Origin:	Original				WGS84 limits	--
Previous Position:	Latitude:		53°30,35'N		Longitude:	04°49,967'W
Position Accuracy:	3 m				Horizontal Datum:	ETRS 1989
Position Last Amended:	3/11/2014				Position Quality:	Surveyed
Position Method:	Differential GPS				Limits:	--
Depth:	54 m				Water Depth:	69 m
Depth Method:	Found by multi-beam				Depth Quality:	Least depth known
Height:	--				Drying Height:	--
Vertical Datum:	Approximate lowest astronomical tide					
Water Level Effect:	Always under water/ submerged				Bottom Texture:	Sand
Sonar Signal Strength:	Strong					
Original Sensor:	Acoustic Sensor				Last Sensor:	Acoustic Sensor
Conspic Visual:	NO				Conspic Radar:	NO
Non-Sub Contact:	--				Contact Description:	Entire wreck
Name:	MANCHESTER					
Type:	STEAM SHIP				Flag:	--
Dimensions (m):	LxBxD = 64,6x9,1x4,9				Tonnage:	915 Gross
Cargo:	GENERAL AND MIXED GOODS				Date Sunk:	11/30/1905
Original Detection Year:	1977				Last Detection Year:	2013
Original Source:	Survey Vessel				Last Source:	Divers
Sonar Dimensions:	LxWxH = 90,8x11,4x14,7				Orientation:	150°
Magnetic Anomaly:	Strong				Debris Field:	NIL
Scour Depth:	0 m				Scour Length:	--
Scour Orientation:	--				Markers:	--
General Comments:	UPRIGHT, INTACT, BOWS SE, CLEAR DAMAGE MIDSHIPS STBD					
Circumstances of Loss:						
	EX-GLANWERN 1882. BUILT IN 1892 BY W DOXFORD & SONS, SUNDERLAND. OWNED AT TIME OF LOSS BY WINGA & CO,					
	CHRISTIANA (ACTIES, OF MANCHESTER, MANAGERS). ONE BOILER, COMPOUND EXPANSION ENGINE OF 110NHP,					
	SINGLE SHAFT. PASSAGE PROSGRUND, NORWAY FOR MANCHESTER. SANK FOLLOWING COLLISON WITH SS FREDA.					
Surveying Details:						
*	24.1.77 H4827/75 NEW WRECK LOCATED 10.11.76 IN 533021N, 0044958W [OGB] USING HIFIX/6 [2 LOP]. LEAST E/S					
	DEPTH 56.6MTRS IN GEN DEPTH 65-70MTRS. DCS3 HT 11MTRS. LENGTH 93MTRS. LYING NW/SE. SCOUR W FROM WK. (HMS FOX, HI 54/76). INS AS WK 56MTRS. BR STD.					
**	12.11.04 H100/351/21 12.11.04 WK IS UPRIGHT IN GEN DEPTH 55MTRS. LENGTH 70MTRS, WIDTH 10MTRS.					
	STANDS ABOUT 8MTRS HIGH. HAS BEEN IDENTIFIED BY THE BELL INSCRIBED 'GLANWERN 1882', RECOVERED IN 2002. (A CORKILL, VIA D STEAD, LTR DTD 11.04). NCA.					
***	11.3.14 EXAM'D 6.11.13 IN 5930.388N, 0450.027W [WGD]. LEAST M/B DEPTH 54.30 IN GEN DEPTH 69MTRS. NO SCOUR. LENGTH 90.8MTRS (SIC), WIDTH 11.4MTRS, HT 14.7MTRS. LIES 150/330 DEGS. STRONG MAGNETIC ANOMALY. INTACT, SLIGHT LIST TO STBD. LARGE HOLE MIDSHIPS. (NETSURVEY, HI 1420). BOWS SE. AMEND WK 54MTRS IN REVISED POSN.					

Table 3.1. The wreck card for UKHO 7347 (Contains public sector information, licenced under Open Government Licence v3.0 from UKHO).

dimensional details, stating the wreck is only 70m long and 8m high. It also states that the wreck was identified by the recovery of its bell in 2002. It is probably because of this letter, that the wreck is named without a suffix, because identification by the recovery of the ship's bell is considered an ultimate validation of any shipwreck identity. Finally in 2013 the site was resurveyed using multibeam (shown as "M/B") and it gives dimensions which are closer to those of 1976 and gives additional information about the state of the wreck.

UKHO legacy misattributions impact on archaeological research

Helpfully the UKHO 7347 wreck card gives the actual dimensions of the wreck, as measured by MBES as 90.8m long. It also gives the dimensions of the ship *Manchester*, derived from the *Lloyd's Register of Shipping*. SS *Manchester* was only 64.6m long. This inconsistency would not be of particular concern for charting purposes, but to an archaeologist it raises the question as to the wreck's supposed identity.

In fact, when this wreck site was surveyed as part of the research it was measured on its DTM as being 96m long. This difference in length is explained simply by convention. Archaeology is interested in the dimensions of the actual wreck; UKHO would be more interested in the dimensions of the obstruction. Since around 6m of the wreck has collapsed to the seabed it was not measured. The differences in length are shown in Figure 3.4.

The difference in actual wreck dimensions and the UKHO recorded dimensions of the same object was noted on several sites during the research and in some cases the differences can be substantial. For example, where the wreck is sitting in a large sand ridge or where only a small portion of the wreck is still extant. In the case of UKHO 7347, the lesson to be derived is that the survey details given on any UKHO wreck card may not necessarily coalesce with the dimensions of the archaeological object. They are more likely to represent the dimensions of the seabed obstruction in which the wreck may or may not predominate.

Moreover, and crucially in this particular case, it must be concluded that UKHO 7347 cannot be SS *Manchester*, because it is dimensionally far too large (being approximately 31.4m too long) to be that ship. The wreck at the location of UKHO 7347 must for the purposes of archaeological research be treated as UNKNOWN. The diver attribution must have come from another wreck and for reasons impossible to establish, became associated with this site, when clearly another shipwreck yielded the bell. See Appendix 2 for the full datasheet of this wreck site, showing all analysis through to the conclusion of Phase 2.

As the data from Bangor University's School of Ocean Sciences came in and was processed into DTMs and their dimensions measured, the discrepancy between the actual dimensions of the wrecks and the proposed identities on the UKHO wreck cards was commonly observed. As my 30 years of experience diving wrecks had made me suspect, legacy misattributions are found throughout the database and remain problematic. In each case the wreck had to be flagged in the research as UNKNOWN.

Aside from dimensional inconsistences such as that seen in UKHO 7347, legacy misattributions were seen in other instances. It was occasionally noted that the identity given by UKHO could not be sustained when newer research suggested a differing fate for the ship. For example, UKHO 7335 GLENCONA (PROBABLY) came under question when it was ascertained from *The Shipwreck Index of the British Isles* (Volume 5, (EB) 28/06/1928) that it was known for certain that *Glencona* had in fact drifted into Morecambe Bay and sunk, 60nm from the wreck noted in UKHO 7335. In Morecambe Bay, the wreck had been

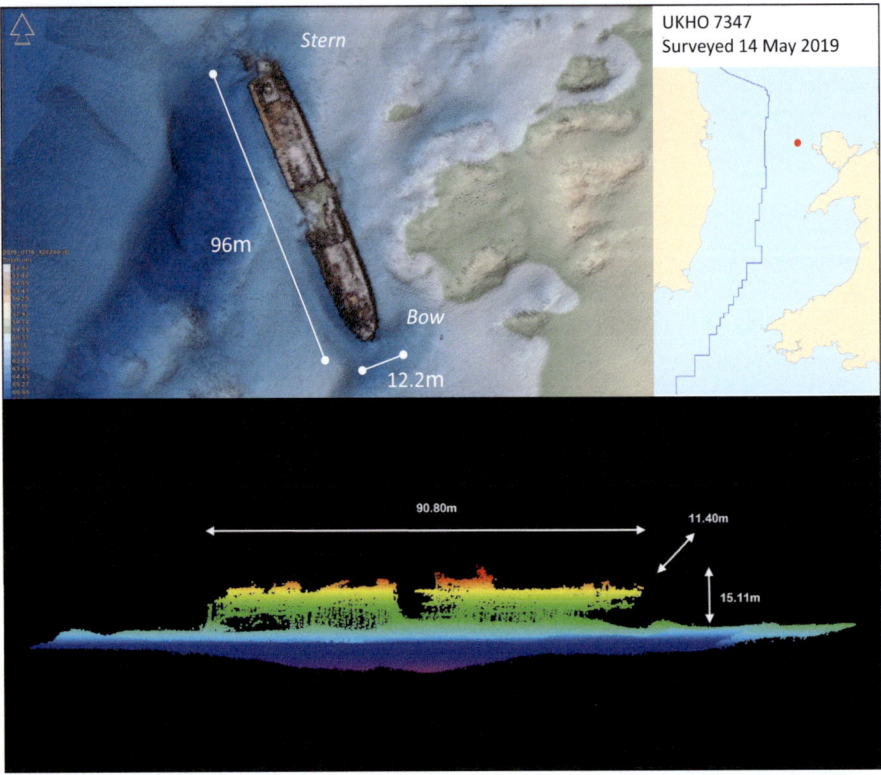

Figure 3.4. Above, the DTM of UKHO 7347 as surveyed by Bangor University's School of Ocean Sciences in May 2019. The stern appears to have collapsed to the seabed, but is present, making the overall length of the wreck approximately 96m. Compare this with the UKHO measurements shown below. In this instance it is the extant portion of the wreck which has been measured, accounting for the difference in lengths (Above: Innes McCartney, Below: UKHO which contains public sector information, licenced under Open Government Licence v3.0 from UKHO).

identified by its physical remains and then dispersed as a navigational hazard. This site too had to be flagged as UNKNOWN, see Appendix 2.

Occasionally too, it was noted on the DTMs that the wrecks manifested features which ruled out the UKHO attribution. For example, UKHO 9955 FRANCOISE DE LILLE (POSSIBLY) came under question because the DTM clearly showed the features of a large (68m long) twin-boiler steamship which has collapsed to the seabed, whereas *Francoise De Lille* was a diesel powered MFV which sunk in 1963. Therefore, the DTM in this case caused the site to be flagged as UNKNOWN, see Appendix 2.

A further five cases listed as CONFIRMED (without the (PROBABLY) or (POSSIBLY) suffix) could not be verified as actually being shipwrecks with CONFIRMED status. One example would be UKHO 9865 CYMRIAN. There is no evidence given on the wreck card to state why a CONFIRMED identity is given. In this case the identity was not changed, but it was flagged as being "(MBES/Archive)", a single definition which replaces both the (POSSIBLY) and (PROBABLY) attributions in all unconfirmed shipwreck identities in this research. In cases such as this, the dimensional, positional, and archival evidence were checked to ensure that the attribution was plausible to the standard of an (MBES/Archive) attribution. If it was, then it was flagged as such. See Chapter 4 for a full definition of the (MBES/Archive) attribution which was devised to create a repeatable process by which all wreck sites could be evaluated and potentially identified to a consistent, repeatable standard.

Research Phase 1: Establishing an accurate list of the UNKNOWN wrecks

As initially intended, it was the aim of the research to use the historical text in conjunction with the survey data to identify the UNKNOWN wrecks in the study area. In order to do this, it was essential to understand which wrecks had been identified in the past and which were UNKNOWN, once legacy misattributions had been eliminated. To do this meant assessing each wreck site individually against its UKHO wreck card. This process was named Phase 1 of the research.

During this phase, a datasheet for each wreck was created and its measured DTM inserted. Comparisons were then made with the UKHO wreck card and any differences which led to a named shipwreck (CONFIRMED, POSSIBLY or PROBABLY) being reverted to another status were noted in the relevant entry field. The criteria used to make any changes to the UKHO name were based mainly on the previous examples given above, with two additions to account for objects which were either not shipwrecks or not identifiable from their remains. The six criteria were:

1. The dimensions of the wreck being inconsistent with the named shipwreck in the UKHO attribution, as given in the *Lloyd's Register of Shipping*, leading to an amendment to UNKNOWN, as seen in the case of UKHO 7347 MANCHESTER,
2. Research into the sinking event ruled out the UKHO attribution, leading to an amendment to UNKNOWN, as seen in the case of UKHO 7336 GLENCONA (PROBABLY),
3. The specific features of the wreck as seen in the survey data ruled out the UKHO attribution, leading to an amendment to UNKNOWN, as seen in the case of UKHO 9955 FRANCOISE DE LILLE (POSSIBLY),
4. The CONFIRMED status given by UKHO could not be verified, but at the same time, the dimensions, position, and archival evidence were consistent with an (MBES/Archive) attribution, as seen in the case of UKHO 9865 CYMRIAN. These cases were denoted as (MBES/Archive), as described later in Chapter 4,
5. Non shipwreck, i.e., where the survey data revealed the object to not be a shipwreck, denoted as NON SHIPWRECK,
6. Unidentifiable, i.e., where the object might be manmade but cannot be identified from the data. For example, this would include piles of stones, which may have once been ballast from a long disintegrated vessel. These were denoted as UNIDENTIFIABLE.

Phase 1 results and the extent of the legacy challenge

When each of the 273 shipwrecks in the study area was processed and assessed against its UKHO wreck card as described, a total of 110 shipwreck sites required amendments to be made. These were based on the six criteria given above. The results are shown in Table 3.2 and in Figure 3.5. For the individual outcomes for each specific shipwreck, see Appendix 1.

By far the most common amendment needed was caused by the dimensions of the shipwreck being inconsistent with the historical dimensions of the named vessel on

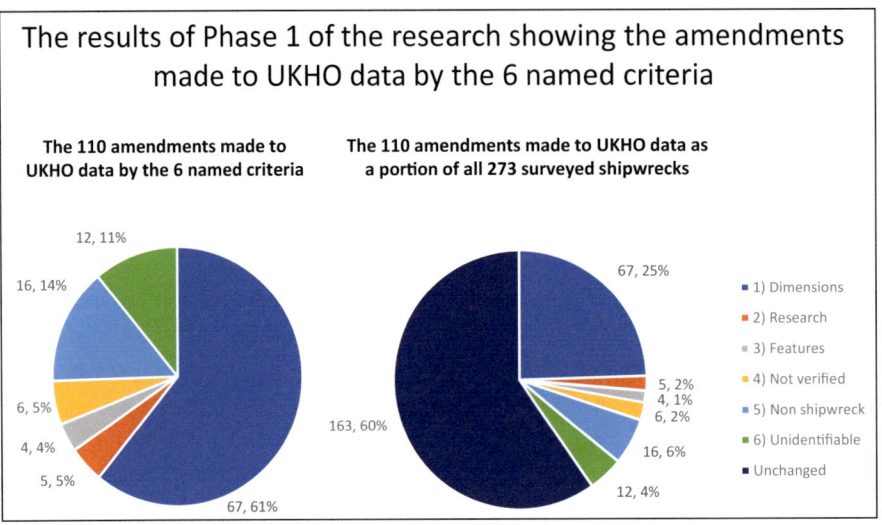

Figure 3.5. The 110 amendments made to the UKHO wreck cards shown by the six criteria used to segment the results. The impact of the results on the full 273 wreck dataset is shown on the right (Innes McCartney).

Criteria	Number of amendments made	Percentages of the 110 amendments	Percentage of all 273 wrecks
1) Dimensions	67	61%	25%
2) Research	5	5%	2%
3) Features	4	4%	1%
4) Not verified	6	5%	2%
5) Non shipwreck	16	15%	6%
6) Unidentifiable	12	11%	4%
Total	110	100%	40%

Table 3.2. The 110 amendments made to the UKHO wreck cards shown by the 6 criteria used to segment the results. The results are shown also as percentages of the full 273 wreck dataset (Innes McCartney).

the UKHO wreck card. All historical ship dimensions were compiled into a list derived from the relevant annual copy of the *Lloyd's Register of Shipping*. In all, 67 shipwrecks were reverted to UNKNOWN because dimensionally they simply could not be the ship named on the wreck card because they were either too large or too small.

These 67 shipwrecks represented a quarter of the 273 in the entire study area, reflecting the still widespread nature of legacy misattributions. The other five criteria made less of an impact but were prevalent in at least four cases. Interestingly 16 wreck sites turned out to be NON SHIPWRECK. A further 12 were UNIDENTIFIABLE. Research overturned five attributions, features another four, and being unable to verify the UKHO CONFIRMED status from known evidence accounted for the final six amendments. In total, 40% of all 273 wrecks examined had to be amended.

Once the 110 amendments were made, the extent to which they affected the total of the wrecks which were CONFIRMED, unconfirmed or UNKNOWN was significant and is

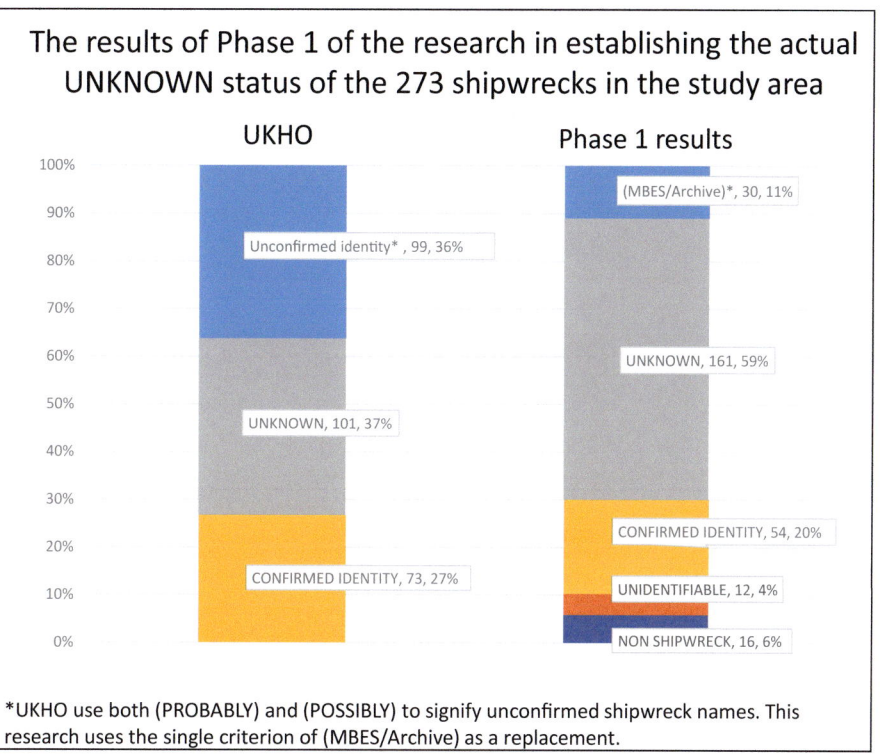

The results of Phase 1 of the research in establishing the actual UNKNOWN status of the 273 shipwrecks in the study area

*UKHO use both (PROBABLY) and (POSSIBLY) to signify unconfirmed shipwreck names. This research uses the single criterion of (MBES/Archive) as a replacement.

Figure 3.6. The attribution changes made during Phase 1 compared to the original UKHO wreck cards shown by number and percentage. The number of UNKNOWN sites grew by 60 to 161 of the overall total of 273 wrecks, accounting for the majority (Innes McCartney).

Status of Shipwreck	UKHO Database number	UKHO Database percentage	Phase 1 results number	Phase 1 results percentage
CONFIRMED IDENTITY	73	27%	54	20%
Unconfirmed identity*	99	36%	30	11%
UNKNOWN	101	37%	161	59%
UNIDENTIFIABLE	0	0%	12	4%
NON SHIPWRECK	0	0%	16	6%
Total	273	100%	273	100%

Table 3.3. The attribution changes made during Phase 1 compared to the original UKHO wreck cards shown by number and percentage. The number of UNKNOWN sites grew by 22% of the overall total of 273 wrecks, finally accounting for 59% of them. *UKHO use both (PROBABLY) and (POSSIBLY) to signify unconfirmed shipwreck names. This research uses the single convention of (MBES/Archive). See Chapter 4 for full description. (Innes McCartney).

shown in Table 3.3 and in Figure 3.6. The number of sites which were CONFIRMED fell by 19. Unconfirmed cases fell by 69, (25% of the 273 wrecks), UNIDENTIFIABLE and NON SHIPWRECK increased by 12 and 16 respectively, where there were of course none beforehand.

The impact and significance of this portion of the research lies in the increase in sites which are in fact UNKNOWN and were previously obscured as such by legacy

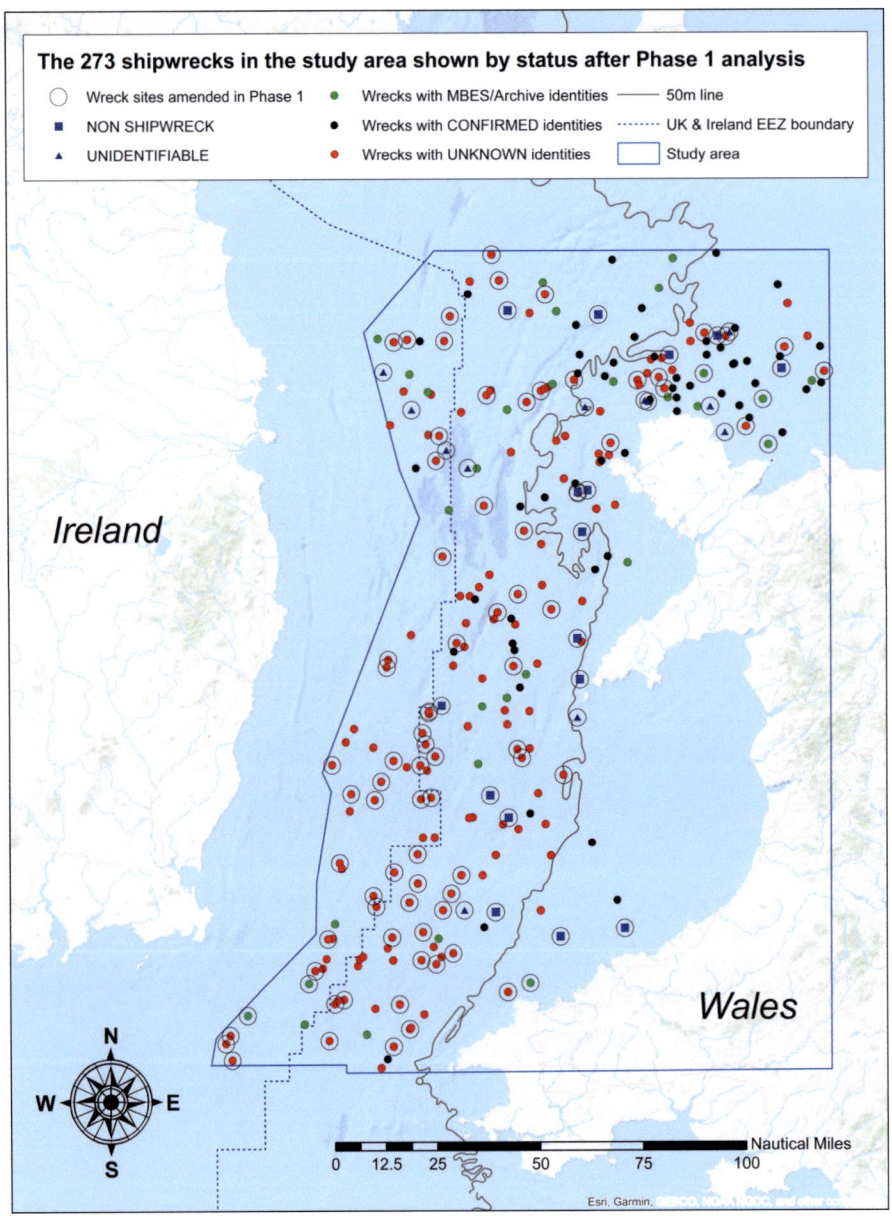

Figure 3.7. The attribution status amendments made during Phase 1 compared to the original UKHO wreck cards. The 110 amendments made are circled (Innes McCartney. Contains public sector information, licenced under Open Government Licence v3.0 from UKHO).

misattributions. They grew from 101 to 161 cases, an increase of 22% of the overall wrecks. Importantly this meant that 59% of the shipwrecks in the study area were actually UNKNOWN. At the beginning of this exercise, the UKHO database suggested that the figure was only 37% of the total. This is the extent to which UKHO legacy misattributions have continued to obscure the actual number of UNKNOWN wrecks. It has, unintentionally, given an optimistic impression. This was of little surprise to me, nor would it be to many other divers who have explored shipwrecks from the 1970s onwards. However, it significantly increased the workload of Phase 2, as an attempt had to be made to reconcile each of these cases to the historical text, as described in Chapter 4.

The results of the Phase 1 analysis are shown spatially in Figure 3.7. The wreck sites which had their status changed are circled for ease of comparison to Figure 3.3. It will be noted that the largest cluster of amended sites lies in the central and southern portion of the study area. These are in the most remote and deepest areas of the study. Conversely, the 54 CONFIRMED sites remain concentrated inside the 50m depth contour line, showing the impact recreational divers have had on the accurate identification of shipwrecks. Where the proximity of shallower wreck sites coincides with higher human population densities, a higher preponderance of identified wrecks sites would be expected, as seen here.

Phase 1 conclusions

Undoubtedly the key finding to emerge from Phase 1 was the actual extent of the legacy misattributions and its impact on the consequently increased number of UNKNOWN wreck sites, they caused once they had been expunged from the data. Whereas it was originally presumed to be 37% of the wrecks, it is in reality 59%. The conclusion to be drawn is that we know far less about the identities of the UK's shipwrecks than the UKHO database may lead us to believe. There is no reason why this legacy issue would not impact the entire 9,845 shipwrecks in the UK EEZ in exactly the same way.

It will be recalled from Chapter 1 that for safe navigation of *Prince Madog*, shipwrecks from 21m and deeper depth only were surveyed. Removing shipwrecks in 20m or shallower from the UKHO database for the UK EEZ reveals that there are 6,239 offshore shipwrecks recorded in the UK EEZ. This would have been the number of shipwrecks surveyed had the study area been extended to its fullest possible extent. By extrapolation therefore, 59% of those wrecks could also be UNKNOWN. This means that 3,681 of the offshore wrecks in the UK EEZ fall into that category. Therefore, up to 1,373 of them can be expected to have legacy misattributions, obscuring their real identities. The UK's shipwrecks cannot be properly inventorised until these are all corrected.

The conclusion drawn when looking at Figure 3.7 is that the number of changes which Phase 1 analysis would have made, had it not been for the impact of diver identifications of shipwrecks. It would have been much higher. In effect, by identifying a wreck, the diver will have in most cases already caused the site to have been amended, so that no change would be needed during Phase 1 analysis. Therefore, what the large cluster of UNKNOWN wreck sites in the southern region of the study area represents is probably what the entire UK EEZ would look like if divers had not reported their discoveries to UKHO over the past decades.

To extend the analogy of the undersea landscape being better understood by the use of MBES surveys in a comparable role to that of aerial photography on land, it is the recreational diver who is comparable to the contribution made by metal detectorists and amateur archaeology societies. By responsibly enjoying their pastimes, they have brought clarity where it would not be expected to exist otherwise. Until now there has been an absence of professional archaeological fieldwork on any of the 273 wreck sites in the survey data. UKHO and amateur divers have created all we know.

It is also important note that the UKHO database has been incorporated to varying degrees into the databases of the UK's heritage agencies, such as its acquisition by English

Heritage in 1992. As a result, the legacy misattributions seemingly found their way, unfiltered into the national heritage record and there is currently no means of correcting it (or even knowing which sites it affects) unless each site is surveyed with high resolution MBES and subjected to Phase 1 analysis.

Finally, and importantly, I would like to reiterate at this point that there is no expressed or implied criticism of UKHO in these results. They simply and only reflect the different requirements of hydrographic data collection for charting and navigational purposes and the archaeological requirement of this research. That a legacy issue about shipwreck names from the 1970's exists would seemingly be of little consequence outside archaeology.

Crucially, the very fact that UKHO has been responsible for finding, cataloguing, and assiduously collecting data from all sources on each of the wrecks for over 100 years makes it a priceless resource when studying the UK's shipwrecks. Its absence would have made this research impossible. The object lesson for users its database is to firstly understand its history, purpose, and processes before exploiting its unique, inestimable value.

References

Morris, R. O., 1995. *Charts and Surveys in Peace and War: The History of the RN Hydrographic Service* 1919-1970. London: HMSO.

Roy, A., 2018a. Email to the author. 21 November 2018.

Roy, A., 2018b. Email to the author. 5 December 2018.

Larn, B. & Larn, R., 2000. *Shipwreck Index of the British Isles, Volume 5: West Coast & Wales*. Redhill: Lloyd's Register.

Lloyd's Register of Shipping (various dates). *Lloyd's register of Shipping*. London: Lloyd's Register of Shipping.

The National Archives (1944). *Wrecks and other permanent non-submarine contacts in British inshore waters*. ADM 219/166: London.

UKHO 2021. *United Kingdom British Isles hydrographic survey status to April 2021*. [online] Available at:<https://assets.publishing.service.gov.uk/government/uploads/system/uploads/attachment_data/file/978845/Q6090_2021_March.pdf> [Accessed 30 September 2021].

UKHO. *7335 GLENCONA (PROBABLY)*. Taunton: UKHO.

UKHO. *7347 MANCHESTER*. Taunton: UKHO.

UKHO. *9865 CYMRIAN*. Taunton: UKHO.

UKHO. *9955 FRANCOISE DE LILLE (POSSIBLY)*. Taunton: UKHO.

wrecksite.eu. 2021. WRECK WRAK EPAVE WRACK PECIO. [online] Available at: <https://wrecksite.eu/wrecksite.aspx> [Accessed 23 November 2021].

4

The Research Phase 2:
Identifying the unknown shipwrecks

Introduction

With the stated purpose of the research being to identify shipwrecks in the study area, it was essential to pinpoint which shipwrecks were in fact not specifically known. This was achieved in Phase 1 by assessing each wreck DTM against the UKHO database, as described in Chapter 3. It will be recalled that when Phase 1 was completed, there were 161 shipwrecks which had been classified as UNKNOWN. The heart of *Echoes from the Deep* is Phase 2. It aimed to see how many of these wrecks could be identified by comparing their survey details with the historic text. When Phase 2 was complete, 129 shipwrecks had been identified using the (MBES/Archive) criterion.

It will be also recalled from Chapter 3 that 30 shipwrecks assessed during Phase 1 were also given the (MBES/Archive) criterion. These were cases where the UKHO identity given was not considered CONFIRMED or UNKNOWN but was consistent with the (MBES/Archive) criterion (which replaces the UKHO (PROBABLY) and (POSSIBLY) suffixes in this research) with the wreck in question being dimensionally, positionally and archivally consistent with the name given. The (MBES/Archive) criterion is fully described in this chapter, along with the case studies which illustrate the process in practice.

The (MBES/Archive) criterion, its design and purpose

The criteria used to assess the validity of a shipwreck's identity as used by UKHO (described in Chapter 3) includes CONFIRMED (usually denoted as just the ship name without a following status in parentheses) for shipwrecks which are undoubtedly identified, the parenthetical denotations of (POSSIBLY) and (PROBABLY) for those with suggested identities, and UNKNOWN for unidentified shipwrecks. Clearly in the cases of CONFIRMED and UNKNOWN, the criteria could stay as they are. A CONFIRMED shipwreck in this research is the same as the UKHO attribution. In reality the only CONFIRMED identities which are given in this research (as shown in Appendix 1) are those which had previously been attributed by UKHO and validated during Phase 1 of this research. This is because without physical interaction with the wrecks, there is no means of knowing that any attributions made during Phase 2 of this research is absolutely correct beyond doubt. UNKNOWN also remains the same as the UKHO criteria for unidentified cases.

Phase 2 of the research was designed to be a systematic and repeatable process which was consistent in its application to each specific shipwreck. Clearly the attributions of (PROBABLY) and (POSSIBLY) used by UKHO would need to be replaced with a single new criterion. This would specifically show that each shipwreck had been assessed in accordance with Phase 2. The criterion devised to replace (PROBABLY) and (POSSIBLY) was termed "(MBES/Archive)". It denotes the fact that a shipwreck has been surveyed by MBES and its DTM and pointcloud has been examined for archaeological details and its dimensions recorded. Then, these details have been compared to the historical text in order to identify the shipwreck. Therefore, the shipwrecks in the datasheets in Appendix 2 and in the case studies which follow later in this chapter show the (MBES/Archive) criterion next to its given identity.

Importantly the 32 shipwrecks which remained unidentified at the end of this research are shown as UNKNOWN in Appendix 1, but it should be noted that they too had been assessed using the (MBES/Archive) process, but that no plausible identity could be found at the time. Therefore, the (MBES/Archive) research process had only two possible outcomes. First, the wreck was identified to (MBES/Archive) standard or second, it remained UNKNOWN.

The (MBES/Archive) assessment process

The assessment process was designed to utilise all of the possible information which could be captured from the survey data and also from each archivally recorded shipping loss. These data were treated as separate and distinct during the fieldwork and archival research so that they could be compared against each other once both datasets had been completed and Phase 2 was ready to start.

The data derived from the surveyed wrecks fell broadly into four categories:

1. the actual dimensions of the shipwreck as measured in the DTM or the survey pointcloud,
2. the position where the wreck lies,
3. the orientation of the wreck, including its possible heading when sunk,
4. any other specific archaeological details of the wreck which could be derived from the survey data. For example, the location of the engine and boilers, the likely area of damage which led the vessel to sink, and the dimensions of specific features, such as the forecastle.

The data which could be derived during the archival research on each named shipping loss fell broadly into three categories:

1. the recorded dimensions of each ship lost, usually derived from *Lloyd's Register of Shipping*, or from ship plans,
2. the reported position where the shipping loss occurred,
3. a textual description of the loss, from survivors, or other witnesses, such as the KTB from the attacking U-boat. This included textual evidence of the damage the ship sustained, how it sank, its heading etc.

Comparing shipwrecks with historical dimensions, positions, and textual accounts

By comparing the data available from both the survey and archival research sources, three distinct criteria emerged where the data from both sources could be productively compared. They were:

1. the dimensions of the shipwreck in comparison with the archivally recorded dimensions of the lost ships,
2. the position of the surveyed wreck in comparison to the archivally reported loss of the ships,
3. the orientation of the wreck and its archaeological features in comparison with the archivally derived details of the ships and the textual accounts of their sinking.

Therefore, for an UNKNOWN shipwreck to be given a specific identity resulting from the (MBES/Archive) process, it had to be dimensionally, positionally and archivally consistent with the historical text. All three specific criteria needed to be consistent for an attribution to be made. Any lesser combination would not meet the criteria for an (MBES/Archive) attribution. During Phase 2, as the cases were resolved, one by one, many which were positionally or dimensionally promising were discounted if they could not satisfactorily meet all three criteria. For example, a wreck in the right position could be the wrong dimensions, or a dimensionally and positionally correct wreck could be wrong because its machinery was not in the right location, or a wreck with a neat torpedo hole in the area of the ship described by the survivors could be wrong because the wreck was dimensionally too large or too small.

After some trial and error to establish the best means of comparing the survey and archive datasets, a consistent process was devised:

1. the surveyed shipwreck's dimensions were checked against the database of lost ships dimensions to find candidate vessels that were similar,
2. these were then checked to ensure they were positionally located within a plausible distance from the reported position of the sinking in the historical text. Any cases that were not positionally consistent were discarded as candidates,
3. the dimensionally and positionally consistent wrecks were then compared against the historical text, specifically textual descriptions of the shipping loss and technical details of the vessels to ensure as much as was possible from the sources available the match was correct. Any obvious discrepancy led to the candidate wreck being rejected.

As Phase 2 was being conducted, it became clear that the process worked well. All of the 129 (MBES/Archive) attributions which were made followed this process and in each case the name given was dimensionally, positionally and archivally consistent with the surveyed wreck. The process drew in all of the data available from the surveyed wrecks and from the historical text and utilised it in a consistent and repeated manner in each case. By using these three separate criteria it was anticipated that it would, eliminate random but incorrect connections which may have occurred by making matches based solely on position and dimensions. How this process worked is shown on each shipwreck in Appendix 2. More detailed examples are given in the case studies which follow later in this chapter.

The shipwreck datasheets

The primary purpose of the shipwreck datasheets was twofold:

1. to capture the survey details of each wreck and to visually present each shipwreck details as seen on their DTMs and to spatially show the approximate positions of the wrecks within the study area,
2. to show how the wrecks were assessed through Phase 1 and Phase 2 of the research. The details presented are intended to allow the reader to understand how the identity of each shipwreck was derived by describing the key contributory factors from the Phase 1 and Phase 2 research processes.

In practice, a datasheet was opened for each of the 273 shipwrecks surveyed during the research and these are all retained in the project archive. In practicality, to prevent this monograph being too large for publication only the 129 datasheets covering the wrecks which were identified by the (MBES/Archive) process are presented in Appendix 2. However, the process by which the other 144 shipwrecks were determined to be CONFIRMED, UNKNOWN, NON SHIPWRECK or UNIDENTIFIABLE can be easily traced by looking at Appendix 1, where results of each phase of the research are shown all 273 cases.

The 129 datasheets shown in Appendix 2 have been sorted by their specific UKHO database number. The title header shows the specific UKHO number and identity attributed to the wreck after the datasheet has been completed fully. In the cases of the datasheets in Appendix 2, this is the name of the wreck as derived by the (MBES/Archive) process carried out during Phase 2. Below the title header is a combined DTM and map. Below this are eight specific data fields pertaining to the wreck:

1. UKHO No & Name: The exact details from UKHO without amendment,
2. Phase 1 Analysis: The end result of the wreck assessment during Phase1,
3. Position WGS84: The position derived from the UKHO wreck card expressed in WGS84,
4. Water Depth: The depth of water at Chart Datum, as given by UKHO,
5. Type of Vessel: If the wreck was CONFIRMED or otherwise identified then the type of vessel was entered. If the wreck was UNKNOWN, but it was clear what it was (i.e., a submarine) then the vessel type was also recorded,
6. Surveyed Length: The length of the actual shipwreck as derived from the survey DTM,
7. Date of Loss: The date on which the ship sank,
8. Circumstances: The circumstances of the loss, e.g., "Torpedoed by U103".

The next section on the datasheet, "Phase 2: Method of wreck identification" serves to explain why a UNKNOWN shipwreck has been given an identity during Phase 2 and the reasons given for the dimensional, positional, and archival consistency are described. There are four fields:

1. Positional Analysis: The difference between the position of the surveyed wreck and the reported sinking position(s) from the historical text is given here. Any comment on the relative proximity of the position is given when it pertains to the process of identifying the wreck. Also, if the position given in archive is at variance from the position of the surveyed wreck, but there is a plausible explanation it will be given here,

2. Dimensional Analysis: The dimensions and constructional details of the shipwreck and the attributed shipping loss are described in this section. Also, any obvious discrepancies are explained, e.g., the when the wreck is foreshortened by the loss of the bow in the torpedo strike,

3. Archival Analysis: The textual evidence derived from the archival research and used to assist in the identification of the wreck is given here. It can come from survivors accounts, official reports, the *Shipwreck Index of the British Isles*, Admiralty sources, newspapers, *Shipping Casualties* etc. Interpretation as to why the evidence points to the identification of the wreck is given when needed,

4. Phase 2 Conclusion: The conclusion of the Phase 2 assessment is given here, along with validation that the wreck is dimensionally, positionally, and archivally consistent with the name given.

The final section on the datasheet, "Phase 1 UKHO Analysis:" contains five fields:

1. Original UKHO Name: The UKHO number and given name of the wreck at the start of the research. This is the same as field 1) under the DTM and served as a check that the UKHO data had been transposed to the correct datasheet,

2. Archival Details: The archival details of the loss of the ship named by UKHO are given here. This will be from the sources as referenced. It will also include any interpretation of the historical text which either confirms or conflicts with the UKHO interpretation,

3. MBES Analysis: This field contains the basic details and analysis of the DTM and point-cloud of the wreck. This will generally include the wreck's physical and geographical orientation and its dimensions. Any other specific features which appear in the DTM and pointcloud may be mentioned if they are considered of utility in identifying the wreck, e.g., presence of boilers, number of holds etc. The condition of the wreck manner in which it appears to have collapsed will also be noted if considered of relevance. If this evidence was obviously in conflict with the UKHO attribution it would be entered here,

4. Additional Evidence: This field contains any specific evidence from the UKHO wreck card which may help the wreck identification. Most commonly the date when the wreck was first located by UKHO is recorded. This can be important in eliminating ships which sunk after the date which the wreck was found,

5. Phase 1 Conclusion: The conclusion of the Phase 1 assessment is given here, along with the reason for the given attribution.

As part of the archival research, a digital folder of research material was opened for each ship reported sunk in the study area. The evidential basis behind any archival interpretation was captured in these files, or in the archival evidence given in datasheets and are source-referenced.

Case studies

By studying the datasheets in Appendix 2, the reader will become familiar with the (MBES/Archive) criterion. However, in order to elucidate the process and how it worked in practice, nine case studies follow. These have been selected to show how the process worked across the historical timeline. They were also selected to describe the range of different archaeology which exists in the survey dataset and the range of historical sources consulted in order to derive the identities described. The case studies are descriptive and are intended to give the historical background and common types of incidents seen. For specific information on the positions of the wrecks, water depth etc. please refer to the datasheets of each wreck in Appendix 2.

Case study 1: Peacetime collisions: SS Waesland sunk in 1902

Collisions at night, especially in poor visibility were all too common in the age before radar and there are a number recorded archivally in the study area. A steamship, even when moving at a low speed (as was expected in fog etc.) was quite capable of slicing another in half or creating fatal damage. One of the most notable of these instances saw the loss of SS *Waesland* on 5 March 1902, after a collision with SS *Harmonides* with the loss of two lives.

Waesland was outward bound from Liverpool to Philadelphia and had experienced fog since departure. After several slow hours of progress, the *Harmonides* appeared out of nowhere and with no time to manoeuvre, *Harmonides* struck the ship on the port side, rebounding and striking it again. It was obvious *Waesland* was going to sink, and evacuation of the crew and passengers began. The sea was calm and all except two were saved. *Waesland* sank around 40 minutes after the collision. *Harmonides* which had been enroute from the USA did not sink and was able to make Liverpool under its own power 28 hours later.

The identification of this particular wreck site was not without its pitfalls, which were mainly due to the long history in service during which it had been lengthened. This ship itself was in service for 35 years by the time it was lost. It had originally been built as SS *Russia* for the British & North American Royal Mail Steam Packet Company (renamed The Cunard Steamship Company in 1878) and was the last Cunarder built with a clipper bow. This can be clearly seen in images of the ship. As built, it was 109.1m long (*Lloyd's Register* 1874-75). It remained in service with Cunard until 1880 when it was sold to the Red Star Line of Belgium. Under its new ownership the ship was transformed, being lengthened by 23.5m in 1881 to an overall length of 132.6m (*Lloyd's Register* 1880-81) and then later re-engined from compound to triple-expansion with the introduction of a pair of double-ended boilers (*Lloyd's Register* 1901-02). The change in the ship's appearance during its life can be seen in Figure 4.1.

With the actual length of the vessel when it sank now ascertained and entered into the shipping loss database, it could be compared to the surveyed wrecks to find candidates of similar dimensions. In fact, *Waesland* was the fifth longest of the UNKNOWN wrecks in the study area and there were only a few potential candidates in the survey dataset which could be this wreck.

A positional match was required next. Accurate positions for losses up to 1914 and again in the inter-war years were the most challenging to locate and verify. Unlike wartime conditions, where Lloyd's and the Admiralty often reported the losses of ships in several sources, this is often not the case in peacetime. For example, *Wreck Returns* for 1902 only gives the general area "*off Anglesey*".

The incident received widespread newspaper coverage. The research retrieved entries in the *New York Times, The Times* of London, and the weekly illustrated paper *The Sphere*. From these sources the textual basis of the evidence used to identify the wreck was

Figure 4.1. SS *Waesland* as originally built for Cunard in 1867 as SS *Russia* (above) and as it appeared around the time it was lost in 1902 (below). Note the difference in length and an additional mast (wrecksite.eu).

compiled. These sources did not give an indication as to where the collision actually took place. However, *Shipping Casualties* (BOT 1901-2, Appendix C, p145) provided positional details. It gives the collision point as "*About 15 miles SE by ¼ E of North Arklow Lightship, St Georges Channel*". It was the only plottable position found during the research. But in this instance, one surveyed shipwreck lies in the locality, being 7.6nm south of this position. This wreck is UKHO 9957, as shown in Figure 4.2.

The UKHO wreck card, 9957 KIRKBY (PROBABLY) had been amended to UNKNOWN status during Phase 1 because it was evident from the DTM that the wreck was far too long to be SS *Kirkby*, which was 96m long (*Lloyd's Register* 1915-16). It was also noted that *Kirkby* had been sunk by a torpedo striking the bows (TNA ADM 137/2959), so dimensionally and archivally the wreck was clearly another ship.

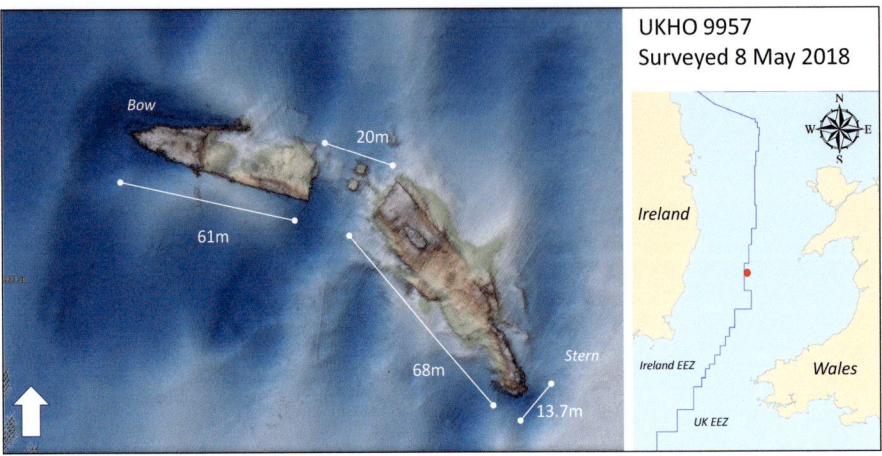

Figure 4.2. The UKHO wreck No 9957 KIRKBY (PROBABLY), amended to UNKNOWN during Phase 1 and attributed during Phase 2 to be SS *Waesland* (MBES/Archive) (Innes McCartney).

Archivally it was reported that *Waesland* had been hit on the port side. The wreck could be interpreted as having been hit in that way from its broken orientation on the seabed. The wreck also can be seen to manifest what appears to be a pair of boilers and one engine. It is also of note that the area where the ship clearly split in half as it sank is where the extension had been added in 1881. So, the conclusion reached was that this wreck should be SS WAESLAND (MBES/Archive), as recorded on its datasheet with the evidence used noted in the relevant fields. The wreck being dimensionally, positionally and archivally consistent with that ship.

Case study 2: Sunk without trace: SS Formby & SS Coningbeg torpedoed in 1917

In a conflict at sea full of the tragedies of war there has always been something particularly poignant about the loss of the steamers SS *Formby* and SS *Coningbeg* in December 1917. The loss of both ships in the space of two days with their entire 83 person complements of crew and passengers had a particularly tragic effect on the town of Waterford, Ireland. A disaster fund was set up to assist the families and it is said that nearly every family in the town knew somebody who was lost.

This event has lived long in the collective memory of the town. In 1997 a memorial was set up at Waterford Quay and opened by President Mary Robinson to commemorate those lost. The relatives of those who died were in attendance. This tragedy was known to me since that time, so that when this research started, I was particularly keen to see if it was possible to locate the potential remains of these two ships to help in a small way to bring some measure of closure to this notably tragic event.

Both vessels were owned by the Clyde Shipping Company and were employed on the Liverpool to Waterford route, carrying goods, livestock, and passengers. They were both sunk after leaving Liverpool to return to Waterford. *Formby* departed on 15 December and was sunk by *U62* that same evening. Its failure to arrive promptly (as it was regularly reputed to do so) was apparently put down to the very bad weather at the time. *Coningbeg* departed Liverpool on 17 December and met the same fate at the hands of the same U-boat.

S.S. "Formby" and "Coningbeg," in the Waterford and Liverpool thrice-weekly Direct Service. Average speed 14½ knots. Refrigerated Holds for Perishable Traffic. Excellent accommodation on upper deck amidships for first-class passengers.

Figure 4.3. SS *Formby* and SS *Coningbeg* were very similar vessels which ran the Liverpool to Waterford route. Both were sunk in a 48-hour period in December 1917 (wrecksite.eu).

The attribution to *U62* sinking both ships was made by the official German historian Arno Spindler (1941, p396). Both ships were sailing independently when lost. The KTB of *U62* provides textual details which make it clear both vessels sank very quickly, with little possibility of anyone surviving. In fact, the circumstances of the loss of both ships was unknown to the British, but the clear suspicion was that they had been sunk by war causes (TNA ADM 137/1361).

Dimensionally both ships were almost identical. *Formby* was 82.34m x 11.02m and *Coningbeg* was 82.44m x 10.99m (*Lloyd's Register* 1917-18). Within the dataset of ships surveyed in the study area there were a number of shipwrecks with similar dimensions, but the number of candidates was whittled down by their proximities to the reported positions of both attacks.

With the absence of any descriptive archival data from the British side to assist in identifying these two wrecks, it meant that the U-boat KTB (NARA PG61627) had to be relied upon to provide the descriptive evidence from the historical text. This was supplemented by the commander's autobiography which describes the patrol in the Irish Sea that claimed these ships (Hashagen 1931, p244-253). Fortunately, the KTB is a detailed one, giving both positional and textual descriptions of both events. The positional data is given using Quadratkarte references.

The KTB describes that *Formby* was seen to be hit in the engine room and sank in around four minutes. The wreck with the best dimensional and positional fit turned out

Figure 4.4. The wrecks which are dimensionally, positionally and, archivally consistent with being *Coningbeg* (UKHO 10004, left) and *Formby* (UKHO 10017, right). In both instances the wrecks were listed by UKHO as UNKNOWN. They both manifest damage consistent with the description given in *U62*'s KTB (Innes McCartney).

to be UKHO 10017, which lies around 5.6nm northeast of from *U62*'s reported position. In this instance the archival evidence that the ship was hit in the engine room matches what is seen on the DTM, as shown in Figure 4.4, where there is a large hole on the starboard side just behind the engine. Therefore UKHO 10017 has been inserted as FORMBY (MBES/Archive) as being dimensionally, positionally, and archivally consistent with that ship.

In the case of *Coningbeg*, the ship was seen to be hit amidships, burst into flames, break in the middle, and sink in three minutes (NARA PG61627). This description closely matches that given in the U-boat commander's later autobiography (Hashagen 1931, p244-253), suggesting he may have had access to the KTB when he wrote the book.

The fatal attack on *Coningbeg* took place further south. Again, there was one good candidate wreck in the survey dataset which matched the ship's dimensions and was positionally close to the attack position given in *U62*'s KTB. UKHO 10004 plots 4.2nm northwest of the position given in *U62*'s KTB. Archivally the evidence that the ship broke amidships is strongly represented in the survey DTM, as shown in Figure 4.4. The wreck can be seen to have been heavily damaged in the central region, accounting for the rapid sinking. Therefore UKHO 10004 has been named CONINGBEG (MBES/Archive) as being dimensionally, positionally, and archivally consistent with that ship. Both wrecks were given the UNKNOWN attribution by UKHO and therefore remained as UNKNOWN after the Phase 1 analysis.

These two ships are examples of the perils faced by ships sailing independently during WW1. Even when armed they were comparatively easy prey for U-boats. The patrols in the study area by *UC65* and *U38* were notably destructive in this regard. By late 1917 independent sailings had been much curtailed and those that did so generally sailed at night, unilluminated. But this afforded only small protection if a U-boat was in the area. Had it been known for certain that a U-boat had claimed *Formby*, *Coningbeg* would not have set sail.

S. S. "Mesaba"   Atlantic Transport Co.

Photo # NH 61498   Torpedoed S.S. City of Glasgow, 1918

Figure 4.5. The steamships SS *Mesaba* (top) and SS *City of Glasgow* (below) were sunk in a convoy attack by *UB118*. In this instance a rare series of photographs showing *City of Glasgow* sinking was taken from one of the escort ships, USS *Beale* (Contemporary postcard & U.S. Naval Historical Centre).

Case study 3: Convoy actions: SS Mesaba & SS City of Glasgow torpedoed in 1918

The seemingly unstoppable destruction of independently sailing cargo ships by U-boats in the opening months of 1917, including the loss of over 10 ships a day in the 2nd quarter (Gibson & Prendergast 1931, p338) was countered by the adoption in July 1917 of the convoy system. This drastically reduced the losses and at a stroke reversed the course of the war at sea 1914-1918. Less than 1% of ships in convoy were lost to enemy action. In

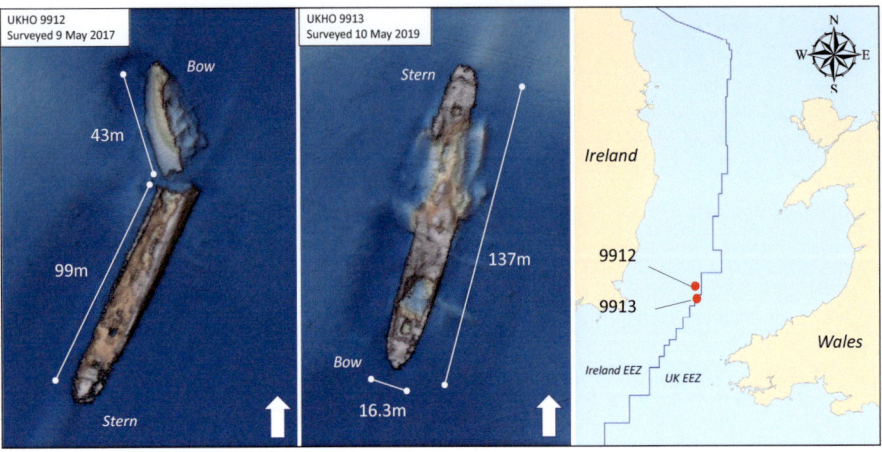

Figure 4.6. The wrecks of *Mesaba* (left) and *City of Glasgow* (right) and their locations. The damage seen on both closely resembles that described and seen in the photograph of *City of Glasgow*. During Phase 1 it was ascertained that the UKHO legacy attributions were in reverse (Innes McCartney).

the coastal waters around the UK and Ireland, the archaeological remains of ships sunk in convoy actions in WW1 are relatively uncommon, making the incident of the loss of SS *Mesaba* & SS *City of Glasgow* on 1 September 1918 of particular interest.

It should also be noted that *Mesaba* is also historically notable as one of the ships in the North Atlantic on the night in 1912 when RMS *Titanic* sank. *Mesaba* sent a radio message directly to *Titanic* warning of the presence of icebergs. The message was received but never passed to *Titanic's* bridge.

Both ships were sailing in the combined convoy OE21 and OL32. They were part of OE21, outward bound from Manchester to Montreal in ballast (TNA ADM 137/1502). The action in which both ships were sunk by *UB118* is well documented from both the allied and German side. The attribution to *UB118* being the U-boat which sank both ships was made by the official German historian Arno Spindler (1966, p317). *UB118's* KTB describes the attack in brief and gives a position in lat/long (NARA PG61873). Both ships were struck by one torpedo in a simultaneous attack, with the ships being hit at the same time.

Dimensionally *Mesaba* was 146.94m x 15.90m and *City of Glasgow* was 135.02m x 16.30m (*Lloyd's Register* 1918-19). Subsequent to Phase 1 analysis, *Mesaba* was the 2nd largest UNKNOWN loss in the study area and *City of Glasgow* the 4th.

In the case of *Mesaba*, the ship was hit on the starboard side around the area of the bridge. It rapidly listed and sank by the bow in nine minutes in a steep dive with its propellers still rotating. Due to its size and the detailed record of the loss of this ship, it was not difficult to pick it out of the survey dataset. The wreck, which turned out to be UKHO 9912 lies 0.7nm from the position plotted from *War Losses* (1990). In this instance the U-boat position from its KTB plots 3.7nm northwest of the wreck. The DTM can be seen in Figure 4.6. It will be noted that the wreck is broken half. This probably happened on impact with the seabed, compounded by the damage caused by the torpedo in the same area of the ship. Interestingly the mean course of the convoy was 200deg, although it was zigzagging. It appears that the ship either turned though 180 degrees as it sank or toppled when the bow hit the seabed. The stern appears in the pointcloud to be to the south, with the after deckhouse seemingly present.

In the case of *City of Glasgow*, the ship was hit on the starboard side amidships. Its back broke but it took an hour to sink. It was photographed by USS *Beale* at this time (see Figure 4.5). Similar to *Mesaba*, this wreck was not difficult to pick out of the survey dataset. Dimensionally and positionally it could only really be UKHO 9913. The wreck lies 5nm south of the position given by *UB118* and 2.75nm south of the position given in *Lloyd's War Losses* (1990). It will be noted that the obvious area of damage seen in the DTM (Figure 4.6) is the same area where ship's back was broken, as seen in the image in Figure 4.5.

Prior to Phase 1 amendment, UKHO listed *Mesaba* as UKHO 9913 MESABA (PROBABLY) and *City of Glasgow* as UKHO 9912 CITY OF GLASGOW (PROBABLY). Both of these legacy attributions were amended to UNKNOWN during Phase 1 analysis because the dimensions of the surveyed shipwrecks did not conform with the ship's actual details. UKHO 9912 was too long and UKHO 9913 too short (see Appendix 2). Despite accurate archival positions, the UKHO legacy attributions were in fact the reverse of what turned out to be correct once the (MBES/Archive) process was complete.

It is concluded that dimensionally, positionally, and archivally UKHO 9912 is MESABA (MBES/Archive) and UKHO 9913 is CITY OF GLASGOW (MBES/Archive). Also, interesting now is the fact this notable single convoy action of 1918 straddles an international boundary which did not exist in WW1, with one shipwreck in the Irish EEZ and one in the UK EEZ.

Case study 4: Wartime accidents: SS Burutu sunk in collision in 1918

Wartime did not mitigate the annual rate of marine accidents, which are just as much a feature of maritime life in war as in peace. In 1921 the wartime edition of *Shipping Casualties* was published, covering 1914-1918. This document shows that 915 UK-registered ships were totally lost by marine casualty in WW1 (BOT 1921, iv).

One such accident occurred in the final weeks of WW1 when the last three ships of the Inward bound convoy HL50, under escort and heading for Liverpool (the others had already departed for other UK ports) ran into the outbound convoy OL37 just before midnight. The results of the two unilluminated (wartime standing orders) convoys passing through each other at night was a collision between the outward bound SS *City of Calcutta* and the inward bound SS *Burutu*.

The most detailed textual account found during the research came from the ship's owners. The Elder Dempster Company produced a history of its wartime service with details of the 36 ships it lost in WW1 (three are wrecks in the study area). The circumstances of the loss of *Burutu* are described within it as the ship being hit on the port side between No 1 & 2 hatches at an oblique angle, with both ships briefly going side on, before reversing apart (Elder Dempster 1921, p45-49). *Burutu* was fatally damaged and sank within minutes. The loss of life was tragically high with 148 passengers and crew lost.

Dimensionally, *Burutu* was a liner of 109.72m x 13.46m (*Lloyd's Register* 1917-18). Within the survey dataset there were a number of potential wrecks which broadly fitted these dimensions. The challenge was to archivally place the collision accurately in the project GIS of shipping losses. Whereas the losses in WW1 to enemy action were recorded through different Admiralty reporting systems and can usually be traced from the German side as well, losses which occurred in wartime but were accidental fall outside of these sources and are, like peacetime incidents more challenging to research.

Figure 4.7. SS *Burutu* of the Elder Dempster Line was sunk in a collision with heavy loss of life in the final weeks of WW1 (wrecksite.eu).

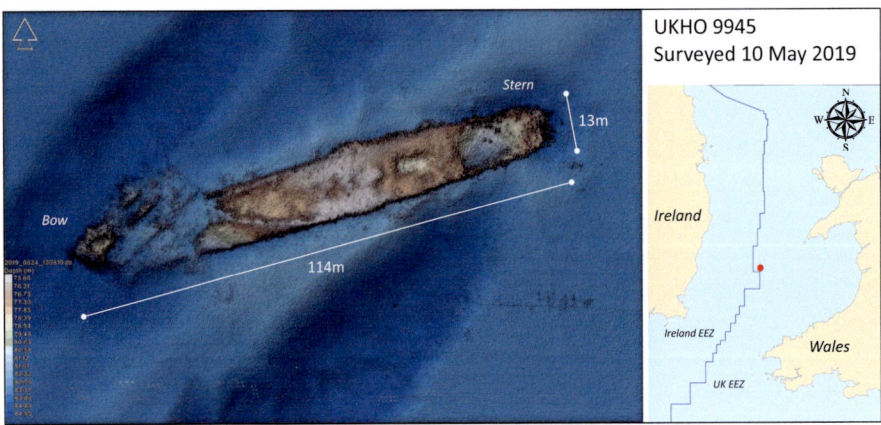

Figure 4.8. The wreck, which is dimensionally, positionally, and archivally the most consistent with *Burutu* turned out to be UKHO 9945. This wreck had been amended to UNKNOWN status during Phase 1 (Innes McCartney).

Positionally *Shipping Casualties* (BOT 1921, p39) gives very brief details and provided a vague position of "*25 miles west of Fishguard*". This was the only position captured during the research phase and is the same as the position given in the *Shipwreck Index* (Volume 5, section EK). The *Wreck Returns* (1918, Qtr.4 p6 table (f)) gave an even more vague position of "*North of Padstow*". However, as described in Chapter 2, it was always desirable to find two alternative positions for any shipping loss. In this instance, as with a few others, further research was needed.

The alternative positional sources came from the National Archives, and due to the accident happening to wartime convoys, the reports of these convoys in the Admiralty

Figure 4.9. The paddle steamer HMS *Mercury* was utilised as an auxiliary minesweeper in 1940. It sank on Christmas Day 1940 after contacting a mine it was sweeping (contemporary postcard).

records held much more accurate positional data. The convoy records provided a pair of positions which plotted close together but were both 37nm north of the position in *Shipping Casualties* (BOT 1921, p39). With the position now archivally fixed, the candidate wrecks in the area around the position given in the convoy reports were examined.

The wreck which dimensionally, positionally, and archivally was the most consistent with *Burutu* was UKHO 9945, as shown in Figure 4.8. Dimensionally the wreck is consistent on width. Its length is slightly extended, probably due to the damage sustained to the bow during the collision and the sinking. Positionally the wreck plots 1.4nm southwest of the position given in the OL37 record (TNA ADM 137/2618) and 5.3nm south of the HL50 record (TNA ADM 137/2657). Archivally, the damage to the wreck is in the region of the fore holds, as described, and could be interpreted as being a result of being struck on the port side, as described. Further archival evidence emerged by consulting the plans of this ship photographed at the Lloyd's Register Foundation, Heritage & Education Centre. They show that the ship had four holds, two forward and two aft. The after ones are clearly discernible in the DTM.

UKHO 9945 had originally been named as SPENSER (POSSIBLY). This UKHO legacy attribution was amended to UNKNOWN during the Phase 1 research due to the fact that this wreck was dimensionally too small to be that ship. Moreover, SS *Spenser* had five holds and was hit aft by a torpedo. These features are not evident on the DTM of UKHO 9945.

Case study 5: Royal Navy wrecks: HMS Mercury, sunk in 1940

Much of the war at sea 1939-45 in home waters relates to minelaying and minesweeping. Mines were laid by both sides and sweeping up enemy fields and older allied ones was the endless duty of the Royal Navy's minesweepers. One such example was HMS *Mercury*, an

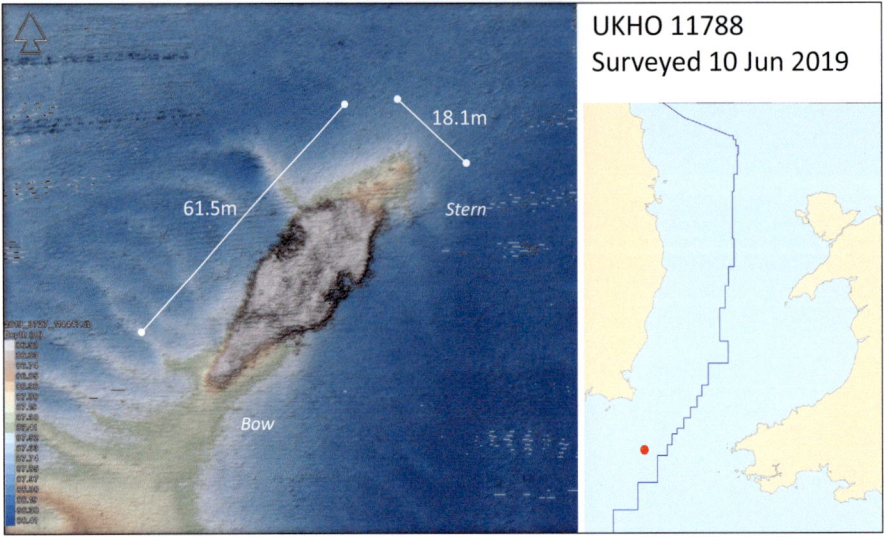

Figure 4.10. The wreck which dimensionally, positionally, and archivally is consistent with HMS *Mercury* as seen in the survey DTM turned out to be UKHO 11788. It had been originally listed as a submarine (Innes McCartney).

auxiliary paddle steamer which succumbed to a mine whilst sweeping. It was built in 1934 for the London Midland Scottish Railway as an excursion passenger steamer. It primarily worked the Greenock, Gourock, and Wemyss Bay route, see Figure 4.9.

Dimensionally, HMS *Mercury* was 68m x 9.2m (*Lloyd's Register* 1939-40). The ship was a "boxed in" paddle steamer, so that the width recorded by Lloyd's was the width of the hull, not including the boxing around the paddle wheels. In the survey dataset, there was only one candidate wreck which was positionally in the right area and dimensionally consistent with HMS *Mercury*. The wreck, UKHO 11788 also conveniently resembles an upside down boxed-in paddle steamer, see Figure 4.10.

Therefore, it was concluded that this must be the wreck of HMS *Mercury*. The original UKHO legacy attribution which was amended during Phase 1 was UKHO 11788 UC33, listed without a (POSSIBLY) or (PROBABLY) suffix, meaning that for some unknown reason this wreck was deemed CONFIRMED. This was easily reverted to UNKNOWN as it was quite obvious the wreck was not a submarine. In fact, *UC33* was later identified during Phase 2 as being UKHO 11798 (See Appendix 2). The U-boat wrecks are addressed in case study 9.

Case study 6: Air attacks: MV Rotula bombed in 1941

The character of the war at sea in the Irish Sea in WW2 is different from WW1. The U-boats only made brief incursions in the last year of the war and for much of the time the area was largely unmolested. However, there were occasional incursions by the German Air Force laying mines, bombing, and torpedoing shipping. One such ship attacked in this manner was the large petrol tanker MV *Rotula*, see Figure 4.11.

A report of the loss of the ship by the master states that, on 1 March 1941, MV *Rotula* was inbound from Halifax NS with a cargo of aviation spirit. The convoy had already reached the Clyde and then been split up into its home destinations. The ship continued south, but now with only a trawler for escort. Traversing the Irish Sea, it was attacked by

Figure 4.11. MV *Rotula* was the 3rd largest unlocated ship (after Phase 1) in the entire study area. It sank after being bombed, burning for three days before being sunk by gunfire (wrecksite.eu).

Figure 4.12. The wreck of MV *Rotula* is dimensionally, positionally, and archivally consistent with UKHO 9925. It is broken into two sections along a line 95m from the stern (Innes McCartney).

three aircraft. A bomb struck aft, and the ship caught fire. The crew fled to the forecastle and were saved by the master, Capt. Kris releasing the anchors which swung the ship around causing the wind to blow the encroaching inferno away from them.

The survivors were picked by the escort's boats. From a crew of 49, 16 had died. The ship remained on a level keel, but its stern was resting on the seabed, while the fore section continued to burn (TNA ADM 199/2136). Two days later, HMT *Armana* was dispatched to sink the ship, which was still burning furiously at that time in a burning sea (TNA ADM 199/2225).

Positionally there was plenty of evidence from archive as to where this ship would be found. In fact, as it turned out HMT *Armana's* reported position for where it sank the burning wreck is only 0.36nm from the wreck site. The position given in *War Losses* (1989) plots 2.9nm northeast of the wreck. Dimensionally, *Rotula* was 146.61m x 18.10m (*Lloyd's Register* 1940-41). Its sheer size made it the 3rd largest wreck in the study area and also the

3rd largest which was UNKNOWN after Phase 1 analysis had been completed. Searching through the survey dataset for a wreck of similar dimensions in the right area yielded only one candidate for this wreck site.

The wreck which turned out to be dimensionally, positionally, and archivally consistent with *Rotula* was UKHO 9925 (see Figure 4.12). In this instance its sheer size, along with detailed positional data meant that it was seemingly an easy case to resolve. The wreck can be seen to be of the correct length, but it has in fact broken into two portions, presumably at the time of the sinking by gunfire from HMT *Armana*. The sea depth is 90m and it is noted that the stern portion is 95m long. The two portions of the wreck are 175m apart.

Prior to Phase 1, the UKHO legacy attribution was UKHO 9925 LANDONIA (PROBABLY). SS *Landonia* was 91.70m x 13.59m (*Lloyd's Register* 1918-19) and therefore easily discounted as being this wreck due to it being dimensionally far too small.

Case study 7: Royal Navy wrecks: LCT 326 lost in 1943

As would be expected, wartime marine accidents continued throughout WW2, as was the case with the loss the Landing Craft Tank *(LCT) 326*. It was a Mk III LCT designed to land armoured vehicles during amphibious operations. This type of vessel was built in large numbers in the last years of WW2. *LCT 326* was built in Middlesbrough and launched in April 1942, see Figure 4.13. The following year it was lost with all 14 of its crew on a transit cruise through the Irish Sea.

After WW2 it was listed by the Admiralty as having been lost to *"weather or mine off Isle of Man"* (Admiralty 1947, p53). Similar to the case of HMS *Mercury*, additional research was needed at the National Archives to find detailed archival data. In the case of *LCT 326*, the 7th LCT Flotilla, under the watch of HMS *Cotilion* had made slow progress south from Troon, on its way to Appledore due to heavy weather (TNA ADM 217/13).

On the evening of 1 February 1943, *Cotilion* made a check on the flotilla and saw *LCT 326* was still with the convoy. That was the last time it was observed. Fortunately for the research, the position at which this check was made was recorded as 53 00; 04 58W, just to the northwest of Bardsey Island. It was not seen again and noticed to be missing when the convoy arrived in Appledore, but this meant that *LCT 326* must have sunk to south of this position. That was the extent of the positional and archival data which could be derived from archive. It was not ideal but could be potentially compensated for by the unique design of the ship, if seen on a DTM.

Dimensionally the Mk III LCT was 58.2m x 9.4m (see Figure 4.13). Although there were a number of wrecks in the survey dataset with similar dimensions, the wreck which stood out as a candidate for being *LCT 326* was UKHO 9950 (see Figure 4.14). It had the characteristics which pointed to it potentially being an LCT. This was because the overall length is approximately 61.0m with a width of 10.2m. The stern portion is to the north, and it appears the vessel had its machinery and bridge aft and what appears to be a discernible ramp at the bows. It was noted that the shipwreck is in two parts which are 133m apart.

Positionally all that can be verified is that the wreck lies in the track the 7th LCT Flotilla would have been taking as it transited south. If this is in fact *LCT 326* , then clearly it foundered by breaking in half in heavy weather. In this case, the DTM appears sufficiently detailed to make up for the weaker evidential basis, so it was concluded that on the best evidence available, dimensionally, positionally, and archivally the wreck of UKHO 9950 was consistent with being *LCT 326* .

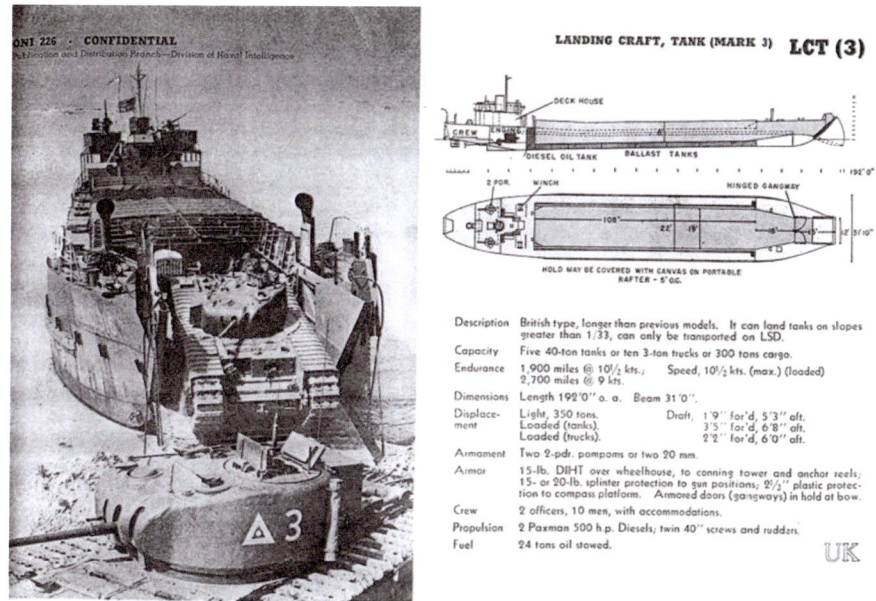

Figure 4.13. The Mk III *LCT 326* was of the type seen in this image. The plan of the vessel is similar to the features seen in the DTM in Figure 4.12 (wrecksite.eu).

Figure 4.14. The wreck, which is dimensionally, positionally and archivally consistent with *LCT 326* . It can be seen that it displays features which appear similar to those seen on the plan in Figure 4.13 (wrecksite.eu).

It is noted that prior to Phase 1 analysis the UKHO legacy attribution was UKHO 9950 FOYLE (POSSIBLY). SS *Foyle* foundered with a cargo of coal on 9 August 1897 (BOT 1897) in a position approximately 7nm southwest of this wreck. It was dimensionally 48.80m x 6.4m (*Lloyd's Register* 1897) and could safely be reverted to UNKNOWN due to the difference in dimensions.

Case study 8: The Inshore Campaign shipwrecks: SS Soreldoc torpedoed in 1945

Incursions by U-boats into the Irish Sea area only got underway during the last phase of the Battle of the Atlantic, which is referred to as the Inshore Campaign. It began on 6 June 1944, D-Day and lasted until the end of the war. The invention of sonar in the interwar years had made operations in shallower coastal waters seemingly too dangerous to be profitable for U-boats, so it was only in a last desperate effort that they made regular incursions into inshore areas. In the Irish Sea, this began in December 1944.

The effectiveness of the U-boats in the study area in WW2 was minimal and the price they paid was high. *War Losses* (1989) records that 11 ships were sunk there by U-boats during the Inshore Campaign. They paid for these successes with the loss of six U-boats confirmed sunk during this research. In contrast *War Losses* (1990) records that the U-boats sank 227 ships in the archival study area in WW1, losing only two U-boats in exchange. The extent of the U-boat wrecks will be addressed in the last case study.

One of the ships lost in WW2 to U-boats was SS *Soreldoc*. Built by Swan Hunter in Newcastle in 1928 as a "Lake Steamer" for Paterson Steamships Ltd (*Lloyd's Register* 1928-29), it was transferred to the US War Shipping Administration in 1943 (*Lloyd's Register* 1942-43). For wrecks in UK waters, the ship has some very distinctive features because it was built for the inland waterways of Canada and the USA. These can be seen in Figure 4.15 and in the ship plans held at the Lloyd's Register Foundation, Heritage & Education Centre (LRF-PUN-004512-004519-0198-P). Dimensionally the ship was 77.10m x 13.10m, with a bridge on the forecastle and machinery aft (*Lloyd's Register* 1944-45). This ship is notably wide for its length, when compared to the other ships in the archival study area database.

Archivally, SS *Soreldoc* was torpedoed and sunk on 28 February 1945 by *U775* (Rohwer 1999, p191). The ship broke in two and sank in four minutes, taking 15 crew to their deaths (Hocking 1969, Vol 2 p655) and 21 survivors were picked by the trawler *Loyal Star* and were landed at Milford Haven (TNA ADM 199/1443).

Positionally, there were a number of sources which gave similar positions to the one from *War Losses* (1989) which had been plotted in the GIS during the initial research. The survey data provided one excellent candidate for *Soreldoc*. It was dimensionally consistent in length and importantly was measured wider than normal. Moreover, it was archivally consistent, being broken into two portions.

This wreck turned out to be UKHO 9892 (see Figure 4.16). It lies 4.2nm south of the plotted archival position. The DTM seems to show that the bow section is to the south and the stern section, possibly upside down is to the north. The plan of the ship shows that it had seven cargo hatches. Hatches 1-3 and hatch 7 can be seen in the DTM, 4-6 were seemingly at the location of the torpedo strike.

Figure 4.15. The steamship SS *Soreldoc*, with is distinctive lines was built in the UK to work on the Great Lakes. It was sunk by *U775* in February 1945, breaking in two and sinking in four minutes with 15 of 36 crew. It was identified during Phase 2 (wrecksite.eu).

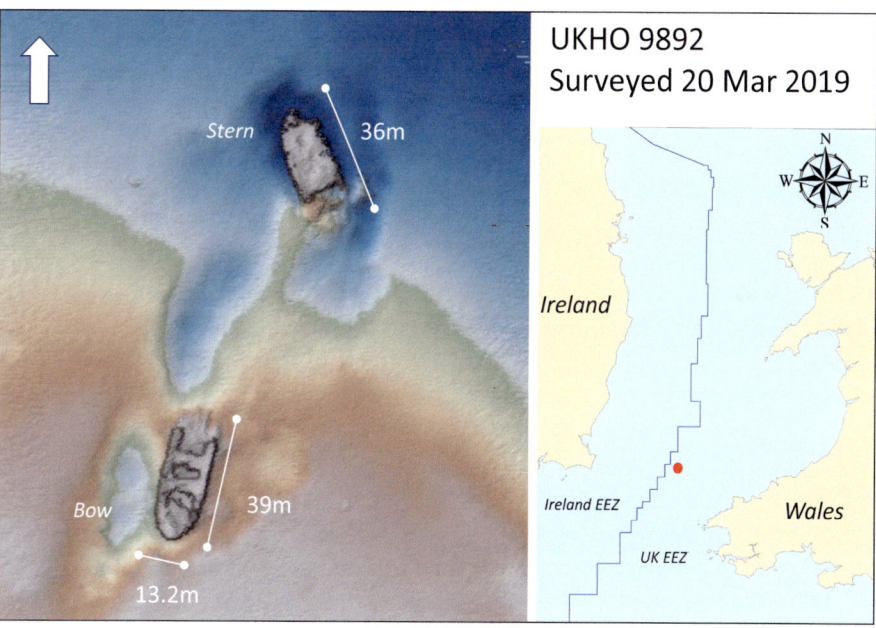

Figure 4.16. The wreck, which is dimensionally, positionally, and archivally consistent with SS *Soreldoc*. The ship broke in two when it was torpedoed. Its original UKHO legacy attribution UKHO 9892 EMPIRE PANTHER (PROBABLY) was amended in Phase 1 to UNKNOWN due to this wreck being far too small (Innes McCartney).

The UKHO legacy attribution was UKHO 9892 EMPIRE PANTHER (PROBABLY). During Phase 1 analysis this was amended to UNKNOWN. SS *Empire Panther* was mined in 1943 at a position plotted from *War Losses* (1989) to be 12nm northeast of this wreck. It was a large ship of 125.0m x 16.6m (*Lloyd's Register* 1943-44) and therefore easily discounted, because this wreck is far too small.

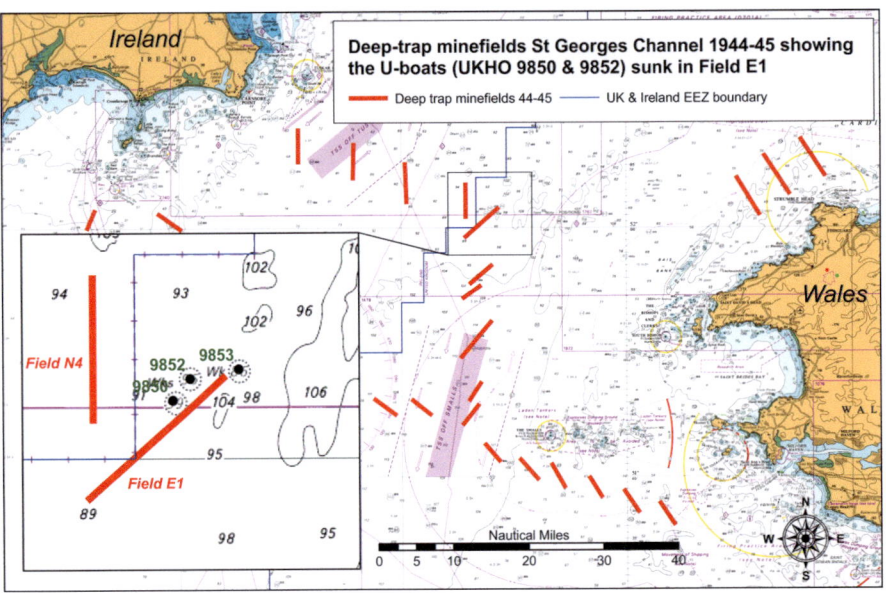

Figure 4.17. The location of the pair of U-boat wrecks (UKHO 9850 and 9852) found to have been lost in deep-trap minefield E1 in the St Georges Channel in 1945. Whereas one U-boat was known to have been sunk in the area, the other was a surprise find (Innes McCartney).

Case study 9: The submarine wrecks and the Inshore Campaign deep-trap minefield

Some of the shipwrecks in the study area had distinctive, if not unique features discernible on their DTMs, for example HMS *Mercury*, SS *Soreldoc*, and *LCT 326* . Equally discernible are submarine wrecks. There were nine submarines in total in the survey dataset. Of these, during Phase 1, four appeared to be correctly attributed by UKHO, being *U87* (UKHO 9989 & 58682), *U246* (UKHO 7373), HMSM *H5* (UKHO 7147) and *U1024* (UKHO 7377). The other five submarine wrecks (all U-boats) were either listed by UKHO as UNKNOWN or by legacy attribution, associated with wrecks which turned out to be ships and were consequentially amended to UNKNOWN during Phase 1.

The challenge in attempting to identify these cases lies in the validity of the historical text. Archival data on U-boat losses is generally plentiful, but it is not always reliable. The issues relating to why this occurred are explored in detail in McCartney (2014). It is sufficient to say here that unless a U-boat was known to be destroyed by unassailable evidence, such as the presence of survivors in the water, the recorded fates were often little more than guesswork and have been found by archaeological surveys to have been correct, as often as not. Unassailable evidence led to "A Grade" recorded sinkings. Any other grade should be considered unreliable. Many were made after wartime to account for U-boats which disappeared without trace and were sometimes made on the scantest of evidence. Problematically, in WW1 there is also evidence that some Admiralty attributions to U-boats destroyed were tacitly fictitious.

Of the five U-boat wrecks identified during Phase 2, three were "A Grade" cases which were found to be dimensionally, positionally, and archivally consistent as: *UC33* (identified at UKHO 11798), *U1051* (identified at UKHO 7106) and *U1302* (identified at UKHO 9911). Fuller details are given in the relevant datasheets in Appendix 2.

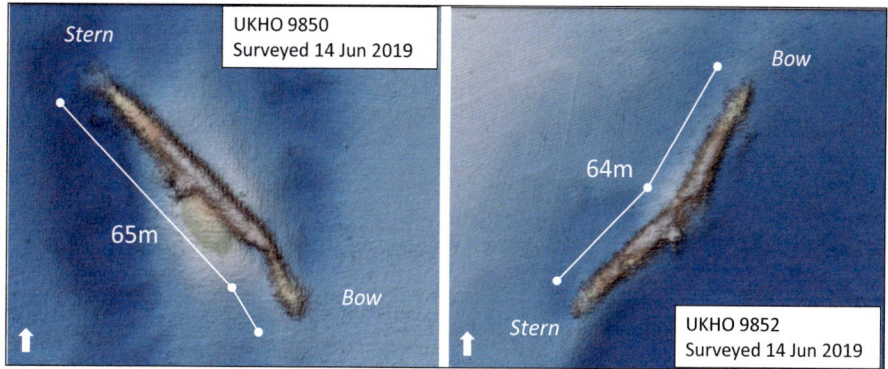

Figure 4.18. The U-boats sunk in Field E1. Both have sustained significant damage consistent with a mine strike. UKHO 9850 can be seen to be heading outbound and UKHO 9852 inbound. They are dimensionally consistent with Inshore Campaign era U-boats (Innes McCartney).

Another U-boat, *U242* was only identified archivally after WW2, although the location of its destruction was witnessed at the time. The final case was a mystery U-boat, with no archival evidence to place it where it was located and its discovery during the fieldwork was a complete surprise. As it turned out, both of these had been mined. Figure 4.17 shows the location of the wrecks in proximity to the nearby deep-trap minefields laid during the Inshore Campaign.

The deep-trap minefields were laid from August 1944 to counter the threat posed by the Inshore Campaign. The plan was to deploy anti-shipping mines laid deep in the water column, allowing ships to pass over without hindrance, but submarines would run into them. The fields laid as shown in Figure 4.17 were part of Operation "CH". The field "E1" which accounted for both U-boats was laid on 23 January 1945 (MOD 1977, Vol 1 p213).

The DTMs of the U-boat wrecks can be seen in Figure 4.18. It is evident in both cases that the type of damage sustained is consistent with what would be expected from a mine hit. Dimensionally the wrecks are consistent with being of the Type VII U-boat as used during the Inshore Campaign. Damage of the type seen is immediately fatal, so the orientations of the wrecks most are probably similar to the course the U-boats were making when they struck the mines. Hence UKHO 9850 can be seen to have been outbound from the central Irish Sea and UKHO 9852 inbound.

From January 1945 onwards, the German historical text covering U-boat operations was not part of the Castle Tambach capture (see Chapter 2) and little now survives from the German side. Allied records on U-boat movements are also vague, as the U-boats had largely abandoned radio use giving few insights to allied codebreakers. Therefore, against a backdrop of weak historical text, reliably pinpointing which U-boats were in any specific area on any specific day is in practice impossible.

Nevertheless, some archival evidence does occasionally give insights into the identities of U-boat wrecks of this period, as was the case with *U242*. It was established after WW2 that a U-boat seen to strike a mine on 5 April 1945 in the proximity of these two wrecks was *U242*. Only two U-boats were known to be allocated to the Irish Sea area. The other *U1024* was known to have been sunk and is recorded at UKHO 7377 (see Appendix 1), so it was logical to presume *U242* was mined here (NHB FDS 476/97). The wrecks of UKHO 9852 and 9850 appear to possibly have originally been detected on 5 May 1945 by Escort

Group 14 (EG14) and attacked (TNA ADM 199/1786). In its report it is stated that EG14 was of the view there were possibly two U-boat wrecks present. This is in fact the case as the two U-boat wrecks are only 0.5nm apart.

During Phase 2 it was concluded that UKHO 9852 was most likely to be *U242*. The wreck lies 0.9nm from the position of the sighted mining of *U242* on 5 April 1945 (TNA ADM 199/1786). *U242* is not known to have had any successes in the Irish Sea and therefore on the thinnest of evidence it was concluded that it most likely would be the inbound U-boat, destroyed before it entered its operational zone. It was listed by UKHO as UNKNOWN and therefore remained so until Phase 2. On the thinnest of evidence, it was concluded that dimensionally, positionally, and archivally the wreck is UKHO 9852 U242 (MBES/Archive).

The question which remained was what is the identity of the other U-boat? A possible clue came from the fact that all the U-boats recorded as destroyed in the study area have been accounted for during this research, except for *U1172*. Archivally, this U-boat was listed post war as being destroyed off the Arklow Bank on 27 January 1945. The actual attack report was assessed as "B Grade" (probably sunk) and therefore is quite possibly incorrect.

In fact, the attack on the U-boat developed after two ships (*Solör* and *Ruben Dario*) had been hit by torpedoes (TNA ADM 199/1786). The loss of these ships has been attributed to *U825*, which survived the patrol and reported being heavily attacked after attacking the ships. The Royal Navy's Naval Historical Branch concluded in 2007 that it was entirely speculative that *U1172* was destroyed at this time (NHB FDS D/NHB/22/2(849). In the survey dataset there is no submarine wreck within a 10nm radius of the reported position of the destruction of *U1172* of 52 24N: 05 42W. Either *U1172* has not been found as a wreck, or more likely it was not present, and it was *U825* which was attacked, but survived to relate what happened.

It is known that *U1172* operated in the Irish Sea where it is credited with one ship sunk (UKHO 7361 VIGSNES) on 23 January, the day the minefield "E1" was being laid. It failed to return to base and there is no evidence as to what really happened to it. Therefore, it seems quite probable that it could be UKHO 9850. The wreck points outbound from the area, as it would have been if destroyed there. Of note is the fact that the minefield was laid just prior to *U1172*'s presumed departure from the area.

Again, on very thin evidence it is concluded that dimensionally and archivally it is UKHO 9850 *U1172* (MBES/Archive). Positionally, the minefield has to suffice. The attribution has been made in part based on my prior experience with U-boats of this period but is only slightly less speculative than the post war attribution made by the Admiralty. At least in this case there is the benefit of an actual, verified U-boat wreck to base the speculation upon, albeit now somewhat esoteric, as its actual identity is almost impossible to ever know for certain now.

The UKHO legacy attribution in this case is UKHO 9850 ETHEL (without a (POSSIBLY) or (PROBABLY) suffix, meaning that for some unknown reason this wreck was assumed to be CONFIRMED). This was reverted to UNKNOWN during Phase 1 as it was obviously not a ship.

Finally, it should be noted that the efficacy of the deep-trap minefields has been shown in recent years to have been significantly greater than presumed at the end of WW2. Records in 1946 showed that it was believed that only three U-boats had been lost in these fields, *U242* (as described), *U275* and *U260*. In fact, there are now a further six U-boats lying in the areas of anti-U-boat minefields, *U1172*, as described and *U325*, *U1021*, *U480*, *U683* and *U400*. Whereas the actual identities of these wrecks will remain permanently uncertain in most cases, their actual physical presence shows that the relative efficacy of these fields was in fact 300% greater than presumed in 1945 (McCartney 2014, p307-310).

Drawing the Phase 2 process to a conclusion

While it was not anticipated that every one of the shipping losses would be reconciled to a surveyed wreck site during Phase 2, it was expected that a high proportion would be. The question was, how many? The Phase 2 process took several months complete. As the case studies show, some shipwreck identifications could be resolved relatively easily. For example, it is difficult to conceive that the wrecks now identified as 9912 MESABA (MBES/Archive) and 9913 CITY OF GLASGOW (MBES/Archive) in case study 3 above, could in fact be anything else but those ships. Other cases, such as 9945 BURUTU (MBES/Archive) in the case study 4 above, involved further research and archival visits to acquire the data needed for resolution.

In practice the list of lengths of the shipwrecks in the database was combed through several times and the more readily obvious matches by dimension were assessed, resolved, and written up as the process moved along. After each pass through the lists, the number of candidate shipwrecks from survey and archive declined so that the data pool was smaller, and the reconciliations became easier to spot and to assess. There were on occasion amendments and changes made to previous attributions as the evidence in each new case was examined.

However, at some point it was evident that the process could go no further and would have to be terminated before the basic tenets of the (MBES/Archive) methodology were no longer being observed. The decision taken was to attempt to identify each DTM as far could be done from the evidence available and conversely that every archivally recorded shipwreck should be given a chance to be assessed against a DTM. But in order to qualify for the (MBES/Archive) attribution, some degree of all three of dimensional, positional, and archival evidential criteria was essential in each case.

As the number of wrecks left to assess declined, quite often the remainder were the cases with weaker evidential bases, especially in terms of textual descriptions of the sinking and the positions where the ships were lost. The final 14 Phase 2 assessments that were made in accordance with the (MBES/Archive) methodology were based on what could be termed sub-optimal positional evidence. In these cases, only an area, such as "Caernarfon Bay", or a route such as "Holyhead to Dublin" were archivally recorded and the areas had to substitute for more accurate positions. *LCT 326* (see case study 7) is an example. However, by this time there were quite often only a few unattributed surveyed wrecks left in these areas and matches could be seen and were made according to the methodology.

However, as the evidence in the remaining cases became increasingly thin, a line had to be drawn beyond which attributions would not conform to the methodology. The final attribution made was 10015 GLYNWEN (MBES/Archive). The details in the datasheet, as shown in Appendix 2 and in Figure 4.19 show that the evidence available at this time had reached the lowest acceptable level for an (MBES/Archive) attribution.

The ship is listed in *Wreck Returns* (1940, Qtr. 4 p9) as having sunk in the Irish Sea after striking a submerged object whilst sailing from Workington to Devonport. This appears to conclusively place it in the study area. Of the unidentified surveyed wrecks still left by this time, only one candidate was dimensionally consistent with SS *Glynwen* and on the ship's likely route. This wreck also manifested a layout which was visually similar with the plans of *Glynwen* held at the Lloyd's Register Foundation, Heritage & Education Centre (LRF-PUN-W399-0009-P). Of particular note is the layout of the holds, with a notably larger hold astern of the superstructure. The ship was well-decked with the quarterdeck being

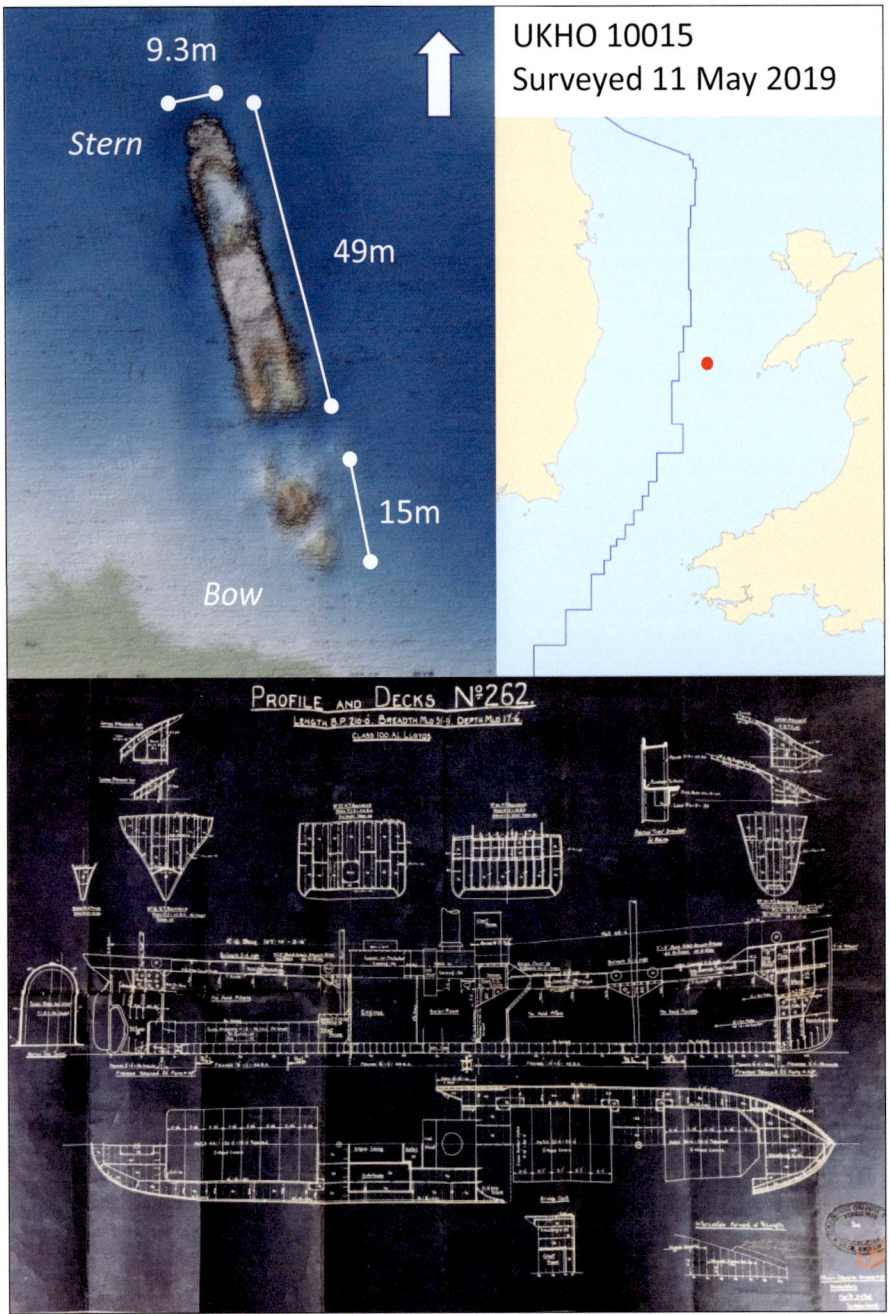

Figure 4.19. UKHO 10015 which by the thinnest of acceptable evidence was given the attribution 10015 GLYNWEN (MBES/Archive). In this instance there is dimensional, positional, and archival consistency, although the positional evidence is weak. The layout of the holds can be seen in the ship plans (Innes McCartney & Lloyd's Register Foundation, Heritage & Education Centre).

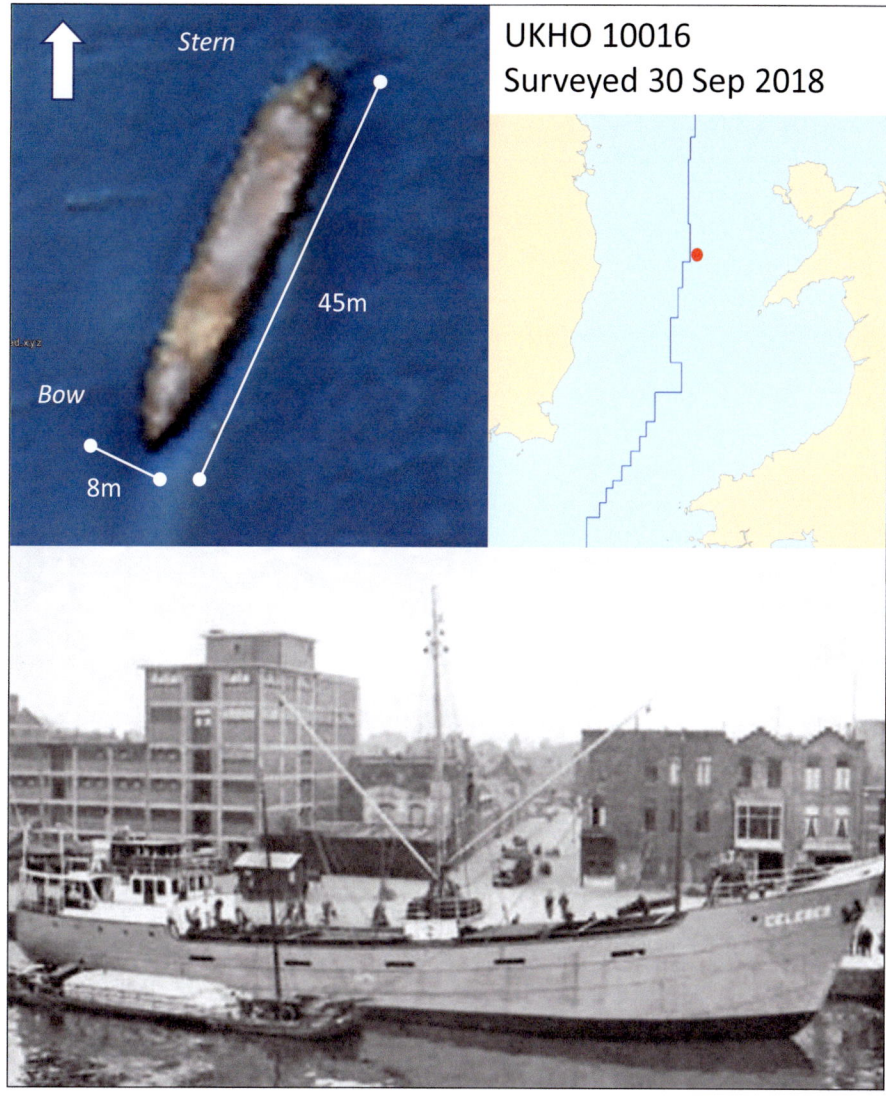

Figure 4.20. The surveyed wreck UKHO 10016 bears some resemblance to the missing MV *Celebes*, but there was not enough evidence available for an (MBES/Archive) attribution to be made. In this case, along with 31 other wrecks, it remains UNKNOWN (Innes McCartney & wrecksite.eu).

33m long (*Lloyd's Register* 1940-41) and it was measured at exactly that length, from the stern to what appears to be the forward end of the superstructure on the DTM. The ship was also on the correct heading.

Therefore, by the definitions of the (MBES/Archive) process the wreck was dimensionally, positionally and archivally consistent with *Glynwen*, but only just. The point to be made in this example, is that the process delineated for an (MBES/Archive) attribution has been followed and that this was the outcome of that process. Positionally the evidence is weak, but there was positional evidence and therefore an attribution could be attempted. This was the final attribution made.

To illustrate the dividing line between an acceptable (MBES/Archive) attribution and a case with some merits, but not enough evidence to complete an attribution, Figure 4.20 shows UKHO 10016. In this instance the wreck is consistent with the ship MV *Celebes*. This Dutch ship disappeared on a voyage from Liverpool to Falmouth, taking it through the study area. It was listed as missing in *Wreck Returns* (1941 Qtr. 2 p8). It is also listed in *War Losses* (1991, p1292) as a missing ship, lost to war causes.

In this case the dimensions of *Celebes*, 47.6m x 8.1m (*Lloyd's Register* 1940-41) are very similar to the dimensions of UKHO 10016 and even some features in the DTM look similar. Also, the wreck is on the expected route and heading in the direction one might expect to see on the DTM. However, while this evidence is consistent, it is not enough for an attribution because the positional evidence is too vague, and the wreck could easily be outside the study area. Therefore, on this occasion the wreck remains as 10016 UNKNOWN. However, on the datasheet it is noted that the wreck could be *Celebes*, but there is not enough evidence for a (MBES/Archive) attribution to be made.

Phase 2 results

The final 129 shipwrecks which were given (MBES/Archive) attributions have all been subject to the same process and assessed against similar evidence in all cases. The case studies and the datasheets in Appendix 2 describe this process in all the cases resolved and readers can see how the identifications were made.

The impact of Phase 2 on the research can be seen in Figure 4.21, which spatially shows the distribution of the UNKNOWN wrecks identified during Phase 2. By comparing this map with the similar one compiled after Phase 1 (see Figure 3.7), the extent to which Phase 2 has transformed our knowledge about these shipwrecks is readily apparent. Phase 2 has placed identities on 80% of all of the 161 UNKNOWN wrecks in the study area.

There appears to be some spatial bias in the results towards the areas further offshore and in water deeper than 50m, but this is a consequence of what was already known of the shallower, more coastal sites, as seen at the end of Phase 1. The area with the highest density of shipwrecks is to the north of Anglesey. In this area, five cases have been resolved. The more remote cases appear evenly spread.

After Phase 2 was completed, there was a remainder of 32 UNKNOWN wrecks which could not be identified. There does also appear to be a cluster of seven still UNKNOWN shipwrecks to the north of Anglesey, with four of them being close together. They lie at the centre of the busy Dublin-Holyhead-Liverpool seaway. It is the area most densely packed with shipwrecks, so it is perhaps not surprising.

The remaining unattributed shipwrecks and archival cases in the study area

It should be of no surprise noting that once the assessment process had run its course there would still be some unresolved archival cases and UNKNOWN wrecks. The sea will always hold some of its mysteries. The case of SS *Chanticleer* (see Chapter 1) is a salutary reminder that ships can simply disappear in places where they may not be expected to be found. In the final reckoning there were 21 shipwrecks archivally recorded in the historical text as lost in the study area which could not be matched to the survey DTMs. There were a further 13 shipwrecks identified during the archival research as missing cases which in theory could have passed through the study area

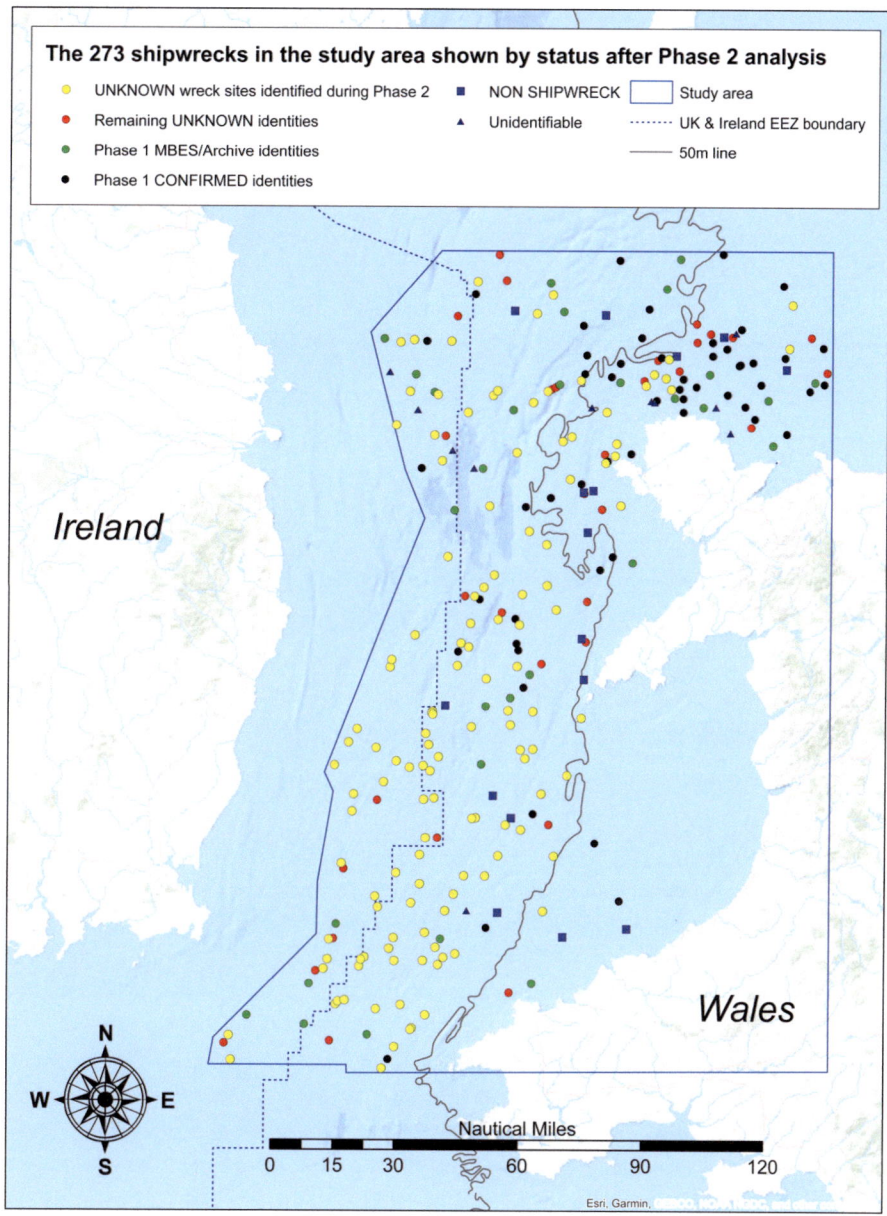

Figure 4.21. The results of Phase 2 on the 273 shipwrecks surveyed. By comparing this map with Figure 3.7, the impact of Phase 2 becomes apparent. It has transformed what is known of these shipwrecks and created a detailed shipwreck inventory (Innes McCartney).

and sunk within it. For example, one of these is *Celebes*, as previously described. So, in total 34 archivally recorded shipwrecks which either should or could be in the study area were not accounted for during Phase 2.

It can be little more than a coincidence that the number of surveyed shipwrecks which could not be matched to the historical text was actually 32; almost identical to the number of shipwrecks in the historical text which could not be matched to the surveyed

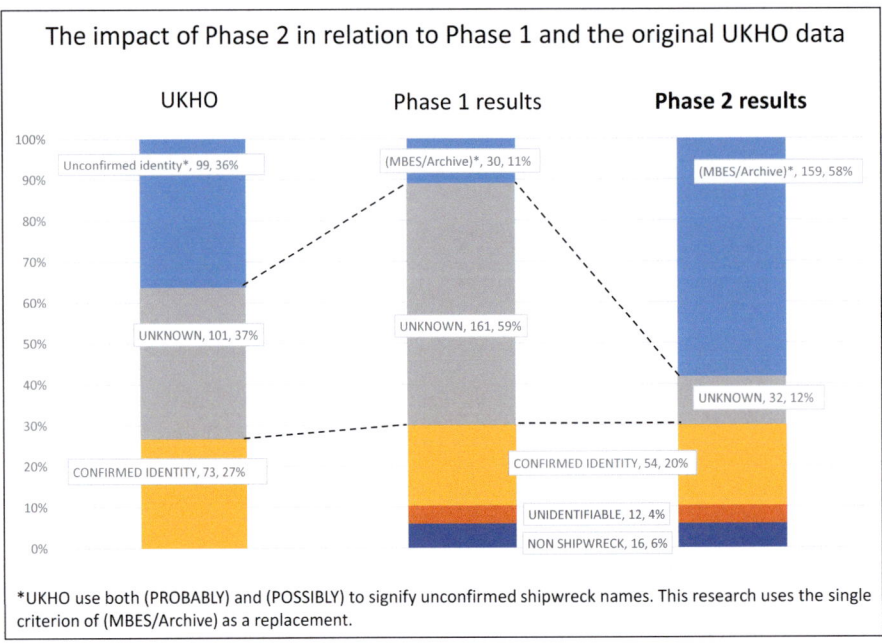

Figure 4.22. The impact of the research through both Phase 1 and Phase 2, showing the reduction of the number of UNKNOWN wrecks in the study area from 59% to 12% (Innes McCartney).

shipwrecks. In reality no shipwreck research project, in my experience would normally expect to create such similar numbers between datasets based on archaeological remains compared to the historical text, but that is how the research came to look once the archival research phase, Phase 1, and Phase 2 had been completed.

However, due to the arbitrary nature of the boundaries placed over a seascape of events which happened long before, it is impossible to know exactly how many shipwrecks should be within the study area. There are several cases on the edges of the study area which could, in theory at least have sunk within it, and conversely the reverse is equally possible. It is also a moot point as to whether every shipwreck has now been found by UKHO surveys. The answer is impossible to know for certain, but it appears that if there are any wrecks still to be found, their numbers in the UK EEZ, at least are relatively limited now and have had a minimal, if any effect on the research.

Phase 2 conclusions

Figure 4.22 statistically shows the impact Phase 2 has had on the overall research. This graph presents the data seen in Figure 3.6 and adds the Phase 2 results to it. The overall percentage of UNKNOWN wrecks has fallen from 59% to 12%. Clearly the numbers of CONFIRMED, NON SHIPWRECK and UNIDENTIFIABLE have not changed since Phase 1. It shows that now completed, the research has established that overall, 78% of the wrecks surveyed have now been identified, based on the strict criteria for a CONFIRMED attribution (Phase 1) or the (MBES/Archive) methodology. When the UNIDENTIFIABLE and NON-SHIPWRECK wreck sites are discounted, because they cannot be identified, the actual overall percentage of identified wrecks is 87% (being 213 of the 245 identifiable

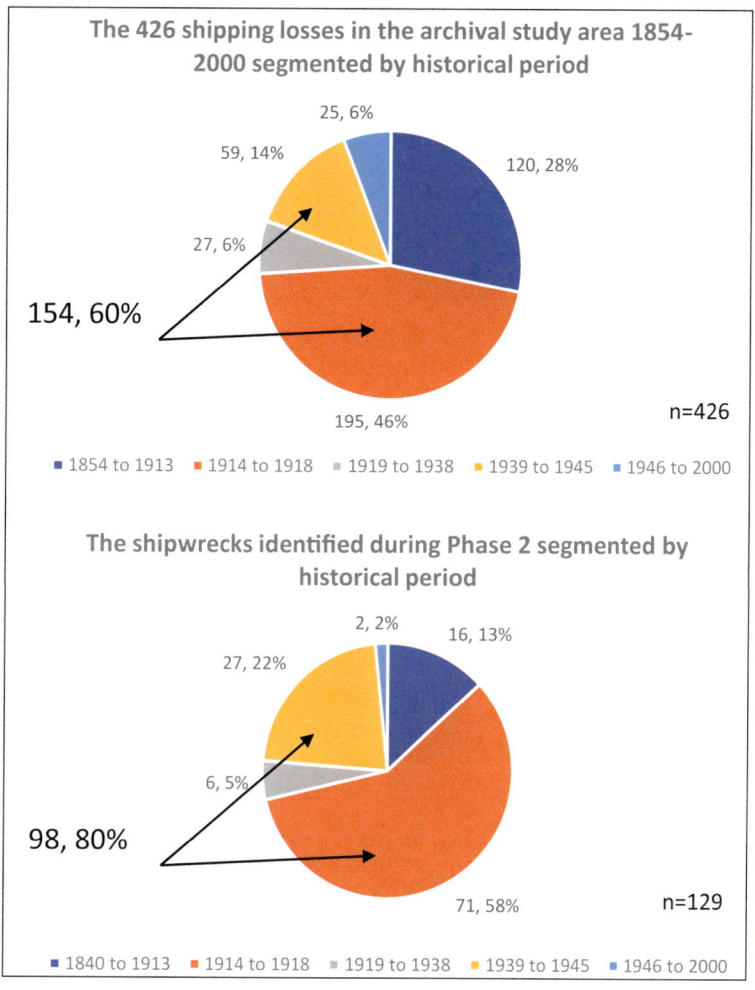

Figure 4.23. The chronological profiles of the historical text (above) and the 129 shipwrecks identified in Phase 2 (below). The world wars remain the dominant feature (Innes McCartney).

wrecks), leaving only 32 UNKNOWN. This exceeds my initial expectations when the research began in 2018. Also, it should be noted that Phase 2 identified 80% (129 out of 161) of the UNKNOWN wrecks.

Moreover, Figure 4.22 clearly shows the necessity for the Phase 1 research. By expunging all of the UKHO legacy attributions from the data and correctly deriving the number of UNKNOWN wrecks, it created a level basis upon which to begin Phase 2. In was essential to chronologically process both phases one after the other in order to accurately derive the final results of the research. It is likely that this represents the highest quota of identified shipwrecks in any stretch of water of comparable and size with a similar shipwreck density anywhere in the world.

The chronological profile of the Phase 2 identifications compared to the historical text

In Chapter 2 the chronological profile of the 426 shipwrecks recorded in the historical text as sunk offshore in the archive study area was presented in Figure 2.2. It will be recalled that it showed that 60% of the ships recorded sunk were from the world wars. The pie chart from Figure 2.2 is shown again in Figure 4.23 alongside the pie chart showing the chronological profile of the 129 shipwrecks identified during Phase 2. It can be seen that the percentage of inter-war wrecks is similar in both profiles, however, there are some notable differences between the two.

There is a reduced percentage of post WW2 wrecks (6% to 2%) which is explained by the fact that positional recording of the losses has improved during this time and the UKHO database has subsequently proven to be far more accurate in pinpointing these wrecks and legacy issues are hardly present. Therefore, the pool of wrecks from this period left to identify during Phase 2 was only two wrecks.

The percentage of pre-WW1 wrecks located during Phase 2 is less than half (13% compared to 28%) that of the chronological timeline of the historical text. This is primarily due to a lower evidential basis in some of these cases which has made pinpointing them a challenge. Of the 21 ships recorded as lost in the study area and not located during Phase 2, six (29%) of them come from this period. Establishing valid positional evidence was the greatest challenge faced in these cases.

Conversely the higher evidential basis commonly seen in the wartime losses to enemy action meant that these cases were more likely to be resolved, (80% against the 58% of the historic text) because good, dimensional, positional, and archival evidence was available. Of the 21 ships recorded as lost in the study area and not located during Phase 2, 14 (66%) of them come from this period. Prior to WW1, the majority of shipping losses occurred on inshore channels and by stranding. The offshore areas saw increased populations of shipwrecks due to wartime losses to enemy action, explaining their dominance.

Shipwreck dimensions

Of the three strands of evidence used to make Phase 2 attributions, the accurately recorded dimensions of the shipwreck DTMs were essential in order to reconcile them with the historically recorded shipping losses. Early on during Phase 2 it was recognised that when the historical losses database was sorted by ship's length that there were lower populations of the dimensionally longer and shorter ships. Figure 4.24 shows the normal distribution curve for the ships archivally recorded as lost in the study area by their length.

As seen in the case studies, there was very high confidence that such cases as SS *Mesaba* and MV *Rotula*, were likely to be accurate when resolved. The normal distribution curve shows why this would be the case, because those wrecks are among the most dimensionally rare in the study area. As Phase 2 progressed and relative dimensional rarity fell off, the other strands of evidence became crucial in making the attributions.

The higher population of shipping losses around 70m in length had an impact on the amount of time and effort required to assess these cases. This was especially so in areas where they tended to form clusters. In these instances, it was sometimes a challenge to fix each recorded shipping loss to the most likely candidate wreck. It was in these cases that positional and textual evidence was so important in being able to differentiate them. All the available evidence had to be carefully considered in each case in order to identify the wrecks according to the (MBES/Archive) methodology.

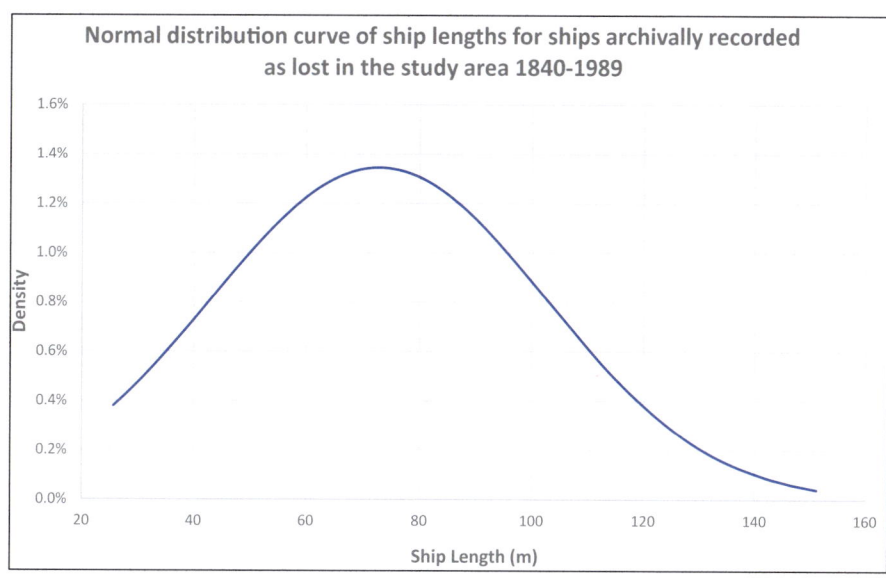

Figure 4.24. The normal distribution curve of the dimensional length of the ships archivally recorded as lost in the study area (Innes McCartney).

	Wrecks identified by (MBES/Archive) during Phase 2 (nm)	CONFIRMED wrecks assessed during Phase 1 (nm)
Mean	5.2	4.5
Median	4.5	2.2
Mode	5.6	0.1
Range	19.6	34.9

Table 4.1. The distances between archivally recorded positions of shipping losses and the actual locations of the wrecks as seen in the CONFIRMED cases from Phase 1 and the 129 outcomes from Phase 2 (Innes McCartney).

Shipwreck positions

One aspect of my previous work prior to this project which encouraged me to believe that the research could yield success, was my experience with the application of loss positions from the historical text to the archaeological remains of ships. Positional recording by dead reckoning navigation during the age of steam could be very accurate. For example, when in 2014 we found and surveyed 22 of the ships sunk during the Battle of Jutland 1916, it was noted that, when witnessed to sink, the wrecks were located on average only 3.5nm from the positions recorded (McCartney 2016, p245). It was of particular interest to see how the archivally recorded positions of merchant ships sunk would compare to the Jutland result.

Table 4.1 shows the distances between the historic positions and the actual wrecks in the cases identified in Phase 2 and the wrecks with CONFIRMED identities, as assessed during Phase 1. The mean distance is 4.5nm for CONFIRMED and 5.2nm for Phase 2. This equates very similarly to what was seen at Jutland and is very encouraging when considering the overall accuracy of the research. However, there are differences between the two which require explanation.

The median and mode are considerably lower in the CONFIRMED cases. The mode is so low among the CONFIRMED cases primarily because in these instances the ship sank and was immediately located. These are predominately cases of ships which have sunk in the post-WW2 era, including modern trawlers etc. Similar cases have had the effect of lowering the median as well.

Also of interest is the range. In the case of the CONFIRMED shipwrecks, the distance of 34.9nm relates to the loss of SV *Loweswater* in 1894. The position in *Shipping Casualties* (BOT 1894, p130) was speculative, being: *"supposed on West Hoyle Bank, Liverpool Bay"*. The wreck was identified by the recovery of its bell in 1991 from UKHO 7366 and plots approximately 35nm west of this position.

The interesting aspect of the difference in range in this instance is that during Phase 2 such a positional discrepancy would most likely not have permitted an attribution to have been considered. This has potential implications for the Phase 2 results overall. It is anticipated that the 21 shipping losses and some of the 13 missing cases which could not be identified as wrecks may be ones where such positional discrepancies lie. Some could be related to the 32 surveyed wrecks which are still UNKNOWN. The case of *Celebes* is relevant in this regard, although in that instance the positional evidence from archive was too vague to plot. This is only conjecture. As previously stated, the sea will always hold some mysteries. Until the wrecks are identified and then the seemingly implausible positional discrepancies can be explained.

Evidence from the historical text

The 3rd string of evidence used to make the Phase 2 attributions was based on textual descriptions from archival research. When seeking evidence of this type, any source pertaining to a shipping loss which could potentially bring evidence to each case was collected. Figure 4.25 shows the percentages of the evidence from archive considered to have been the most important in each of the 129 cases resolved during Phase 2.

The decision as to what was the most important evidence in each case is based on what is cited in the datasheets. It is evident that when available, accounts of survivors have proven to yield the types of evidence which is most important in this type of work and of the highest priority to locate archivally. The type of evidence yielded by survivors would include some indication of the damage sustained by the ship and importantly its location in the ship's structure. It could also include a description of how the ship sank, whether it broke into pierces etc. Beyond survivors, other witness accounts would be found in the newspapers, books and, U-boat KTBs.

Other evidence of utility if survivor evidence is not forthcoming is spread among whatever else may be available in each instance. In reality the U-boat KTBs show they witnessed events which were often very similar to the survivor accounts, but they did also provide evidence in cases where there were no survivors. In peacetime, newspaper articles proved the most useful, as did *Shipping Casualties* and *Wreck Returns*.

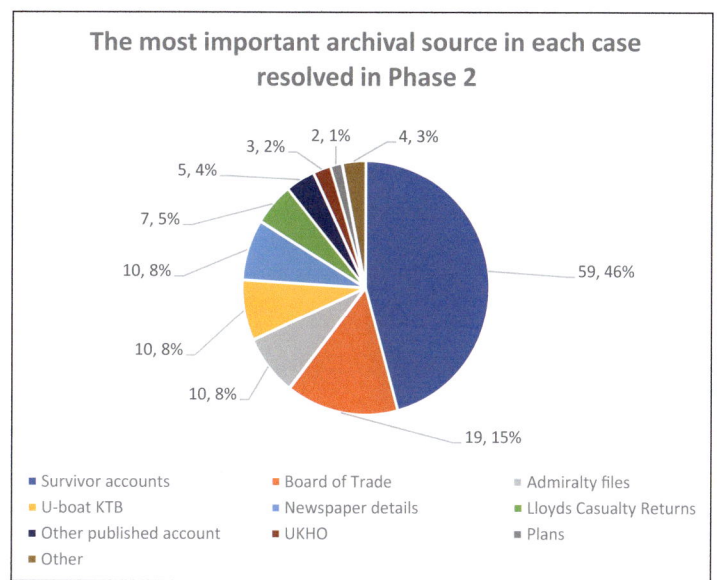

Figure 4.25. The archival evidence considered most important in the resolution of the 129 cases in Phase 2 (Innes McCartney).

Accuracy of the Phase 2 results

Overall, the outcomes of the research offer a major step forward from the shipwreck identities given in the UKHO database. They have made a transformative step change to our knowledge of shipping losses in the area and identified up to 87% of them. Time will ultimately tell how accurate the results of Phase 2 will prove to be. As the nine case studies in this chapter and the 129 datasheets in Appendix 2 show, some cases cannot in reality be anything other than the result shown and they will inevitably prove to be as described.

However, there is little doubt that some degree of adjustment may be needed in the future, as the shipwrecks are physically identified to the status of being CONFIRMED cases. Any adjustment needed will be a simple task, because every shipwreck has been measured and its dimensions and the reason for its attribution being made has been recorded, so the actual number of dimensionally consistent candidates in any given area, which could plausibly be amended is limited.

The primary reason for publishing all of the datasheets is so that readers can assess the evidence in each case. They are all slightly different and have varying aspects of emphasis across the three strings of evidence: dimensions, positions, and archival descriptions. In replacing the (POSSIBLY) and (PROBABLY) UKHO criteria with a single (MBES/Archive) criterion, it is acknowledged that some outcomes will seem to be more "probable" than others, but at the same time it seeks not to pigeonhole each case, but to let the available evidence speak for itself. The full transparency of the results allows readers to see how the attributions were made.

Finally, it should be clearly noted for readers who would seek to use this publication as a shipwreck guide, that it is reiterated it was not devised for that purpose and to do so may lead to disappointment. The outcomes are not guaranteed. They are not my opinions, but the outcome of a specifically designed research process. They are the results of a work of applied archaeological theory. The ultimate outcomes in each case are the results of the three distinct layers of research; archival, Phase 1 and Phase 2, validated using three distinct evidential bases: shipwreck dimensions, shipwreck positions and the historical text. They are optimum outcomes the research process generated from the evidence available.

References

Admiralty, 1947. *Ships of the Royal Navy statement of losses during the Second World War*. London: HMSO.

Board of Trade, Marine Department (various dates). *Return of Shipping Casualties and Loss of Life*. London: HMSO.

Elder Dempster & Co. Ltd., 1921. *The Elder Dempster Fleet in the War 1914-1918*. Liverpool: The Elder Dempster & Co. Ltd.

Gibson, R. H. and Prendergast, M., 1931. *The German Submarine War 1914-1918*. London: John Constable Ltd.

Hashagen, E., 1931. *The Log of a U-boat Commander or U-boats Westward 1914-1918*. London: Putnam.

Hocking, C., 1969. *Dictionary of Disasters at Sea During the Age of Steam: Including sailing ships and ships of war lost in action 1824-1962. Vol II, M to Z*. London: Lloyd's Register of Shipping.

Larn, B. & Larn, R., 2000. *Shipwreck Index of the British Isles, Volume 5: West Coast & Wales*. Redhill: Lloyd's Register.

Larn, B. & Larn, R., 2002. *Shipwreck Index of Ireland: Volume 6 – All Ireland. Part of the Shipwreck Index of the British Isles series*. Redhill: Lloyd's Register.

Lloyd's Register Foundation, Heritage & Education Centre (various dates) *Report of Survey: SS Burutu*. London: Lloyd's Register.

Lloyd's Register Foundation, Heritage & Education Centre (various dates) *Plan of SS Soreldoc LRF-PUN-004512-004519-0198-P*. London. Lloyd's Register.

Lloyd's Register Foundation, Heritage & Education Centre (various dates) *Plan of SS Glynwen LRF-PUN-W399-0009-P*. London: Lloyd's Register.

Lloyd's of London, 1989. *Lloyd's War Losses: The Second World War Volume 1: British, Allied and Neutral Merchant Vessels Destroyed by War Causes*. London: Lloyd's of London Press.

Lloyd's of London, 1990. *Lloyd's War Losses: The First World War: Casualties to Shipping Through Enemy Causes 1914-1918*. London: Lloyd's of London Press.

Lloyd's of London, 1991. *Lloyd's War Losses: The Second World War Volume 2. Statistics Showing Monthly Losses of British, Allied and Neutral Merchant Vessels*. London: Lloyd's of London Press.

Lloyd's Register of Shipping (various dates). *Lloyd's Register Wreck Returns*. London: Lloyd's Register of Shipping.

Lloyd's Register of Shipping (various dates). *Lloyd's Register of Shipping*. London: Lloyd's Register of Shipping.

Ministry of Defence, Director of Naval Warfare, 1973. *British Mining Operations 1939-1945, Volume 1. Naval Staff History BR1736(56)(1)*. Portsmouth.

Ministry of Defence, Director of Naval Warfare, 1977. *British Mining Operations 1939-1945, Volume 2. Naval Staff History BR1736(56)(2)*. Portsmouth.

McCartney, I., 2014. *The Maritime Archaeology of a Modern Conflict: Comparing the archaeology of German submarine wrecks to the historical text*. New York: Routledge.

McCartney, I., 2016. *Jutland 1916: The Archaeology of a Naval Battlefield*: London: Bloomsbury.

National Archives and Records Administration (various dates), 1984. *Microfilmed Records of the German Navy Roll No T1022 PG61626 35*. Washington DC: NARA.

National Archives and Records Administration (various dates), 1984. *Microfilmed Records of the German Navy Roll No T1022 PG61873 42*. Washington DC: NARA.

Naval Historical Branch, (various dates). *Foreign Documents Section Files, Loss of U683, with reference also to the loss of U1208, U1276, U927, U275, U1195, U246, U1169, and U242 FDS 476/97*. Portsmouth: Naval Historical Branch, Ministry of Defence.

Naval Historical Branch, (various dates). *Foreign Documents Section Files The destruction of U1051 and U1172 FDS D/NHB/22/2(849)*. Portsmouth: Naval Historical Branch, Ministry of Defence.

Rohwer, J., 1999. *Axis Submarine Successes of World War Two*. London: Greenhill.

Spindler, A., 1941. *Der Handelskrieg mit U-Booten, Bd.4, Januar bis Dezember 1917*. Berlin: Verlag von Mittler & Sohn.

Spindler, A., 1966. *Der Handelskrieg mit U-Booten, Bd 5, Januar bis November 1918*. Berlin: Verlag von Mittler & Sohn.

The National Archives (various dates). *British Merchant Vessels sunk and captured by the enemy, August 1914 to December 1915. ADM 137/2959*. London.

The National Archives (various dates). *Irish Sea: Various, 1917. ADM 137/1361*. London.

The National Archives (various dates). *Convoys: OL21 – OL41 ADM 137/2618*. London.

The National Archives (various dates). *Precis of reports of Homeward Convoys, May 1917 November 1918 ADM 137/2657*. London.

The National Archives (various dates). *South West Approach: German Submarines, 1-15 September 1918. ADM 137/1502*. London.

The National Archives (various dates). *Court Martial of Temporary Lieutenant B Palmer RNVR for the hazard and loss of HMS MERCURY ADM 156/216*. London.

The National Archives (various dates). *Portland, Portsmouth, and Western Approaches: War Diaries ADM 199/1443*. London.

The National Archives (various dates). *Director of Torpedo, Anti-Submarine and Mine Warfare Division: proceedings of U-Boat Assessment Committee, reassessments, and identity of U-boats. ADM 199/1786*. London.

The National Archives (various dates). *Survivors' Report: Merchant Vessels Indexed. ADM 199/2136*. London.

The National Archives (various dates). *HMS COTILION: report on passage of 7th LCT Flotilla and loss of LCT 326 . ADM 217/13*. London.

The National Archives (various dates). *War diary summaries: situation reports ADM 199/2225*. London.

The New York Times. The Waesland Disaster. *The New York Times*, 8 March 1902. New York: The New York Times.

The Times. The Sinking of the Waesland. *The Times*, 8 March 1902. London: The Times.

The Sphere. Sinking of the Atlantic Liner "Waesland" *The Sphere*, 15 March 1902. London: London Illustrated Newspapers.

5

Conclusions, impact, and significance

Introduction

Whereas the conclusions of the three strings of the research, archive, Phase 1, and Phase 2 are given at the end of their relevant chapters, this short chapter serves to draw together the overall results of this unique interdisciplinary study. It is intended to clarify for the reader both the practical impact of the research and its significance for wider applications in the real world. It will answer the question why is this research is important and prescient to archaeology, heritage, and marine sciences in 2022 and beyond?

In Chapter 1 I used the analogy of how aerial photography changed landscape studies forever and that MBES is offering the same epochal transformation to studies of the undersea landscape and to shipwreck archaeology. Conversely, it is increasing the need to understand the distribution and status of the shipwrecks being uncovered. This is crucial for environmental, legislative, and cultural reasons. The challenge of managing myriad shipwrecks on potentially vast scales is an emerging one. A nation may know where the shipwrecks are, but only ever partially knows what they are. This no longer needs to be the case. Shipwrecks can be identified in large swathes and then appropriately managed through targeted responses in means that were impossible up to now.

Looking back at the original research proposal, the previous chapters confirm that its aims and objectives have been more than achieved:

1. A globally important stock of MBES surveyed shipwrecks has been inventorised and provably linked to specific shipping losses represented in the historical text,
2. The development from scratch of a methodology to carry out the identification of the shipwrecks using MBES in combination with historical text has been successfully completed,
3. As the case studies and Phase 2 (Chapter 4) show, this methodology has created a finer-grained understanding of the extent, distribution, and identities of the 273 ships surveyed. The ships now exist in place, alongside time,
4. The research has clearly signposted the way by which many thousands of shipwrecks in any given region can be identified using the new methodology,
5. The research is easily applied to any offshore area which has been sufficiently surveyed, and has an archivally recorded history of shipping losses,
6. The research has provably shown that a low-cost (by the standards of maritime operations) approach to inventorising large numbers of shipwrecks in any given region now exists.

With the project successfully completed, its impact and significance can now be fully assessed.

Archival research phase

The results of Chapter 2 show that the historical text in the UK relating to offshore iron and steel hulled shipping losses is very comprehensive. In order to reconcile them with the shipwrecks a spatial database in GIS had to be developed. Due to the fact that the UK has no national database of historic shipping losses, it had to be built from scratch.

When completed, it was established that at least 426 shipwrecks fell into the iron and steel offshore category. It was noted that ships sunk during both world wars make a dominant 60% of these losses, mainly being from WW1. The UK-based records were then enhanced by databasing the records of the German navy from WW1, namely its U-boats which sank the greatest proportion of all of the ships lost.

In practice the archival research phase looked for and nearly always found positional data which was a key component of the triumvirate of evidence (dimensions, positions, and textual descriptions) used to identify the wrecks in Phase 2. The fact that by the end of the research 87% of the wrecks were named is in no small part down to this alone. Moreover, the fact that the mean distance between the historical text and matching archaeology is 5.2nm is of notable impact. Its significance is that it clearly demonstrates that the surviving records of shipping losses from the industrial age are perfectly capable of yielding the crucial positional evidence needed for this type of research.

Likewise, survivor descriptions which proved equally valuable as part of the "description" portion of the evidential triumvirate are proven to be more than plentiful. Where missing, the historical text yielded alternative evidence from other sources such as ship plans etc. In summary then, there is no shortage of archival evidence pertaining to shipwrecks lost at least around the UK and Ireland. Significantly, the historical text is fully capable of delivering the evidence required to roll out this research around the UK and its neighbouring nations. No claim is made here as to how this would work in other global regions with historic texts of differing quality or extent, although a recent limited study in Greece suggests similar approaches are applicable (Geraga et al, 2020).

Archaeology Phase 1 and the UKHO legacy challenge

The overarching lesson for users of the UKHO shipwreck database is to understand its history, purpose, and processes before using it for archaeological research. This is because as the results of Chapter 3 show that the UKHO legacy misattributions of the past significantly impact what is actually known of the UK's underwater cultural heritage. In the study area it affected 110 of the 273 wrecks we surveyed. This was a combination of 82 shipwrecks and 28 other seabed objects. By the end of the process the number of completely UNKNOWN shipwrecks had increased to 161 from 101.

In other words, whereas the quotient of UNKNOWN shipwrecks was originally presumed to be 37% of the dataset, it was in reality 59%. The conclusion to be drawn is that we know far less about the identities of the UK's shipwrecks than the current UKHO database may lead us to believe. By extrapolation therefore if this result is extended across the entire UK EEZ, it would be expected that 1,373 shipwrecks are currently misattributed. How the historical misattributions were made in the first instance is described in Chapter 3.

It was also noted that at one time, this figure would have been much higher, but that the reporting of finds by recreational divers, mainly in waters less than 50m depth has steadily improved the accuracy of shipwreck attributions in these shallower waters. It is the wrecks beyond 50m where the legacy misattributions still dominate.

Moreover, the UKHO attributions are to varying degrees currently incorporated in the UK heritage agency's own databases. For example, the NMR (English Heritage), adopted the UKHO database in 1992, hence legacy misattributions have seemingly found their way into the national heritage record and there is currently no means of correcting it (or even knowing which sites are affected) unless each site is subject to Phase 1 analysis. In the UK it means that meaningful inventorisation cannot go ahead until these issues are weeded out, because they will persist in obscuring the real identities of the wrecks until this is done.

However, in Ireland it appears that this is already happening. The study area straddles the EEZ border between Ireland and Wales. In Ireland INFOMAR manages the MBES inventory of shipwrecks. In the UK, the UKHO manages its surveys for charting purposes, but heritage and archaeology stand alone and distinct from it. The national differences are notable and significant.

Ireland: INFOMAR Shipwrecks Viewer

The way the shipwreck datasets are being managed in Ireland can be partially gleaned from examining the INFOMAR data online. INFOMAR is a DECC-funded joint programme between the Geological Survey Ireland and the Marine Institute, surveying Ireland's unmapped marine territory and creating a range of integrated mapping products of the physical, chemical, and biological features of the seabed (About Us | Infomar, 2021).

Within the MBES survey dataset of this research are 54 wrecks (20% of the total surveyed) which are in the Irish EEZ. They lie in the sliver of water between Ireland's 12 mile limit (where we did not have permission to survey) and its EEZ border with the UK and can be seen in the maps in this publication (e.g., Figure 1.5). Some of the 54 shipwrecks are visible on INFOMAR's superb shipwreck viewer portal (Shipwrecks Viewer | Infomar, 2021). Although now seemingly diverged, its shipwreck database was originally based on the UKHO one, so some legacy would be expected to remain, including UKHO legacy misattributions, which INFOMAR appears to have begun correcting.

Looking at the data INFOMAR provides, it appears that it has surveyed 10 the 54 wrecks during its own MBES fieldwork, presenting high resolution DTMs on its shipwreck viewer. Looking at these 10 sites, it appears to have carried out some degree of validation comparable to Phase 1 of this research, although from the limited descriptions given, it is difficult to discern a consistent pattern in its approach. There is no evidence of an attempt to identify any of the wrecks reverted to UNKNOWN, so nothing comparable to Phase 2 of this research is seen in the small sample.

Of the 10 cases, seven are identical to the outcomes of this research and can be seen to be absolutely consistent with it in Appendix 1; they are: UKHO 7039, 9865, 11805, 12135, 7023, 9870, 11796. A further site (UKHO 7091, see Appendix 2) was identified during Phase 2 and therefore will differ from INFOMAR's UNKNOWN attribution, which up to the end of Phase 1 was just as consistent with this research.

The two primary differences in wreck identity relate to UKHO 11788 and 11798 (see Appendix 2). The case of 11788 has been described in case study 5 in Chapter 4 and was identified during Phase 2 as 11788 HMS MERCURY (MBES/Archive). The historic UKHO attribution to this site being the submarine UC33 was obviously wrong as the wreck clearly is not a submarine. During Phase 2 this made UC33 available to be another wreck. Coincidentally, UKHO 11798 is in fact a submarine and in Phase 2 was identified as the correct resting place of UC33 (see Appendix 2).

INFOMAR had, consistent with this research, reverted 11798 to UNKNOWN presumably based on its dimensional discrepancy with the UKHO attribution. Therefore only 11788 is divergent with this research at the Phase 1 stage and it is odd because at Phase 1 it should have been reverted to UNKNOWN because the wreck is actually twice the length of a UCII class submarine. Why this inconsistency arose is difficult to discern from the evidence available on the shipwreck viewer.

However, the conclusion drawn from this very small dataset is that, where evident to it, INFOMAR has amended the obvious dimensional differences caused by the UHKO legacy attributions it inherited, in effect carrying out a validation exercise comparable to Phase 1. However, it has not gone further and attempted any widespread reconciliation with the historic text comparable to Phase 2.

Importantly if the validation seen in the seven sites above was consistent across all its shipwrecks it would be very well situated to undertake a Phase 2 type research project in the future. In fact, it could be a world leader in this field. It was very encouraging to see another organisation's MBES dataset being subjected to similar validation of the historic UKHO data, and I do hope the other 44 cases in our MBES dataset in Irish waters will be of utility to INFOMAR. INFOMAR is much further advanced in its inventorisation of shipwrecks than anything seen in the UK, which increasingly lags far behind.

Wales: U-Boat Project, 1914-18: Commemorating the War at Sea

The current situation in Wales and in the broader UK could not be more different from Ireland. UKHO is a separate and distinct organisation from the UK heritage agencies of England, Scotland, and Wales. The agencies have in the past adopted the UKHO data in slightly different ways into their own databases. The UKHO wreck cards (legacy attributions included) were accepted as written, there being no other option because traditionally there has not been any means of validating them, until now. Historically in Wales the Royal Commission on the Ancient and Historical Monuments of Wales (RCAHMW) has not used MBES data in anything more than site-specific instances and it is educative to see what happened when it attempted a broader shipwreck study based in part on MBES data.

The U-Boat Project, 1914-18: Commemorating the War at Sea was a RCAHMW run project, funded by the Heritage Lottery Fund. It *"commemorates the Welsh experience of the Great War at sea and the underwater sites that form its legacy"* and it is still hosted online, where the MBES data is presented alongside its analysis (Prosiect Llongau-U 1914-18 | U-boat Project Wales 1914-18, 2021).

The project website presents MBES models and interpretation of 18 shipwrecks it says were lost in WW1, 15 of which are inside the study area and can be compared to the research. In fact, the U-Boat Project used exactly the same MBES pointcloud data from Bangor University's School of Ocean Sciences as was used in this research. In 10 of the 15 cases (see Appendix 1, UKHO 7147, 7331, 7332, 7334, 7363, 7430, 9977, 9983, 9985 and 9989 (&68682)) the project showcased shipwrecks which UKHO attribute as being CONFIRMED. All of these cases are consistent with this research. One further case (UKHO 9984, see Appendix 1) listed as a (POSSIBLY) by UKHO, the project also showcased.

A further two sites are clearly UKHO legacy misattributions based on their dimensions (9987 and 9976 (&10020), see Appendix 2). Had basic validation of the UKHO data been carried out to a pre-planned process, then the attributions clearly would not have been made. On another site (UKHO 9975, see Appendix 2), it appears somebody noticed that the UKHO legacy attribution was dimensionally inconsistent with the name given. But no alternative is suggested, and in fact Phase 2 analysis in this research shows that the wreck appears most likely to have been sunk in WW2 (Mystery wreck – Prosiect Llongau-U 1914-18 | U-boat Project Wales 1914-18, 2021).

Finally, the last wreck (UKHO 9970, see Appendix 2) has been subjected to an attempt to identify its original UKHO UNKNOWN status, comparable to Phase 2. In this instance RCAHMW concluded that the wreck was a timber SV, *Walpas* which sank after burning to the waterline (WALPAS – Prosiect Llongau-U 1914-18 | U-boat Project Wales 1914-18, 2021). However, the wreck shown is upright, intact, clearly made of metal and showing no signs of burning to the waterline. As this research has shown such a detailed project requires careful methodical work, carried out to a prescribed process. Without the benefit of a fully built archival and spatial database covering the entire historic timeline of the industrial age to correctly identify the UNKNOWN wrecks, the result in this case was inevitable.

It is interesting to see the *U-Boat Project, 1914-18* has been accepted without qualification elsewhere. *The Welsh Wreck Web Research Project – 2020 / 2021* (www.madu.org.uk Page 4.42 – Research Project – 2020, 2021) was an amateur volunteer-based desktop project, sponsored and published by the Nautical Archaeology Society (NAS) (a UK-based charity) and the Malvern Archaeological Diving Unit (MADU) (a Wales-based archaeology consultancy). Again, in this instance no validation seemingly took place. Consequently, the same mistakes made by RCAHMW were simply duplicated (see UKHO 9987, 9976 & 10020, and 9970).

The overarching conclusion is that the UKHO legacy issue appears to be better understood in Ireland. In the UK to date, nothing has been done to address it, because it has not been perceived as a problem or perhaps, even where it has been, there is still no means of correcting it. MBES has yet to be used in any wide scale shipwreck inventorisation project in the UK. But significantly, the research has unequivocally pointed to the challenge UKHO legacy misattributions present. Their erasure via a Phase 1 assessment process is a prerequisite to any serious attempt to inventorise the UK's shipwrecks.

Archaeology Phase 2: towards a national shipwreck inventory

The results of Chapter 4 show that the overall number of UNKNOWN wrecks has fallen from 161 (59% of all 273 wrecks surveyed) to 32 (12% of all 273 wrecks surveyed). The actual overall percentage of sites which could be potentially identified has reached 87% (being 213 of the 245 identifiable wrecks), leaving only 32 UNKNOWN sites left. By extrapolation therefore, if rolled out across the UK EEZ up to 2,945 shipwrecks could be identified by replicating this research.

Moreover, looking specifically that the 161 shipwrecks established as UNKNOWN at the end of Phase 1, 80% (129 out of 161) were identified during Phase 2. Crucially, the actual identities of 62 (48%) of the 129 wrecks identified during Phase 2 were originally obscured by UKHO legacy misattributions and would never have been identified if Phase 1 research had not amended each case first.

Wartime losses accounted for 80% of the Phase 2 identifications, showing that the detailed reporting of losses compiled in wartime has resulted in a rich archival record which has yielded the highest chance of success when utilised to identify wrecks. In fact, the high evidential emphasis placed on deriving reported sinking positions from the historical text paid off well, with the mean distance between an identified wreck and its archival sinking position being only 5.2nm. This has a significant impact on how this research could be rolled out in other areas, as it gives a considerable degree of confidence that wrecks can in many cases be located near to the position where they were recorded lost. This is especially the case where evidence is gathered from friend and foe.

Whereas Phase 1 created the level playing field from which the research could be used to identify the UNKNOWN wrecks, it was Phase 2 which lay at the heart of the research. It has confirmed that MBES is an excellent tool for recording and analysing the details of any shipwreck. It is the key, in company with archival research, which is unlocking the means by which shipwreck inventories can be created on almost any geographic scale. Importantly, the truly multidisciplinary nature of this research is revealed when its impact on marine sciences is fully realised.

Traditionally in the UK, an archaeology project manager on a specific shipwreck might call in a specialist consultant to do a MBES survey, but this work is not widespread. To date MBES has not been employed to inventorise widescale groupings of UK shipwrecks. This research shows that MBES is not this year's trendy archaeology gizmo, it is the key to cutting the gordian knot of inventorising the UK's 6,239 offshore shipwrecks and many more worldwide. The concept has been tested and proven beyond doubt in this research.

Moreover, a further aspect of Phase 2 is its significant impact on the broader field of marine sciences. The key impact has been the consistent ability of this research to provide dating for the shipwrecks. As described in Chapter 1 the wrecks serve as proxies for artificially emplaced static objects on the seabed, but in order to fully realise their utility in this regard, it is essential to know when they sank. The research has been able to provide sinking dates for 87% of the wrecks in Bangor University's School of Ocean Sciences MBES dataset and has thereby provided its scientists with an extremely valuable resource in a world-leading first. The multidisciplinary and far-reaching impact and significance of Phase 2 inventorisation across its key subject areas is now described.

Pollution & environment

It is axiomatic that the extent of any potential undersea environmental problem, such as leaks from sunken tankers, cannot be properly assessed beyond the desktop unless the wrecks causing the problem are found. For marine scientists, knowing the identity of any potentially polluting wreck means that the environmental effects of its cargo can be investigated. Leaking cargoes can then be traced and analysed in the seabed around the wreck.

During Phase 2 of the research, four tankers were identified in the MBES dataset which had previously not been known wreck sites and significantly the identities of three of them had been obscured by UKHO legacy misattributions. The four are UKHO 7347 BRITISH VISCOUNT (MBES/Archive), 9925 ROTULA (MBES/Archive), 9975 RFA OSAGE (MBES/Archive), and 10013 RFA VITOL (MBES/Archive), see Appendix 2).

Three others were known sites, UKHO 7063, 7334, and 11805, see Appendix 1. Overall, the research has more than doubled the number of these potentially highly polluting wrecks in the study area. Of note are the 2 RFAs, VITOL is an Admiralty oiler, which was carrying heavy fuel oil for the fleet and OSAGE, which was thought sunk in the Irish EEZ, is in fact it seems, a UK problem.

Climate change & renewable energy

The effect of the marine environment on static objects can be better assessed if a datable shipwreck is present because it offers the means to refine models of computational fluid dynamics (Quinn et al, 2018). The effects of tide on shipwrecks has been shown not to be temporally linear in the Irish Sea, but knowing when scouring effects slow down, or even if they are permanent through time is extremely useful. For example, craters around one shipwreck seemingly caused by depth charges are still visible over 70 years later, showing that even strong tidal effects appear in some circumstances to be unable to infill certain seabed substrates, even many decades later. This can have significant impact in the fight against climate change, because with this knowledge, the placement of static objects such as windfarms and tidal generators can be more dependably situated and anchored.

Importantly, climate change has impacts on the marine ecology and the man made materials situated in the marine environment. Shipwrecks can act as proxies, if the specific types of iron and steel of the wreck's construction are known for certain. The effects of increased acidification and warming etc on known materials can be temporally measured with higher degrees of accuracy and predictability.

Marine life and & diversity

Marine life diversity and change can be very usefully studied on shipwrecks (Paxton et al, 2019). They are artificial reefs, the wildflower meadows of the undersea world compared to the mowed lawns, perceived to make up so much of the rest of the seabed where it is exploited as fisheries. Clearly knowing the date any wreck sank plays an important role in being able ascertain the temporal scale of the extent of life seen on each example and can help understand how quickly invasive species can settle into new territories.

Seabed geology & sediment research

Targeted regular surveys can be strategically managed to derive data from the best sites for research into how structures modify the surrounding seabed environment as they degrade (Majcher et al, 2021). All shipwrecks undergo modification due to continued exposure to marine processes, such as corrosion, sediment movement, storm activity etc. Crucially, the temporal nature of this change requires dated shipwrecks to develop accurate modelling.

Every type of shipwreck from submarines to tankers have their unique 'erodibility' profile and knowing a vessels identity helps to understand how it has changed the seabed surrounding it over time. Bangor University's School of Ocean Sciences has learned that this type of work is very well suited to MBES survey and has demonstrated that sonar surveys produce better results than visual inspections of wreck scours (which is all but impossible in much of the Irish Sea due to limited visibility), but these are only useful if the wreck is a known and datable object.

By understanding a known shipwreck's construction and knowing the way in which a ship was built (often synonymous with the way it which shipwrecks degrade), its status relative to the surrounding marine environment can improve fine-scale sediment transport models, and these in turn can provide insights into seabed geology, sub-surface stratigraphy and even the geotechnical properties of seabed material within the affected adjacent area. This can be extended when coupled with other surveying techniques such as sidescan sonar and sub-bottom profiling (Majcher et al 2020).

Bathymetric research

Shipwrecks can serve as high-resolution bathymetric spot samples, from which site-to-site comparisons can be accurately measured as each wreck site can effectively be used as a 'ground control point' for testing, calibrating, and improving national bathymetric survey grids and low-resolution, large-scale bathymetric data sets used in the development of numerical models.

Archaeology & heritage

History books and heritage websites will highlight the shipwrecks of the past which are of greater historical note, for whatever reason. But who is going to find them? The case studies clearly demonstrate that the research will pick up the notable just as efficiently as the mundane. Case study 2 revealed the locations of the potentially quite sensitive maritime graves of SS *Formby* and SS *Coningbeg*.

Case study 3 revealed the location of the historic, RMS *Titanic*-related wreck of SS *Mesaba*. These cases too could draw the attention of sea users for varying legitimate and nefarious purposes. In the study area several wrecks of importance for differing reasons have been found. This research is in the public domain and these ships now have a home in place as well as time. They are living entities, whereas before they resided on the pages of books.

Shipwreck vulnerability & the need for protection

It is not just the historic shipwrecks which are at risk of exploitation. Case study 4 in Chapter 4 showcased the case of the liner *Burutu* which sank in a collision between two convoys in 1918. According to the *Shipwreck Index* (Volume 5, section EK), *Burutu* was carrying a cargo of "*South African produce*" which allegedly included tin and copper. Apparently "*a number of major salvage attempts have been made to recover the more valuable items*".

If salvage companies have in fact located this wreck in the past, they may have done so furtively (i.e., without UKHO knowing of it and noting it as a record on a wreck card or in notices to mariners) and excavated the wreck for its valuable cargo. If so, their activity has left little archival trace beyond the aforementioned comment. However, looking at the DTM of the wreck, it could be interpreted as if hold No4 has been excavated, but that is not certain. If the wreck has in fact not been found in the past, then its cargo puts it immediately at risk.

Of course, vulnerability is not just linked to historic wreck sites and metal theft is not just linked to valuable cargoes. Metal theft is on the rise around the world. Importantly too, it is believed that climate change will cause an upswing in the demand for recyclable steels. Shipwrecks can make excellent donors of plenty of recyclable metals. The higher value sites, such as the ocean liners, all of which have been identified during this research (e.g., UKHO 9896 PORT TOWNSVILLE (MBES/Archive), see Appendix 2) will be at the forefront of a new wave of salvage activity, not seen around the UK for many years. How is this to be monitored if the locations of the vulnerable sites are not known or obscured by the misattributions of the past?

The research identified 243 shipwrecks by name, all of which are at some degree of risk in the future. Significantly, inventorisation projects can name wrecks, allowing for some degree of prioritisation to the monitoring of sites. Uninventorised valuable or famous shipwrecks cannot be monitored, and no desk-based study can save them from destruction. It was the saddest work experience of my life to see this for real with the

Battle of Jutland wrecks. This type of research, where fully introduced, ensures that it need not happen again.

Human casualties & family history

During the research, the databases built up of all of the shipwrecks expected to be found within the study area were designed to serve purposes beyond archaeology and ocean sciences. It is easy forget sometimes that even most commonplace of shipwrecks is likely to be the grave of somebody's ancestor. This is why the recorded numbers of human lives lost, when known was captured during the database building.

In fact, by the end of the research the 243 shipwrecks which had been named were noted to hold the known remains of 2,392 mariners and civilians that had died when the vessels sank. Importantly Phase 1 of the research uncovered the fact that the UKHO legacy misattributions had obscured the resting places for 1,085 of those lives. Phase 2 research was able to locate the wrecks which accounted for 1,272 lives lost. It is also important to note that the survivor accounts contain the names of many who lived through a sinking event. Clearly, these voices resonate with families and historians as they quite often have much to tell us.

If archaeology differs from history in its study of events in place as well as time, then it shares a great deal of affinity with personal and family history. The extent to which this matters can be measured on internet sites such as wrecksite.eu and on the numbers of emails fielded by Bangor University' s School of Ocean Sciences by people asking if certain wrecks have been found because their ancestors were lost on them or survived an event which marked them for life. Inventorising shipwrecks at the national scale creates the potential for so many new human connections to the past to be made in place and in time. This is an important human need and should not be overlooked. Also, personal history is one of the best ways to engage the public with shipwrecks.

Importantly, the compilation of the databases and the accompanying individual ship files has led to the drawing together of a unique array of sources from a range of historical texts. This opens up the possibility for a range of potential historical studies extending beyond family history. For example, by bringing together the survivors accounts of ships torpedoed, and the accounts of the U-boat commanders involved, the interactions between survivors in boats and the U-boats which approached them can be studied from both sides. This reveals occasions of great humanity, bravery, and suffering. The extremely diverse ethnic mix of the merchant crews was also noted in many cases.

Cost of fieldwork and research

The cost of the 273 shipwreck surveys which Bangor University's School of Ocean Sciences carried out, which formed the basis of this research also shows why this type of inventorisation project is attractive compared to the alternatives. It estimated that the highest cost per pointcloud was around GBP800. In other words, the entire dataset collection cost a maximum around GBP220,000. When the additional costs of post-processing and research are added, the entire project cost around GBP500,000 to bring to this point.

This is just a tiny fraction of what it would have cost to survey just a limited number of sites with divers. A rule of thumb is that marine diver surveys are around 120 times more expensive than geophysics. ROV surveys too require vessels with multi-point mooring capability or dynamic positioning and the lowest estimated cost per wreck would still be many times the cost of MBES survey.

Assuming costs were to be the same, the projected cost of rolling out this research across the remaining 5,966 of the UK's uninventorised offshore shipwrecks is GBP11,000,000; or the mere equivalent of 645 yards of new motorway at 2011 estimates (BBC.co.uk | The UK's last, great, expensive, short roads, 2021). This would be significantly reduced with access to MBES datasets, such as the Civil Hydrography Program and wind farm surveys. It seems small money to know the real identities of the UK's shipwrecks for environmental safety, our national history and heritage protection. With so many calls on the public purse, I have doubts the work will go any further.

Although the finances mentioned in relation to this research might appear high to the lay reader, in terms of marine operations they are in fact in the bargain basement. The cost to the UK of extending this work across the entire country is now known. There is simply no other current means of inventorising large numbers of shipwreck sites which is anything close to as cost effective. The processes needed to achieve this have been piloted, tested, and are now defined and are repeatable. This is what the research set out to investigate and its results provably confirmed its viability.

Concluding remarks

This research was a long time coming to fruition. Its gestation as an idea can be traced back to the days in the late 1980s when I first dived on unseen shipwrecks in UK waters. In those halcyon days for wreck diving fondly remembered, it was as if the more of them we saw, the more we realised that many were misattributed. Being able to quantify what it means in practice across 7,500 sqnm of sea has been something of a pressing need for a very long time now. Phase 1 was an essential prerequisite to identifying the wrecks and its results should not be ignored.

The potential for MBES surveys of the wrecks to work as well as it has in this research was made clear to me during my time in the North Sea and Scapa Flow with the Sea War Museum Jutland. The chief surveyor patiently taught me how to use this incredible technology for recording and processing shipwreck data. Once I had learned and seen the results of the broader surveys in the North Sea, the die was cast to apply this experience to the waters of my home country.

This is now done, and it reveals the huge potential for this type of research, extending far beyond just the benefits to archaeology. At the opening of Chapter 1, I described how this research can be seen as analogous to how aerial photography transformed studies of the UK's historic landscape. Its shipwrecks are now receiving similar treatment. This will only improve in scope and quality of resolution in the future, offering enhanced scientific research opportunities, greater environmental and heritage protections, stronger evidence-based wildlife diversity studies, better data for the renewable energy sector, and the elimination of pollutants from our seas. Of course, it also offers similar opportunities to criminals.

The UK is a seafaring nation, an island linked to the sea for all time and all history. Its waters consequently teem with thousands of shipwrecks. Such a vast assemblage could never in reality, be individually inspected, surveyed, and identified by artefact recovery. However, this research project sought to develop an alternative approach by applying archaeological theory to inventorise the wrecks, mitigating the need for widespread ground truthing by using geophysics.

The results seem to show that a means by which the intractable problem of knowing where the nation's stock of offshore underwater cultural heritage actually lies in place and in time has been developed to the extent that names can be given to the majority of the UNKNOWN wrecks. It is hoped that the process by which geophysics can enhance our understanding of the world's shipwreck heritage is a little better illuminated for the hardy pioneers who have chosen to embark on this new epochal adventure.

Dr Innes McCartney
Bournemouth University, December 2021
imccartney@bournemouth.ac.uk

References

BBC News. 2011. The UK's last, great, expensive, short roads. [online] Available at: <https://www.bbc.co.uk/news/magazine-13924687> [Accessed 18 November 2021].

Infomar.ie. 2021. About Us | Infomar. [online] Available at: <https://www.infomar.ie/about-us> [Accessed 19 November 2021].

Infomar.ie. 2021. Shipwrecks Viewer | Infomar. [online] Available at: <https://www.infomar.ie/maps/downloadable-maps/shipwrecks-viewer> [Accessed 15 & 16 November 2021].

Larn, B. & Larn, R., 2000. *Shipwreck Index of the British Isles, Volume 5: West Coast & Wales*. Redhill: Lloyd's Register.

Madu.org.uk. 2021. Page 4.42 – www Research Project – 2020. [online] Available at: <http://www.madu.org.uk/Page%204.42%20-%20www%20Research%20Project%20-%202020.dwt> [Accessed 26 November 2021].

Majcher, J., Plets, R. & Quinn, R. Residual relief modelling: digital elevation enhancement for shipwreck site characterisation. Archaeol Anthropol Sci 12, 122 (2020). https://doi.org/10.1007/s12520-020-01082-6

Majcher, J, Quinn, R, Plets, R, et al. Spatial and temporal variability in geomorphic change at tidally influenced shipwreck sites: The use of time-lapse multibeam data for the assessment of site formation processes. *Geoarchaeology*. 2021; 36: 429- 454. https://doi.org/10.1002/gea.21840

Geraga, M.; Christodoulou, D.; Eleftherakis, D.; Papatheodorou, G.; Fakiris, E.; Dimas, X.; Georgiou, N.; Kordella, S.; Prevenios, M.; Iatrou, M.; Zoura, D.; Kekebanou, S.; Sotiropoulos, M.; Ferentinos, G. Atlas of Shipwrecks in Inner Ionian Sea (Greece): A Remote Sensing Approach. Heritage 2020, 3, 1210-1236. https://doi.org/10.3390/heritage3040067

Paxton AB, Taylor JC, Peterson CH, Fegley SR, Rosman JH (2019) Consistent spatial patterns in multiple trophic levels occur around artificial habitats. Mar Ecol Prog Ser 611:189-202. https://doi.org/10.3354/meps12865

Quinn, R., Smyth, T.A.G. Processes and patterns of flow, erosion, and deposition at shipwreck sites: a computational fluid dynamic simulation. *Archaeol Anthropol Sci* 10, 1429-1442 (2018). https://doi.org/10.1007/s12520-017-0468-7

Uboatproject.wales. 2021. Prosiect Llongau-U 1914-18 | U-boat Project Wales 1914-18. [online] Available at: <https://uboatproject.wales/> [Accessed 16 November 2021].

Uboatproject.wales. 2021. WALPAS – Prosiect Llongau-U 1914-18 | U-boat Project Wales 1914-18. [online] Available at: <https://uboatproject.wales/wrecks/walpas/> [Accessed 17 November 2021].

Uboatproject.wales. 2021. Mystery wreck – Prosiect Llongau-U 1914-18 | U-boat Project Wales 1914-18. [online] Available at: <https://uboatproject.wales/wrecks/rose-marie/> [Accessed 17 November 2021].

Appendix 1

This table lists all 273 shipwrecks featured in the research. It shows their original UKHO identities and how or if these were amended during the research process. The details of the 129 of these wrecks which were identified during Phase 2 are given in Appendix 2.

LIVE UKHO No.	UKHO Name	UKHO Status	Status after Phase 1	Phase 1 status change	Primary Reason for Change	Phase 2 Identification	Phase 2 status change	Final Status
6985	UNKNOWN	UNKNOWN	UNKNOWN	No	N/A	TARBETNESS	Yes	MBES/Archive
6986	EARL OF ELGIN	Confirmed	UNKNOWN	Yes	Dimensions of wreck	CYRENE	Yes	MBES/Archive
6987	UNKNOWN	UNKNOWN	UNKNOWN	No	N/A	EARL OF ELGIN	Yes	MBES/Archive
6992	TRINIDAD	Probably	UNKNOWN	Yes	Dimensions of wreck	MEXICO CITY	Yes	MBES/Archive
6993	BALDERSBY	Probably	MBES/Archive	No	N/A	N/A	No	MBES/Archive
6994	FOSTIMORE	Possibly	UNKNOWN	Yes	Dimensions of wreck	PSS WILLIAM HUSKISSON or MANCHESTER	Yes	MBES/Archive
6999	ENERGY	Possibly	UNKNOWN	Yes	Features of wreck	ALCAZAR	Yes	MBES/Archive
7000	MEXICO CITY	Confirmed	UNKNOWN	Yes	Dimensions of wreck	ESKMERE	Yes	MBES/Archive
7002	LOWTHER RANGE	Probably	MBES/Archive	No	N/A	N/A	No	MBES/Archive
7006	ANTEROS	Possibly	UNIDENTIFIABLE	Yes	Unidentifiable	N/A	No	UNIDENTIFIABLE
7007	UNKNOWN	UNKNOWN	UNKNOWN	No	N/A	TRINIDAD	Yes	MBES/Archive
7010	BALTHAZAR	Confirmed	Confirmed	No	N/A	N/A	No	Confirmed
7021	FAIREARN	Probably	MBES/Archive	No	N/A	N/A	No	MBES/Archive
7023	INNISFALLEN	Possibly	UNKNOWN	Yes	Dimensions of wreck	N/A	No	UNKNOWN
7025	ESKMERE	Possibly	UNKNOWN	Yes	Dimensions of wreck	PENVEARN	Yes	MBES/Archive
7027	PALMELLA	Possibly	MBES/Archive	No	N/A	N/A	No	MBES/Archive
7028	HOLYHEAD	Possibly	UNKNOWN	Yes	Dimensions of wreck	KENMARE	Yes	MBES/Archive
7030	UNKNOWN	UNKNOWN	UNKNOWN	No	N/A	TRIESTE	Yes	MBES/Archive
7031	UNKNOWN	UNKNOWN	UNKNOWN	No	N/A	INNISFALLEN	Yes	MBES/Archive
7032	PEACEFUL STAR	Possibly	UNIDENTIFIABLE	Yes	Unidentifiable	N/A	No	UNIDENTIFIABLE
7033	UNKNOWN	UNKNOWN	UNKNOWN	No	N/A	NORMANDIET	Yes	MBES/Archive
7039	SALAMINIA	Probably	MBES/Archive	No	N/A	N/A	No	MBES/Archive
7043	THE MARQUIS	Possibly	UNKNOWN	Yes	Dimensions of wreck	PEACEFUL STAR	Yes	MBES/Archive
7051	U1051	Confirmed	UNKNOWN	Yes	Dimensions of wreck	N/A	No	UNKNOWN
7056	GLENFORD	Possibly	UNKNOWN	Yes	Dimensions of wreck	HMS MANNERS (Stern section)	Yes	MBES/Archive
7059	KENMARE	Possibly	NON SHIPWRECK	Yes	Geology	N/A	No	NON SHIPWRECK
7063	MAJA	Confirmed	Confirmed	No	N/A	N/A	No	Confirmed
7065	NADEJDA	Possibly	UNKNOWN	Yes	Dimensions of wreck	N/A	No	UNKNOWN
7066	UNKNOWN	UNKNOWN	UNKNOWN	No	N/A	GLENFORD	Yes	MBES/Archive
7070	DALEWOOD	Probably	UNKNOWN	Yes	Dimensions of wreck	N/A	No	UNKNOWN
7086	UNKNOWN	UNKNOWN	UNIDENTIFIABLE	Yes	Unidentifiable	N/A	No	UNIDENTIFIABLE
7087	SHARELGA	Confirmed	Confirmed	No	N/A	N/A	No	Confirmed
7089	UNKNOWN	UNKNOWN	UNIDENTIFIABLE	Yes	Unidentifiable	N/A	No	UNIDENTIFIABLE
7091	UNKNOWN	UNKNOWN	UNKNOWN	No	N/A	ANTEROS	Yes	MBES/Archive
7093	UNKNOWN	UNKNOWN	UNKNOWN	No	N/A	THE MARQUIS	Yes	MBES/Archive
7106	NORMANDIET	Possibly	UNKNOWN	Yes	Dimensions of wreck	U1051	Yes	MBES/Archive
7107	CRESSWELL	Probably	MBES/Archive	No	N/A	N/A	No	MBES/Archive
7127	UNKNOWN	UNKNOWN	UNKNOWN	No	N/A	CRESSADO	Yes	MBES/Archive
7128	UNKNOWN	UNKNOWN	UNKNOWN	No	N/A	HOLYHEAD	Yes	MBES/Archive
7146	UNKNOWN	UNKNOWN	UNKNOWN	No	N/A	PERSEUS	Yes	MBES/Archive

LIVE UKHO No.	UKHO Name	UKHO Status	Status after Phase 1	Phase 1 status change	Primary Reason for Change	Phase 2 Identification	Phase 2 status change	Final Status
7147	HMSM H5	Confirmed	Confirmed	No	N/A	N/A	No	Confirmed
7162	FARFIELD	Confirmed	NON SHIPWRECK	Yes	Geology	N/A	No	NON SHIPWRECK
7163	VIVAR	Confirmed	Confirmed	No	N/A	N/A	No	Confirmed
7173	SLIEVE BLOOM	Confirmed	UNKNOWN	Yes	Dimensions of wreck	NORMANDY COAST	Yes	MBES/Archive
7178	ROANOKE	Confirmed	Confirmed	No	N/A	N/A	No	Confirmed
7228	MAARTEN CORNELIS	Confirmed	Confirmed	No	N/A	N/A	No	Confirmed
7261	ORIA	Confirmed	Confirmed	No	N/A	N/A	No	Confirmed
7262	RIVER HUMBER	Possibly	UNKNOWN	Yes	Dimensions of wreck	MARQUESS OF BUTE	Yes	MBES/Archive
7279	UNKNOWN	UNKNOWN	UNKNOWN	No	N/A	SLIEVE BLOOM	Yes	MBES/Archive
7293	LADY WINDSOR	Probably	UNKNOWN	Yes	Features of wreck	KINKORTH	Yes	MBES/Archive
7329	UNKNOWN	UNKNOWN	UNIDENTIFIABLE	Yes	Unidentifiable	N/A	No	UNIDENTIFIABLE
7330	CAPTAIN MCCLINTOCK	Possibly	MBES/Archive	No	N/A	N/A	No	MBES/Archive
7331	CAMBANK	Confirmed	Confirmed	No	N/A	N/A	No	Confirmed
7332	APAPA	Confirmed	Confirmed	No	N/A	N/A	No	Confirmed
7334	DERBENT	Confirmed	Confirmed	No	N/A	N/A	No	Confirmed
7335	GLENCONA	Probably	UNKNOWN	Yes	Archival evidence	CRESSIDA	Yes	MBES/Archive
7336	BRAGANZA	Confirmed	Confirmed	No	N/A	N/A	No	Confirmed
7338	UNKNOWN	UNKNOWN	UNKNOWN	No	N/A	N/A	No	UNKNOWN
7339	GUILLERMO	Confirmed	Confirmed	No	N/A	N/A	No	Confirmed
7340	UNKNOWN	UNKNOWN	UNKNOWN	No	N/A	N/A	No	UNKNOWN
7341	ADELA	Possibly	MBES/Archive	No	N/A	N/A	No	MBES/Archive
7345	LA ROCHELLE	Possibly	MBES/Archive	No	N/A	N/A	No	MBES/Archive
7346	HMS WESTPHALIA	Possibly	UNKNOWN	Yes	Archival evidence	N/A	No	UNKNOWN
7347	MANCHESTER	Confirmed	UNKNOWN	Yes	Dimensions of wreck	BRITISH VISCOUNT	Yes	MBES/Archive
7348	THEME	Confirmed	Confirmed	No	N/A	N/A	No	Confirmed
7350	RIVER HUMBER	Confirmed	Confirmed	No	N/A	N/A	No	Confirmed
7352	UNKNOWN	UNKNOWN	UNKNOWN	No	N/A	COGNAC	Yes	MBES/Archive
7353	ARNO MENDI	Confirmed	Confirmed	No	N/A	N/A	No	Confirmed
7355	HYACINTH	Confirmed	Confirmed	No	N/A	N/A	No	Confirmed
7360	BIJOU	Confirmed	Confirmed	No	N/A	N/A	No	Confirmed
7361	VIGSNES	Confirmed	Confirmed	No	N/A	N/A	No	Confirmed
7362	WESTWARD HO	Confirmed	Confirmed	No	N/A	N/A	No	Confirmed
7363	CORK	Confirmed	Confirmed	No	N/A	N/A	No	Confirmed
7365	SONIA (sic)	Confirmed	Confirmed	No	N/A	N/A	No	Confirmed
7366	LOWESWATER	Confirmed	Confirmed	No	N/A	N/A	No	Confirmed
7369	BENGOLLYUN	Confirmed	Confirmed	No	N/A	N/A	No	Confirmed
7370	SEAGULL (sic)	Confirmed	Confirmed	No	N/A	N/A	No	Confirmed
7372	CONARGO	Confirmed	Confirmed	No	N/A	N/A	No	Confirmed
7373	U246	Probably	MBES/Archive	No	N/A	N/A	No	MBES/Archive
7374	U242	Probably	UNKNOWN	Yes	Dimensions of wreck	SARPFOS	Yes	MBES/Archive
7377	U1024	Probably	MBES/Archive	No	N/A	N/A	No	MBES/Archive
7379	DUNDALK	Confirmed	Confirmed	No	N/A	N/A	No	Confirmed
7407	MINERVA	Confirmed	Confirmed	No	N/A	N/A	No	Confirmed

LIVE UKHO No.	UKHO Name	UKHO Status	Status after Phase 1	Phase 1 status change	Primary Reason for Change	Phase 2 Identification	Phase 2 status change	Final Status
7415	WERN	Possibly	UNIDENTIFIABLE	Yes	Unidentifiable	N/A	No	UNIDENTIFIABLE
7416	UNKNOWN	UNKNOWN	NON SHIPWRECK	Yes	Geology	N/A	No	NON SHIPWRECK
7418	UNKNOWN	UNKNOWN	UNKNOWN	No	N/A	FOSTIMORE	Yes	MBES/Archive
7429	CONSTANCE	Confirmed	Confirmed	No	N/A	N/A	No	Confirmed
7430	CARTAGENA	Confirmed	Confirmed	No	N/A	N/A	No	Confirmed
7431	UNKNOWN	UNKNOWN	NON SHIPWRECK	Yes	Geology	N/A	No	NON SHIPWRECK
7432	UNKNOWN	UNKNOWN	UNIDENTIFIABLE	Yes	Unidentifiable	N/A	No	UNIDENTIFIABLE
7433	MIRIAM THOMAS	Confirmed	Confirmed	No	N/A	N/A	No	Confirmed
7436	MONA	Confirmed	MBES/Archive	Yes	No definitive proof	N/A	No	MBES/Archive
7437	DELFINA	Confirmed	MBES/Archive	Yes	No definitive proof	N/A	No	MBES/Archive
7438	PACIFIC	Possibly	UNKNOWN	Yes	Dimensions of wreck	N/A	No	UNKNOWN
7439	UNKNOWN	UNKNOWN	UNKNOWN	No	N/A	HILDENA	Yes	MBES/Archive
7441	UNKNOWN	UNKNOWN	UNKNOWN	No	N/A	N/A	No	UNKNOWN
7442	PAMELA	Confirmed	MBES/Archive	Yes	No definitive proof	N/A	No	MBES/Archive
7443	BRITISH VISCOUNT	Possibly	UNKNOWN	Yes	Dimensions of wreck	BENHOLM	Yes	MBES/Archive
7444	LEEDS	Possibly	UNKNOWN	Yes	Archival evidence	HERBERT	Yes	MBES/Archive
7446	LIGHTSTONE (sic)	Probably	MBES/Archive	No	N/A	N/A	No	MBES/Archive
7447	UNKNOWN	UNKNOWN	UNKNOWN	No	N/A	N/A	No	UNKNOWN
7448	CLYTIE	Possibly	UNKNOWN	Yes	Archival evidence	N/A	No	UNKNOWN
7449	UNKNOWN	UNKNOWN	UNKNOWN	No	N/A	N/A	No	UNKNOWN
7452	TIJL UILENSPIEGEL	Confirmed	Confirmed	No	N/A	N/A	No	Confirmed
7454	EMPRESS EUGENIE	Confirmed	Confirmed	No	N/A	N/A	No	Confirmed
7464	NORMANDY COAST	Probably	NON SHIPWRECK	Yes	Geology	N/A	No	NON SHIPWRECK
7466	CAERNARVON BAY LV	Probably	MBES/Archive	No	N/A	N/A	No	MBES/Archive
7467	MTB 539	Confirmed	Confirmed	No	N/A	N/A	No	Confirmed
7491	OPAL	Possibly	MBES/Archive	No	N/A	N/A	No	MBES/Archive
7492	CLWYD	Probably	MBES/Archive	No	N/A	N/A	No	MBES/Archive
7494	TREVEAL	Confirmed	Confirmed	No	N/A	N/A	No	Confirmed
7496	UNKNOWN	UNKNOWN	NON SHIPWRECK	Yes	Geology	N/A	No	NON SHIPWRECK
7516	GHAMBIRA (sic)	Confirmed	Confirmed	No	N/A	N/A	No	Confirmed
7711	STRATHRYE	Confirmed	Confirmed	No	N/A	N/A	No	Confirmed
7946	SEA GULL	Possibly	UNKNOWN	Yes	Dimensions of wreck	CLYTIE	Yes	MBES/Archive
7969	UNKNOWN	UNKNOWN	UNKNOWN	No	N/A	N/A	No	UNKNOWN
8124	ALBANIAN	Confirmed	Confirmed	No	N/A	N/A	No	Confirmed
8140	NYDIA	Probably	MBES/Archive	No	N/A	N/A	No	MBES/Archive
8161	UNKNOWN	UNKNOWN	NON SHIPWRECK	Yes	Geology	N/A	No	NON SHIPWRECK
8162	UNKNOWN	UNKNOWN	UNKNOWN	No	N/A	N/A	No	UNKNOWN
8204	AMLWCH ROSE	Probably	UNKNOWN	Yes	Dimensions of wreck	N/A	No	UNKNOWN
8239	ARDLOUGH	Confirmed	Confirmed	No	N/A	N/A	No	Confirmed
9850	ETHEL	Confirmed	UNKNOWN	Yes	Dimensions of wreck	U1172	Yes	MBES/Archive
9852	UNKNOWN	UNKNOWN	UNKNOWN	No	N/A	U242	Yes	MBES/Archive
9853	U1169	Possibly	UNKNOWN	Yes	Dimensions of wreck	AMIRAL ZEDE	Yes	MBES/Archive

LIVE UKHO No.	UKHO Name	UKHO Status	Status after Phase 1	Phase 1 status change	Primary Reason for Change	Phase 2 Identification	Phase 2 status change	Final Status
9855	MOYALLON	Confirmed	UNKNOWN	Yes	Dimensions of wreck	N/A	No	UNKNOWN
9859	PENTWYN	Possibly	UNKNOWN	Yes	Dimensions of wreck	SAPPHIRE	Yes	MBES/Archive
9864	ST PATRICK	Possibly	UNKNOWN	Yes	Dimensions of wreck	HEENVLIET	Yes	MBES/Archive
9865	CYMRIAN	Confirmed	MBES/Archive	Yes	No definitive proof	N/A	No	MBES/Archive
9866	BARON CARNEGIE	Confirmed	MBES/Archive	Yes	No definitive proof	N/A	No	MBES/Archive
9870	HIGHCLIFFE	Confirmed	UNKNOWN	Yes	Dimensions of wreck	N/A	No	UNKNOWN
9872	UNKNOWN	UNKNOWN	UNKNOWN	No	N/A	HIGHCLIFFE	Yes	MBES/Archive
9873	PORT TOWNSVILLE	Probably	UNKNOWN	Yes	Dimensions of wreck	EMPIRE GUNNER	Yes	MBES/Archive
9876	UNKNOWN	UNKNOWN	UNKNOWN	No	N/A	ISIDORO	Yes	MBES/Archive
9877	EMPIRE GUNNER	Possibly	UNKNOWN	Yes	Dimensions of wreck	THORNFIELD	Yes	MBES/Archive
9880	UNKNOWN	UNKNOWN	UNKNOWN	No	N/A	GLENBY	Yes	MBES/Archive
9881	UNKNOWN	UNKNOWN	UNKNOWN	No	N/A	ALU MENDI	Yes	MBES/Archive
9883	UNKNOWN	UNKNOWN	UNKNOWN	No	N/A	ST PATRICK	Yes	MBES/Archive
9887	UNKNOWN	UNKNOWN	UNKNOWN	No	N/A	THE QUEEN	Yes	MBES/Archive
9891	KASSANGA	Possibly	MBES/Archive	No	N/A	N/A	No	MBES/Archive
9892	EMPIRE PANTHER	Probably	UNKNOWN	Yes	Dimensions of wreck	SORELDOC	Yes	MBES/Archive
9894	TWEED	Possibly	UNKNOWN	Yes	Dimensions of wreck	WHEATFLOWER	Yes	MBES/Archive
9895	EGDA	Possibly	MBES/Archive	No	N/A	N/A	No	MBES/Archive
9896	STRATHNAIRN	Possibly	UNKNOWN	Yes	Dimensions of wreck	PORT TOWNSVILLE	Yes	MBES/Archive
9898	BOSCASTLE	Possibly	UNIDENTIFIABLE	Yes	Unidentifiable	N/A	No	UNIDENTIFIABLE
9902	ETHEL	Confirmed	Confirmed	No	N/A	N/A	No	Confirmed
9903	GRESHAM	Possibly	UNKNOWN	Yes	Dimensions of wreck	BOSCASTLE	Yes	MBES/Archive
9904	UNKNOWN	UNKNOWN	UNKNOWN	No	N/A	GRESHAM	Yes	MBES/Archive
9909	BOLTONHALL	Possibly	UNKNOWN	Yes	Dimensions of wreck	LYCIA	Yes	MBES/Archive
9911	SORELDOC	Possibly	UNKNOWN	Yes	Dimensions of wreck	U1302	Yes	MBES/Archive
9912	CITY OF GLASGOW	Probably	UNKNOWN	Yes	Dimensions of wreck	MESABA	Yes	MBES/Archive
9913	MESABA	Probably	UNKNOWN	Yes	Dimensions of wreck	CITY OF GLASGOW	Yes	MBES/Archive
9914	U1302	Possibly	UNKNOWN	Yes	Dimensions of wreck	KYANITE	Yes	MBES/Archive
9915	ROTULA	Possibly	UNKNOWN	Yes	Dimensions of wreck	VOLTAIRE	Yes	MBES/Archive
9920	PRINS FREDERIK HENDRIK	Possibly	UNKNOWN	Yes	Dimensions of wreck	BOLTONHALL	Yes	MBES/Archive
9921	ARCADIAN	Possibly	UNKNOWN	Yes	Dimensions of wreck	CITY OF DUNDEE	Yes	MBES/Archive
9922	UNKNOWN	UNKNOWN	UNKNOWN	No	N/A	GREENLAND	Yes	MBES/Archive
9923	UNKNOWN	UNKNOWN	UNKNOWN	No	N/A	N/A	No	UNKNOWN
9925	LANDONIA	Probably	UNKNOWN	Yes	Dimensions of wreck	ROTULA	Yes	MBES/Archive
9929	UNKNOWN	UNKNOWN	UNKNOWN	No	N/A	SERULA	Yes	MBES/Archive
9932	UNKNOWN	UNKNOWN	UNKNOWN	No	N/A	STRATHNAIRN	Yes	MBES/Archive
9933	METRIC	Confirmed	Confirmed	No	N/A	N/A	No	Confirmed

LIVE UKHO No.	UKHO Name	UKHO Status	Status after Phase 1	Phase 1 status change	Primary Reason for Change	Phase 2 Identification	Phase 2 status change	Final Status
9934	UNKNOWN	UNKNOWN	NON SHIPWRECK	Yes	Geology	N/A	No	NON SHIPWRECK
9935	UNKNOWN	UNKNOWN	UNKNOWN	No	N/A	SALLAGH	Yes	MBES/Archive
9936	YTURRI BIDE	Possibly	UNKNOWN	Yes	Dimensions of wreck	LANDONIA	Yes	MBES/Archive
9937	ELSENA	Possibly	UNKNOWN	Yes	Dimensions of wreck	NOR	Yes	MBES/Archive
9940	ENNISTOWN	Possibly	UNKNOWN	Yes	Dimensions of wreck	GLORIA	Yes	MBES/Archive
9941	LADOGA	Possibly	UNKNOWN	Yes	Archival evidence	N/A	No	UNKNOWN
9945	SPENSER	Possibly	UNKNOWN	Yes	Dimensions of wreck	BURUTU	Yes	MBES/Archive
9946	ARDGLASS	Possibly	UNKNOWN	Yes	Dimensions of wreck	HMY KETHAILES	Yes	MBES/Archive
9948	FORT RONA	Possibly	MBES/Archive	No	N/A	N/A	No	MBES/Archive
9950	FOYLE	Possibly	UNKNOWN	Yes	Dimensions of wreck	LCT 326	Yes	MBES/Archive
9951	RFA VITOL	Possibly	UNKNOWN	Yes	Dimensions of wreck	ROBERT EGGLETON	Yes	MBES/Archive
9953	UNKNOWN	UNKNOWN	UNKNOWN	No	N/A	BOSCAWEN	Yes	MBES/Archive
9954	BIRCHWOOD	Possibly	UNKNOWN	Yes	Dimensions of wreck	SPENSER	Yes	MBES/Archive
9955	FRANCOISE DE LILLE	Possibly	UNKNOWN	Yes	Features of wreck	HALBERDIER	Yes	MBES/Archive
9957	KIRKBY	Probably	UNKNOWN	Yes	Dimensions of wreck	WAESLAND	Yes	MBES/Archive
9959	GALATEA	Possibly	UNKNOWN	Yes	Dimensions of wreck	KIRKBY	Yes	MBES/Archive
9964	UNKNOWN	UNKNOWN	UNKNOWN	No	N/A	PULTENEY	Yes	MBES/Archive
9966	AMETHYST	Probably	MBES/Archive	No	N/A	N/A	No	MBES/Archive
9970	UNKNOWN	UNKNOWN	UNKNOWN	No	N/A	FRANCOISE DE LILLE	Yes	MBES/Archive
9972	UNKNOWN	UNKNOWN	UNKNOWN	No	N/A	KORSNAES	Yes	MBES/Archive
9975	ROSE MARIE	Probably	UNKNOWN	Yes	Dimensions of wreck	RFA OSAGE	Yes	MBES/Archive
9976	BOSCOWAN	Probably	UNKNOWN	Yes	Dimensions of wreck	EMBLETON	Yes	MBES/Archive
9977	CHELFORD	Confirmed	Confirmed	No	N/A	N/A	No	Confirmed
9978	ABBAS COMBE	Possibly	MBES/Archive	No	N/A	N/A	No	MBES/Archive
9981	PERSEUS	Probably	UNKNOWN	Yes	Features of wreck	KNUT	Yes	MBES/Archive
9983	JANVOLD	Confirmed	Confirmed	No	N/A	N/A	No	Confirmed
9984	DAMAO	Possibly	MBES/Archive	No	N/A	N/A	No	MBES/Archive
9985	ORONSA	Confirmed	Confirmed	No	N/A	N/A	No	Confirmed
9987	AGBERI	Possibly	UNKNOWN	Yes	Dimensions of wreck	SNOWDON RANGE	Yes	MBES/Archive
9988	UNKNOWN	UNKNOWN	UNKNOWN	No	N/A	AGBERI	Yes	MBES/Archive
9989	U87	Confirmed	Confirmed	No	N/A	N/A	No	Confirmed
9990	UNKNOWN	UNKNOWN	UNKNOWN	No	N/A	DERELICT (half with 9994)	Yes	MBES/Archive
9994	DERELICT	Confirmed	Confirmed	No	N/A	N/A	No	Confirmed
9995	MEMPHIAN	Possibly	UNKNOWN	Yes	Dimensions of wreck	PAROS	Yes	MBES/Archive
9996	GRELDON	Possibly	UNKNOWN	Yes	Dimensions of wreck	DROMEDARY	Yes	MBES/Archive
9998	UNKNOWN	UNKNOWN	UNKNOWN	No	N/A	N/A	No	UNKNOWN
10000	UNKNOWN	UNKNOWN	UNKNOWN	No	N/A	N/A	No	UNKNOWN
10001	UNKNOWN	UNKNOWN	UNKNOWN	No	N/A	LADOGA	Yes	MBES/Archive

LIVE UKHO No.	UKHO Name	UKHO Status	Status after Phase 1	Phase 1 status change	Primary Reason for Change	Phase 2 Identification	Phase 2 status change	Final Status
10002	UNKNOWN	UNKNOWN	UNKNOWN	No	N/A	POSEIDON	Yes	MBES/Archive
10003	HALBERDIER	Possibly	UNKNOWN	Yes	Dimensions of wreck	ANT CASSAR	Yes	MBES/Archive
10004	UNKNOWN	UNKNOWN	UNKNOWN	No	N/A	CONINGBEG	Yes	MBES/Archive
10005	UNKNOWN	UNKNOWN	UNKNOWN	No	N/A	GARTHLOCH	Yes	MBES/Archive
10007	UNKNOWN	UNKNOWN	NON SHIPWRECK	yes	Geology	N/A	No	NON SHIPWRECK
10009	CYRENE	Possibly	UNKNOWN	Yes	Dimensions of wreck	N/A	No	UNKNOWN
10010	UNKNOWN	UNKNOWN	UNKNOWN	No	N/A	MEMPHIAN	Yes	MBES/Archive
10011	UNKNOWN	UNKNOWN	UNKNOWN	No	N/A	ROSE MARIE	Yes	MBES/Archive
10012	UNKNOWN	UNKNOWN	UNKNOWN	No	N/A	FORTUNA	Yes	MBES/Archive
10013	UNKNOWN	UNKNOWN	UNKNOWN	No	N/A	RFA VITOL	Yes	MBES/Archive
10014	UNKNOWN	UNKNOWN	UNKNOWN	No	N/A	ARDRI	Yes	MBES/Archive
10015	UNKNOWN	UNKNOWN	UNKNOWN	No	N/A	GLYNWEN	Yes	MBES/Archive
10016	UNKNOWN	UNKNOWN	UNKNOWN	No	N/A	N/A	No	UNKNOWN
10017	UNKNOWN	UNKNOWN	UNKNOWN	No	N/A	FORMBY	Yes	MBES/Archive
10018	WALPAS	Possibly	NON SHIPWRECK	Yes	Geology	N/A	No	NON SHIPWRECK
10019	UNKNOWN	UNKNOWN	UNKNOWN	No	N/A	ONYX	Yes	MBES/Archive
10022	UNKNOWN	UNKNOWN	UNKNOWN	No	N/A	LODANER	Yes	MBES/Archive
10026	UNKNOWN	UNKNOWN	UNKNOWN	No	N/A	GEORGE BEWLEY	Yes	MBES/Archive
10028	UNKNOWN	UNKNOWN	UNKNOWN	No	N/A	HAVNA	Yes	MBES/Archive
10030	MARGARETHA	Confirmed	NON SHIPWRECK	Yes	Geology	N/A	No	NON SHIPWRECK
10031	UNKNOWN	UNKNOWN	UNKNOWN	No	N/A	FERGA	Yes	MBES/Archive
10052	SERULA	Confirmed	NON SHIPWRECK	Yes	Geology	N/A	No	NON SHIPWRECK
10053	SUTTON	Probably	NON SHIPWRECK	Yes	Geology	N/A	No	NON SHIPWRECK
10055	UNKNOWN	UNKNOWN	UNKNOWN	No	N/A	EMPIRE PANTHER	Yes	MBES/Archive
10056	HMS WHIRLWIND (F187)	Confirmed	Confirmed	No	N/A	N/A	No	Confirmed
10064	ROBERT EGGLETON	Possibly	UNKNOWN	Yes	Dimensions of wreck	SOMERSET COAST	Yes	MBES/Archive
10070	UNKNOWN	UNKNOWN	UNKNOWN	No	N/A	N/A	No	UNKNOWN
10088	TARBETNESS	Confirmed	UNKNOWN	Yes	Dimensions of wreck	DUKE OF CONNAUGHT	Yes	MBES/Archive
10094	UNKNOWN	UNKNOWN	UNKNOWN	No	N/A	N/A	No	UNKNOWN
10104	HMS LEWISTON	Confirmed	Confirmed	No	N/A	N/A	No	Confirmed
11788	UC33	Confirmed	UNKNOWN	Yes	Dimensions of wreck	HMS MERCURY	Yes	MBES/Archive
11796	COPELAND	Possibly	UNKNOWN	Yes	Dimensions of wreck	N/A	No	UNKNOWN
11798	THORALF	Possibly	UNKNOWN	Yes	Dimensions of wreck	UC33	Yes	MBES/Archive
11805	HILDEFJORD (sic)	Confirmed	MBES/Archive	Yes	No definitive proof	N/A	No	MBES/Archive
12097	KING EDGAR	Confirmed	Confirmed	No	N/A	N/A	No	Confirmed
12124	SAPPHIRE	Possibly	UNKNOWN	Yes	Dimensions of wreck	CHATHAM	Yes	MBES/Archive
12125	HAVNA	Possibly	UNKNOWN	Yes	Dimensions of wreck	N/A	No	UNKNOWN
12132	UNKNOWN	UNKNOWN	UNKNOWN	No	N/A	BRYNMOR	Yes	MBES/Archive
12135	DARU	Possibly	MBES/Archive	No	N/A	N/A	No	MBES/Archive
12137	UNKNOWN	UNKNOWN	UNKNOWN	No	N/A	NORFOLK COAST	Yes	MBES/Archive
12163	UNKNOWN	UNKNOWN	UNKNOWN	No	N/A	ST FINTAN	Yes	MBES/Archive
12164	BESTWOOD	Possibly	MBES/Archive	No	N/A	N/A	No	MBES/Archive

LIVE UKHO No.	UKHO Name	UKHO Status	Status after Phase 1	Phase 1 status change	Primary Reason for Change	Phase 2 Identification	Phase 2 status change	Final Status
12165	DAN BEARD (AFTER PART)	Possibly	UNKNOWN	Yes	Dimensions of wreck	ABERDEEN	Yes	MBES/Archive
58321	UNKNOWN	UNKNOWN	NON SHIPWRECK	Yes	Geology	N/A	No	NON SHIPWRECK
58341	LIVERPOOL	Confirmed	Confirmed	No	N/A	N/A	No	Confirmed
58454	MARY COLES	Possibly	UNIDENTIFIABLE	Yes	Unidentifiable	N/A	No	UNIDENTIFIABLE
60302	FIN-AR-BED	Confirmed	Confirmed	No	N/A	N/A	No	Confirmed
62529	UNKNOWN	UNKNOWN	UNKNOWN	No	N/A	CASTLEDOBBS	Yes	MBES/Archive
66101	MORNING SPRAY	Confirmed	Confirmed	No	N/A	N/A	No	Confirmed
69684	UNKNOWN	UNKNOWN	UNKNOWN	No	N/A	BIRCHWOOD	Yes	MBES/Archive
69685	UNKNOWN	UNKNOWN	UNKNOWN	No	N/A	MAGGIE	Yes	MBES/Archive
69696	UNKNOWN	UNKNOWN	UNKNOWN	No	N/A	YTURRI BIDE	Yes	MBES/Archive
75674	UNKNOWN	UNKNOWN	UNKNOWN	No	N/A	N/A	No	UNKNOWN
76491	UNKNOWN	UNKNOWN	UNKNOWN	No	N/A	AFTON	Yes	MBES/Archive
78681	SWANLAND	Confirmed	Confirmed	No	N/A	N/A	No	Confirmed
80722	UNKNOWN	UNKNOWN	UNKNOWN	No	N/A	SKERNAHAN	Yes	MBES/Archive
81382	UNKNOWN	UNKNOWN	UNKNOWN	No	N/A	N/A	No	UNKNOWN
81383	UNKNOWN	UNKNOWN	UNKNOWN	No	N/A	YEWS	Yes	MBES/Archive
81384	UNKNOWN	UNKNOWN	UNKNOWN	No	N/A	SAINT OLAF	Yes	MBES/Archive
81385	UNKNOWN	UNKNOWN	UNIDENTIFIABLE	Yes	Unidentifiable	N/A	No	UNIDENTIFIABLE
81386	UNKNOWN	UNKNOWN	UNIDENTIFIABLE	Yes	Unidentifiable	N/A	No	UNIDENTIFIABLE
81387	UNKNOWN	UNKNOWN	UNKNOWN	No	N/A	N/A	No	UNKNOWN
81388	UNKNOWN	UNKNOWN	UNKNOWN	No	N/A	POCHARD	Yes	MBES/Archive
81389	UNKNOWN	UNKNOWN	UNKNOWN	No	N/A	ISLE OF ARRAN	Yes	MBES/Archive
82347	UNKNOWN	UNKNOWN	UNKNOWN	No	N/A	DUNSLEY	Yes	MBES/Archive
82348	UNKNOWN	UNKNOWN	UNIDENTIFIABLE	Yes	Unidentifiable	N/A	No	UNIDENTIFIABLE
83384	UNKNOWN	UNKNOWN	UNKNOWN	No	N/A	PENTWYN	Yes	MBES/Archive
83692	UNKNOWN	UNKNOWN	UNKNOWN	No	N/A	N/A	No	UNKNOWN
83693	UNKNOWN	UNKNOWN	UNKNOWN	No	N/A	KANGAROO	Yes	MBES/Archive
83694	UNKNOWN	UNKNOWN	UNKNOWN	No	N/A	N/A	No	UNKNOWN
83695	UNKNOWN	UNKNOWN	NON SHIPWRECK	Yes	Geology	N/A	No	NON SHIPWRECK
83696	UNKNOWN	UNKNOWN	NON SHIPWRECK	Yes	Geology	N/A	No	NON SHIPWRECK
89263	OLIVER CROMWELL	Confirmed	Confirmed	No	N/A	N/A	No	Confirmed

Appendix 2

The datasheets in this appendix represent the 129 shipwrecks which were identified during Phase 2, using the (MBES/Archive) methodology.

The results of Phase 1 on all 273 wrecks can be found in Appendix 1.

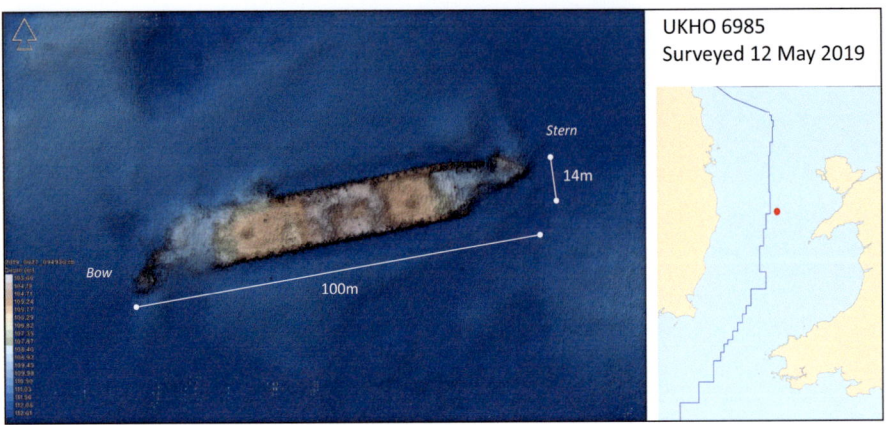

6985 TARBETNESS (MBES/Archive)

Original UKHO: 6985 UNKNOWN	**Phase 1 Analysis:** UNKNOWN
Position WGS84: 53 1.084 N, 5 13.247 W	**Water Depth:** 112m
Type of Vessel: SS Cargo	**Surveyed Length:** 100m (approx)
Date of Loss: 7 Mar 1918	**Circumstances:** Torpedoed by U110

Phase 2: Method of wreck identification:

Positional Analysis:
U110 did not survive the patrol, so the only positions are from the British side. A pair of positions from Irish Sea (TNA ADM 137/1515) plot 1.5-2.1nm east of the wreck. One is from the escort ship which found the survivors and is probably the most accurate.

Dimensional Analysis:
TARBETNESS was 100.8m x 14.6m (*Lloyd's Register* 1916-17) which is consistent with the dimensions of the wreck.

Archival Analysis:
Various survivor accounts in Irish Sea (TNA ADM 137/1515) all state the torpedo struck aft on port side. This is consistent with an area of damage seen on the wreck. Drawings and photos of the ship in the project archive show that it was well-decked fore and aft with No1 & No4 hatches lower than No2 & 3 hatches. This can be discerned clearly in the MBES data.

Phase 2 Conclusion:
TARBETNESS (MBES/Archive)

Phase 1: UKHO Analysis:

Original UKHO Name:
6985 UNKNOWN

Archival details:
N/A

MBES Analysis:
Appears to be an upright cargo vessel of approximately 100m length and 14m width. Central bridge with holds fore and aft, showing collapse, revealing propeller tunnel at the stern portion, pointing east.

Additional Evidence:
N/A

Phase 1 Conclusion:
UNKNOWN

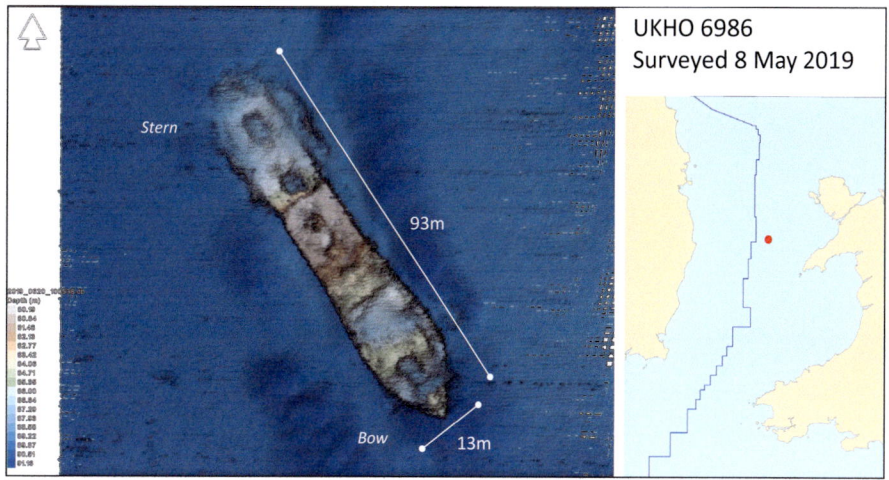

UKHO 6986
Surveyed 8 May 2019

6986 CYRENE (MBES/Archive)

Original UKHO: 6986 EARL OF ELGIN	**Phase 1 Analysis:** UNKNOWN
Position WGS84: 53 0.084 N, 5 3.948 W	**Water Depth:** 91m
Type of Vessel: SS Cargo	**Surveyed Length:** 93m (approx)
Date of Loss: 5 Apr 1918	**Circumstances:** Torpedoed by UB64

Phase 2: Method of wreck identification:

Positional Analysis:
UB64 (NOT UC31 was responsible for this sinking, as described) position for this attack is not recorded accurately in its KTB, but it is clear it was in Caernarfon Bay. Only three persons survived the sinking and only a vague position was given to interrogators but is clear the ship sank in Caernarfon Bay.

Dimensional Analysis:
CYRENE was 92m x 12.8m (*Lloyd's Register* 1917-18). This is consistent with the dimensions of this wreck. CYRENE was spar-decked. Nothing in the MBES data discounts this being the case.

Archival Analysis:
The account of the sinking (TNA ADM 137/2964) gives a time which matches with an attack in this area by UB64 (KTB). The damage (hit in engine room starboard side) and course of ship (given in KTB as 170deg, as per the wreck) both match. Spindler's attribution to UC31 cannot be correct as the timings differ by several hours. The attack reported by UC31 was on the steamship CLAM, which survived (TNA ADM 137/1516). Spindler attributed this to UB64. When substituted for each other, the timings and positions clearly match correctly. UC31 attacked CLAM and UB64 sunk CYRENE.

Phase 2 Conclusion:
CYRENE (MBES/Archive). This wreck is dimensionally, archivally, and positionally consistent with CYRENE.

Phase 1: UKHO Analysis:

Original UKHO Name:
6986 EARL OF ELGIN

Archival details:
EARL OF ELGIN was a steamship of 117.2m long and 15.11m wide (Larn & Larn). Torpedoed 2.5nm from the wreck (*War Losses* 1990).

MBES Analysis:
Appears to be an upright steamer of approx. 93m long and 13m wide. Central bridge with holds fore and aft.

Additional Evidence:
Located in 1969. UKHO give no reason why the (POSSIBLY) or (PROBABLY) suffix is not present. I.e., the wreck was considered CONFIRMED.

Phase 1 Conclusion:
UNKNOWN. The wreck is too small to be EARL OF ELGIN.

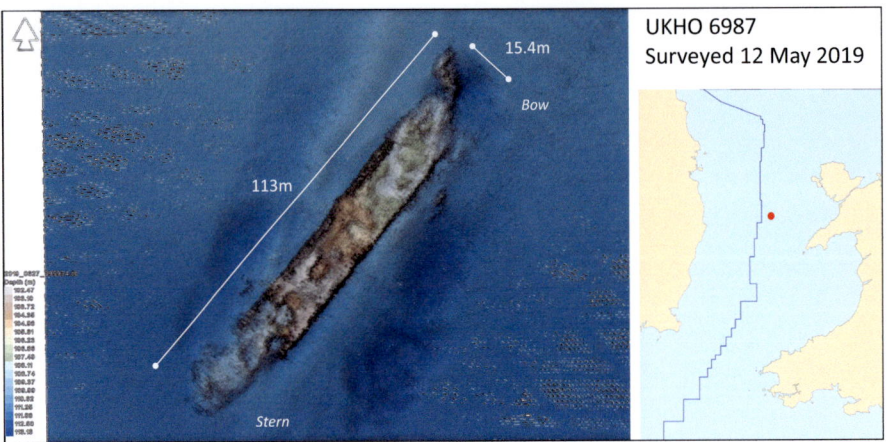

15.4m

Bow

113m

Stern

6987 EARL OF ELGIN (MBES/Archive)

Original UKHO: 6987 UNKNOWN	**Phase 1 Analysis:** UNKNOWN
Position WGS84: 53 2.834 N, 5 10.714 W	**Water Depth:** 112m
Type of Vessel: SS Cargo	**Surveyed Length:** 113m (approx)
Date of Loss: 7 Dec 1917	**Circumstances:** Torpedoed by UC75

Phase 2: Method of wreck identification:

Positional Analysis:
The convoy report (TNA ADM 137/1362) and survivors report for EARL OF ELGIN (TNA ADM 137/2963) both give positions which plot within 3nm of this wreck. UC75's KTB does not give a position for its attack, but it is clear that it took place in the Caernarfon Bay area.

Dimensional Analysis:
EARL OF ELGIN was 117.2m x 15.09m, which is consistent with the overall dimensions of this wreck site. Its overall length is difficult to discern with accuracy.

Archival Analysis:
The ship was torpedoed aft on the starboard side and sank by the stern (TNA ADM/1362). This is consistent with a notable area of damage seen on the wreck aft. Sank in four minutes and was heading NE at the time of the torpedo hit. The wreck points NE.

Phase 2 Conclusion:
EARL OF ELGIN (MBES/Archive)

Phase 1: UKHO Analysis:

Original UKHO Name:
6987 UNKNOWN

Archival details:
N/A

MBES Analysis:
Appears to be an upright cargo vessel of approx. 113m length, although it could be longer. 15.4m wide. Central bridge with holds fore and aft.

Additional Evidence:
Wreck located in 1945 (UKHO).

Phase 1 Conclusion:
UNKNOWN

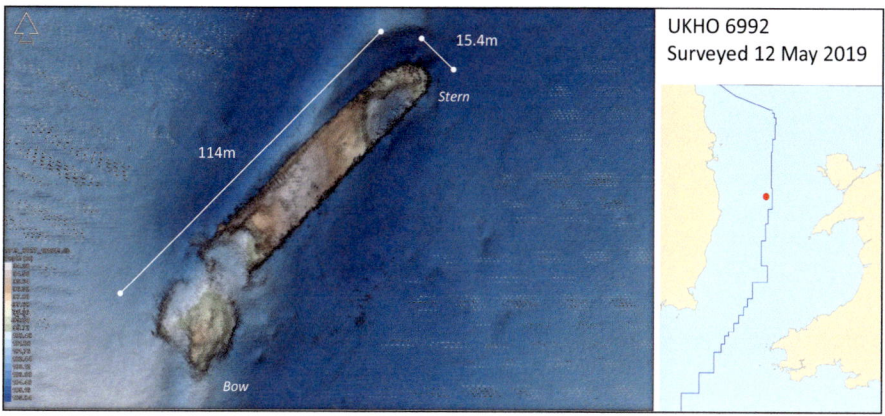

Stern

15.4m

114m

Bow

6992 MEXICO CITY (MBES/Archive)

Original UKHO: 6992 TRINIDAD (PROBABLY)	**Phase 1 Analysis:** UNKNOWN
Position WGS84: 53 5.483 N, 5 22.083 W	**Water Depth:** 101m
Type of Vessel: SS Cargo	**Surveyed Length:** 114m (approx)
Date of Loss: 5 Feb 1918	**Circumstances:** Torpedoed by U101

Phase 2: Method of wreck identification:

Positional Analysis:
U101's KTB position plots 5nm NW of this wreck. The report by SS WAR BRACKEN in Irish Sea (TNA ADM 137/1514) gives the position where four survivors from MEXICO CITY were found. It plots 8.6nm NW of the wreck.

Dimensional Analysis:
MEXICO CITY was 121.9m x 14.5m (*Lloyd's Register* 1917-18). The wreck is a close fit on width but is shorter on length. The fore section looks to have broken off, most likely when the ship impacted the seabed.

Archival Analysis:
The few survivors stated ship was hit in region of No2 hold, probably on starboard side and ship sank at once by the bow (TNA ADM 1514). U101's KTB states the course of the ship was 210deg. This is consistent with the orientation of the wreck. The bell of this ship when named SS NARRUNG is recorded against UKHO 7000 as being recovered in the 1950s, but the location is not known. MEXICO CITY carried copper and may have been salvaged in the past.

Phase 2 Conclusion:
MEXICO CITY (MBES/Archive)

Phase 1: UKHO Analysis:

Original UKHO Name:
6992 TRINIDAD (PROBABLY)

Archival details:
TRINIDAD was 94.48 x 11.30m (Larn & Larn). Although recorded sunk in the area, the U-boat (U101 KTB) position plots 25.7nm to the NE.

MBES Analysis:
Appears to be an upright steamship of approx 114m long and 15.4 m wide.

Additional Evidence:
Wreck located in 1982. Charted as Position Approximate (PA) beforehand (UKHO).

Phase 1 Conclusion:
UNKNOWN. The width and length of the wreck are too big for TRINIDAD.

Bow? Stern?

10.5m

36m

6994 PSS WILLIAM HUSKISSON/MANCHESTER (MBES/Archive)

Original UKHO: 6994 FOSTIMORE (POSSIBLY)	**Phase 1 Analysis:** UNKNOWN
Position WGS84: 53 9.1 N, 5 2.382 W	**Water Depth:** 53m
Type of Vessel: PSS Passenger/Cargo	**Surveyed Length:** 36m (approx)
Date of Loss: Uncertain	**Circumstances:** Foundered

Phase 2: Method of wreck identification:

Positional Analysis:
Two early paddle steamers are recorded as lost off Holyhead. MANCHESTER in 1829 (https://www.liverpool.ac.uk/~cmi/books/earlySS/earlySS.html#MAN) and WILLIAM HUSKISSON on 12 Jan 1840 (*The Times*, Jan 16, 1840 & Jan 25, 1840). In both cases the ships foundered after springing leaks.

Dimensional Analysis:
The dimensions of MANCHESTER are currently not known. WILLIAM HUSKISSON was 40m long (http://www.clydeships.co.uk/view.php?ref=20445).

Archival Analysis:
It is not possible to say what the identity of the wreck might be. The archival records available prior to 1850 are not accurate enough to derive a satisfactory identity. If a paddle steamer, then the wreck is (from sources available) most likely to be MANCHESTER or WILLIAM HUSKISSON.

Phase 2 Conclusion:
PSS WILLIAM HUSKISSON or PSS MANCHESTER (MBES/Archive)

Phase 1: UKHO Analysis:

Original UKHO Name:
6994 FOSTIMORE (POSSIBLY)

Archival details:
FOSTIMORE sank in collision four miles from the Carnarvon Bay LV in 1898 carrying a cargo of slag (*Wreck Returns* 1898). The wreck plots 8.5nm from the location of the LV. FOSTIMORE was a single screw steamship of 39.14m x 6.49m (*Lloyd's Register* 1898-99).

MBES Analysis:
The wreck measures approx. 36m long and appears to be very wide at approx. 10.5m. The centre section of the wreck gives the impression that it could have been a paddle wheeler, but this is based on the evidence from the pointcloud, which could be misleading.

Additional Evidence:
Wreck located in 1982 (UKHO).

Phase 1 Conclusion:
UNKNOWN. Although this wreck is broadly correct on length for FOSTIMORE (identified at 7418), it appears to be nearly twice the width. The cargo of slag does not show on the pointcloud or DTM.

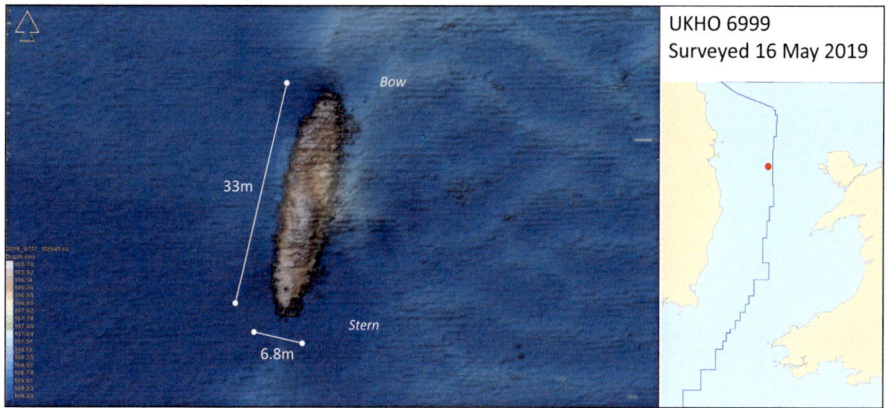

Bow

33m

Stern

6.8m

6999 ALCAZAR (MBES/Archive)

Original UKHO: 6999 ENERGY (POSSIBLY)	Phase 1 Analysis: UNKNOWN
Position WGS84: 53 19.017 N, 5 23.567 W	Water Depth: 110m
Type of Vessel: Steam trawler	Surveyed Length: 33m
Date of Loss: 7 Aug 1938	Circumstances: Collision

Phase 2: Method of wreck identification:

Positional Analysis:
Trawler ALCAZAR was sunk in a collision in fog with CAMBRIA while steering a course NNE from Codling LV. CAMBRIA was on its usual route from Holyhead to Kingstown (Dún Laoghaire). The wreck lies close to the intersection point of these two routes.

Dimensional Analysis:
ALCAZAR was 31.2m x 6.28m (*Lloyd's Register* 1938), which is consistent with the dimensions of the wreck.

Archival Analysis:
ALCAZAR was struck amidships (BOT Wreck Report No 7927). There is a noticeable area of damage on the wreck amidships on the starboard side.

Phase 2 Conclusion:
ALCAZAR (MBES/Archive)

Phase 1: UKHO Analysis:

Original UKHO Name:
6999 ENERGY (POSSIBLY)

Archival details:
ENERGY was an SV reported sunk in 1918 off the Codling Bank. The reported position is 14nm south of the wreck (*War Losses* 1990).

MBES Analysis:
The wreck is upright and appears to be a steam trawler or coaster of 33m length and around 6.8m wide.

Additional Evidence:
Located in 1982 (UKHO).

Phase 1 Conclusion:
UNKNOWN. ENERGY was an SV, and this wreck is clearly a steamship.

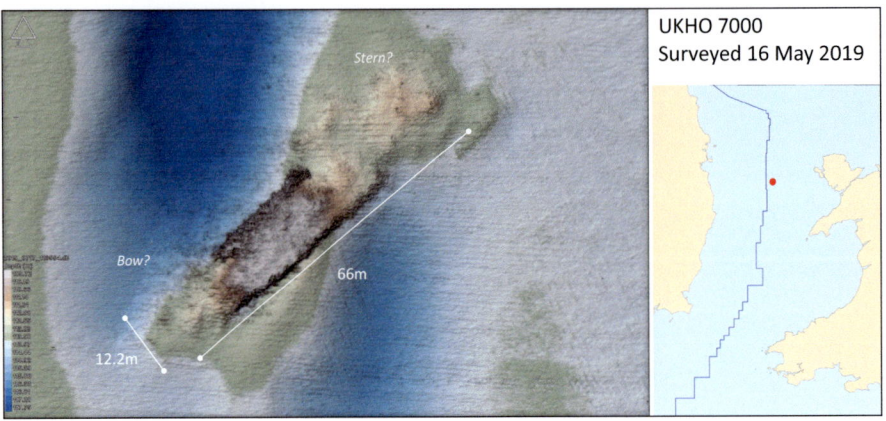

UKHO 7000
Surveyed 16 May 2019

Stern?

Bow?

66m

12.2m

7000 ESKMERE (MBES/Archive)

Original UKHO: 7000 MEXICO CITY	**Phase 1 Analysis:** UNKNOWN
Position WGS84: 53 12.649 N, 5 11.981 W	**Water Depth:** 117m
Type of Vessel: SS Cargo	**Surveyed Length:** 66m (approx)
Date of Loss: 13 Oct 1917	**Circumstances:** Torpedoed by UC75

Phase 2: Method of wreck identification:

Positional Analysis:
The wreck lies 9nm south of the position given by the 2nd mate, one of the few survivors (TNA ADM 137/1362). The position given in UC75's KTB is vague, but the position on its patrol track chart plots 10.6nm north of the wreck.

Dimensional Analysis:
ESKMERE was 87.5m x 12.2m (*Lloyd's Register* 1917). The wreck is very similar in width, but difficult to measure in length.

Archival Analysis:
The wreck appears to be pointing in the general direction ESKMERE was steering when sunk. It sank in three minutes after being torpedoed in the region of Nos 3 & 4 holds. The torpedo is reported to have blown out both sides of the ship, which then sank at once (TNA ADM 137/1362).

Phase 2 Conclusion:
ESKMERE (MBES/Archive). The evidence seems to point to this being ESKMERE, but it is not certain in this case. The wreck is much degraded, and the attack positions are both some miles to the north.

Phase 1: UKHO Analysis:

Original UKHO Name:
7000 MEXICO CITY

Archival details:
MEXICO CITY (ex-NARRUNG) was sunk in 1918 by U101. The position of the wreck plots 19nm from the *War Losses* (1990) position and 10nm from the U-boat (KTB) position.

MBES Analysis:
Cargo ship with bow and stern broken down. The wreck is approximately 66m long and 12.2m wide.

Additional Evidence:
Wreck located in 1945. It is attributed as "*could*" be MEXICO CITY. Not clear on the wreck card why a suffix of (POSSIBLY) or (PROBABLY) is not present. The bell of "NARRUNG" has been recovered but position was not known (UKHO).

Phase 1 Conclusion:
UNKNOWN. The wreck is too small to be MEXICO CITY which was 121m x 14.5m (*Lloyd's Register* 1918-19).

UKHO 7007
Surveyed 11 Jul 2019

7007 TRINIDAD (MBES/Archive)

Original UKHO: 7007 UNKNOWN	**Phase 1 Analysis:** UNKNOWN
Position WGS84: 53 20.198 N, 5 5.432 W	**Water Depth:** 120m
Type of Vessel: SS Cargo	**Surveyed Length:** 84m (approx)
Date of Loss: 7 Mar 1918	**Circumstances:** Torpedoed by U110

Phase 2: Method of wreck identification:

Positional Analysis:
The wreck of TRINIDAD lies 3.2nm north of the position reported in U110's KTB. The British position is vague due the fact the ship sank immediately, leaving only three survivors (TNA ADM 137/1515).

Dimensional Analysis:
TRINIDAD was 94.5m x 11.3m (*Lloyd's Register* 1917). The wreck is identical on width, but difficult to accurately measure on length due to the collapsed bows.

Archival Analysis:
U110's KTB states that it torpedoed a two-funnelled steamer in the bridge area, resulting in the fore section falling away and the ship sinking immediately.
Trinidad had two funnels after it was lengthened in 1893. The drawings in the project archive show its original build with only one funnel. But this website shows photos of the ship as it was in 1918.
https://sailstrait.wordpress.com/2019/02/14/cruising-to-new-york-the-s-s-trinidad/

Phase 2 Conclusion:
TRINIDAD (MBES/Archive)

Phase 1: UKHO Analysis:

Original UKHO Name:
7007 UNKNOWN

Archival details:
N/A

MBES Analysis:
Upright SS cargo with bow section missing or collapsed. 84m long 11m wide.

Additional Evidence:
Located in 1945.

Phase 1 Conclusion:
UNKNOWN

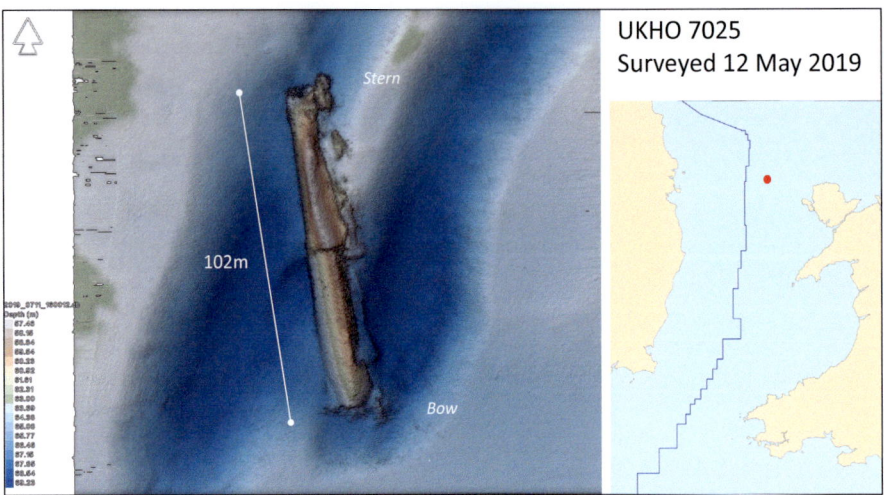

7025 PENVEARN (MBES/Archive)

Original UKHO: 7025 ESKMERE (POSSIBLY)	**Phase 1 Analysis:** UNKNOWN
Position WGS84: 53 27.297 N, 5 1.583 W	**Water Depth:** 62m
Type of Vessel: SS Cargo	**Surveyed Length:** 103m
Date of Loss: 2 Mar 1918	**Circumstances:** Torpedoed by U105

Phase 2: Method of wreck identification:

Positional Analysis:
The wreck lies 10.7nm SW of the U-boat (KTB) position and 11.9nm SW of the position given in *War Losses* (1990). The ship was towed from the position where it was originally hit.

Dimensional Analysis:
PENVEARN was 104.26m x 14.17m. It was a turret-decked ship (*Lloyd's Register* 1916-17). The length and width of the wreck is similar to the *Lloyd's Register* dimensions. The turret feature, if present is difficult to pick out of the pointcloud due to the wreck's orientation.

Archival Analysis:
The ship did not sink initially and was taken in tow (TNA ADM 137/4102) although the direction of tow is not stated. The initial damage being to the thwartship bunker, which was holed both sides of the ship. The wreck is clearly split in the middle, as seen in the DTM. The ship was reported to have been torpedoed again the following morning, with a hit in the bows causing the ship to sink rapidly from forward. This is not mentioned at all in U105's KTB which states the ship simply burned until its destruction was assured. The bow section is only slightly damaged. Note, the ship sank on 2 Mar 1918, not 1 Mar, as generally reported.

Phase 2 Conclusion:
PENVEARN (MBES/Archive). The position, dimensions, and condition of the wreck all favour PENVEARN as the identity. If this attribution is in fact correct, this would make it a rare example of a turret-decked ship.

Phase 1: UKHO Analysis:

Original UKHO Name:
7025 ESKMERE (POSSIBLY)

Archival details:
ESKMERE was a steamship of 87.5m x 12.2m (*Lloyd's Register* 1917-18). It was sunk in 1917 in a position which plots 3.8nm from the wreck (*War Losses* 1990).

MBES Analysis:
Cargo ship of 102m length lying on its port side beam ends. Its orientation means its width is difficult to measure with accuracy but appears to be 13-14m.

Additional Evidence:
Located in 1976 (UKHO).

Phase 1 Conclusion:
UNKNOWN. The wreck is too long to be ESKMERE.

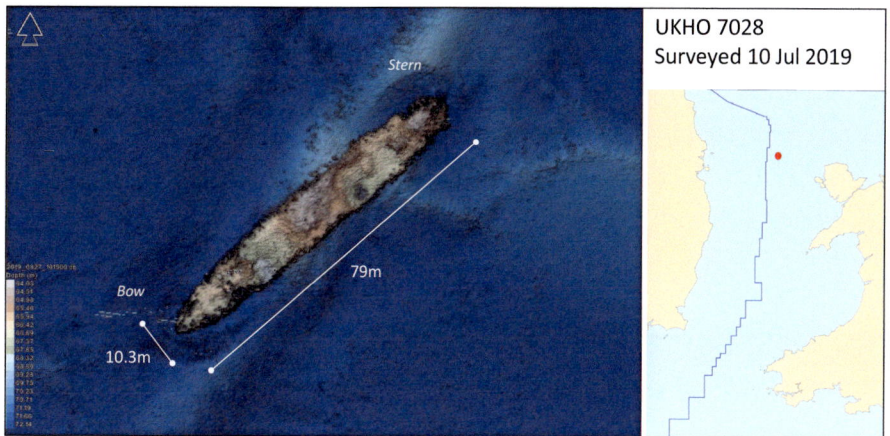

UKHO 7028
Surveyed 10 Jul 2019

7028 KENMARE (MBES/Archive)

Original UKHO: 7028 HOLYHEAD (POSSIBLY)	**Phase 1 Analysis:** UNKNOWN
Position WGS84: 53 28.28 N, 5 11.215 W	**Water Depth:** 75m
Type of Vessel: SS Cargo	**Surveyed Length:** 79m
Date of Loss: 2 Mar 1918	**Circumstances:** Torpedoed by U104

Phase 2: Method of wreck identification:

Positional Analysis:
The U104 KTB position plots 5.6nm north of this wreck. Due to there being few survivors the Allied position which plots 15nm north of the wreck is considered the more inaccurate (TNA ADM 137/1515).

Dimensional Analysis:
KENMARE was 80.6m x 10.9m (*Lloyd's Register* 1917-18). This is consistent with the dimensions of the wreck.

Archival Analysis:
U104's KTB states the ship was struck in the stern by a single torpedo and sank in four minutes. The survivors stated the ship immediately and quickly sank by the stern (TNA ADM 137/1515).

Phase 2 Conclusion:
KENMARE (MBES/Archive)

Phase 1: UKHO Analysis:

Original UKHO Name:
7028 HOLYHEAD (POSSIBLY)

Archival details:
HOLYHEAD sank after a collision with ALHAMBRA on 30 Oct 1883 (Larn & Larn 2000). The position (BOT 1883-4) plots 3.4nm from this wreck site. HOLYHEAD was 91.4m x 10.10m (*Lloyd's Register* 1883).

MBES Analysis:
Cargo vessel upright and pointing SW. 79m long and 10.3m wide.

Additional Evidence:
Located in 1977 (UKHO).

Phase 1 Conclusion:
UNKNOWN. This wreck is too small to be HOLYHEAD.

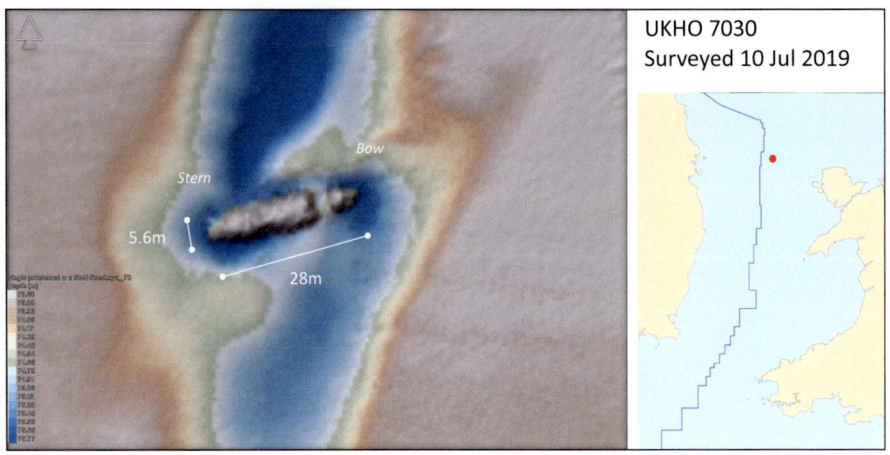

7030 TRIESTE (MBES/Archive)

Original UKHO: 7030 UNKNOWN	**Phase 1 Analysis:** UNKNOWN
Position WGS84: 53 28.947 N, 5 10.198 W	**Water Depth:** 76m
Type of Vessel: SS Tug	**Surveyed Length:** 28m
Date of Loss: 25 Oct 1896	**Circumstances:** Foundered

Phase 2: Method of wreck identification:

Positional Analysis:
Wreck Returns (1896) state TRIESTE sank 20nm NW of Holyhead. The position plots 7.5nm NW of this wreck.

Dimensional Analysis:
TRIESTE was 27.6m x 5.8m (*Lloyd's Register* 1896-7). This is consistent with the dimensions of this wreck.

Archival Analysis:
Sank on transit from Glasgow to Constantinople carrying bunker coal (*Wreck Returns* 1896).

Phase 2 Conclusion:
TRIESTE (MBES/Archive). The wreck is dimensionally, archivally, and positionally consistent with TRIESTE.

Phase 1: UKHO Analysis:

Original UKHO Name:
7030 UNKNOWN

Archival details:
Unknown

MBES Analysis:
SS/MV/FV of 28m x 5.6m.

Additional Evidence:
Located in 1976 (UKHO).

Phase 1 Conclusion:
UNKNOWN

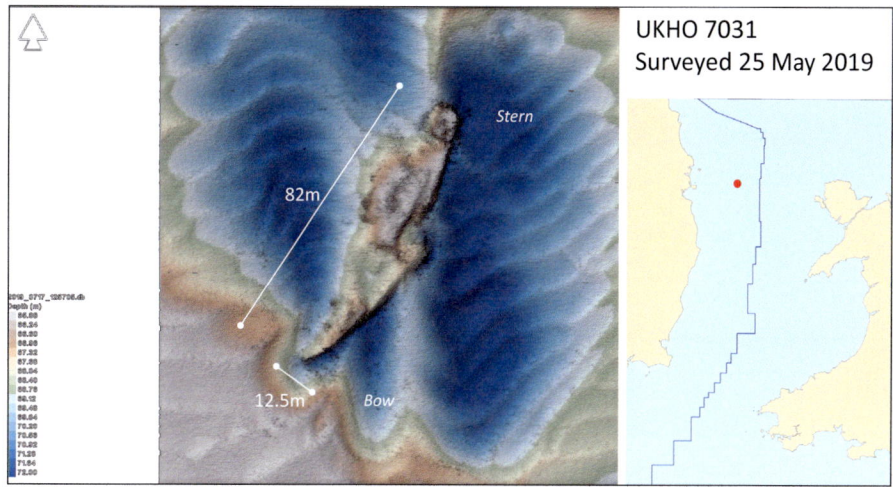

UKHO 7031
Surveyed 25 May 2019

7031 INNISFALLEN (MBES/Archive)

Original UKHO: 7031 UNKNOWN	**Phase 1 Analysis:** UNKNOWN
Position WGS84: 53 24.066 N, 5 34.657 W	**Water Depth:** 70m
Type of Vessel: SS Passenger/Cargo	**Surveyed Length:** 82m
Date of Loss: 23 May 1918	**Circumstances:** Torpedoed by UB64

Phase 2: Method of wreck identification:

Positional Analysis:
UB64 position (KTB) plots 9.6nm SE of this wreck. The master's report of the sinking (TNA ADM 137/1516) gives a position which plots 2.9nm NE of the wreck.

Dimensional Analysis:
INNISFALLEN was 82.9m x 11m (*Lloyd's Register* 1917-18). This is consistent with the dimensions of the wreck. Photos of the ship show it to have been a small ferry/liner with central superstructure. The MBES scan could be interpreted as being similar.

Archival Analysis:
The master stated the ship was hit port side in the stokehold and sank in four minutes. The U-boat then surfaced and approached the survivors but was driven off by HMS KESTREL (TNA ADM 137/1516). UB64 KTB states the ship was hit in the after part. The KTB refers to having to crash dive on the approach of a destroyer.

Phase 2 Conclusion:
INNISFALLEN (MBES/Archive). This wreck is positionally, dimensionally, and archivally consistent with INNISFALLEN.

Phase 1: UKHO Analysis:

Original UKHO Name:
7031 UNKNOWN

Archival details:
N/A

MBES Analysis:
Steamship broken in half amidships. Probably upright with the forward section partially buried along the starboard side. Length is 82m. Width is estimated at approx. 12-13m. Prominent central high point.

Additional Evidence:
First located in 1980 (UKHO).

Phase 1 Conclusion:
UNKNOWN

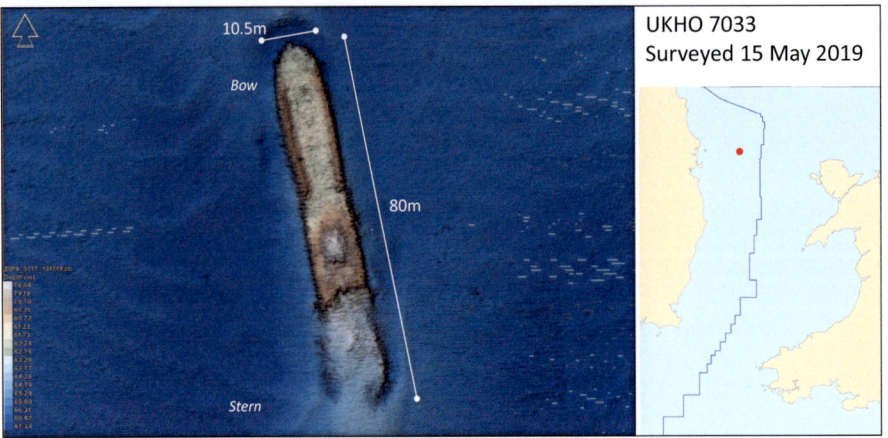

7033 NORMANDIET (MBES/Archive)

Original UKHO: 7033 UNKNOWN	**Phase 1 Analysis:** UNKNOWN
Position WGS84: 53 28.81 N, 5 31.375 W	**Water Depth:** 88m
Type of Vessel: SS Cargo	**Surveyed Length:** 80m
Date of Loss: 21 Apr 1918	**Circumstances:** Torpedoed by U91

Phase 2: Method of wreck identification:

Positional Analysis:
The position in U91's KTB plots 3.8nm south of this wreck. It should also be noted that the position given in *War Losses* (1990) is the same as given by the senior survivor (3rd engineer) and is given as 34nm SW of Holyhead (TNA ADM 137/1516). The position was also given by another survivor as 34nm SW½W of the Chickens (TNA ADM 137/1516). In the light of the fact that all the senior officers perished in the sinking, the survivor depositions appear less accurate. They plot 70nm apart. The nearest plots 15.2nm north of the wreck.

Dimensional Analysis:
NORMANDIET was 81.1m x 11.3m (*Lloyd's Register* 1918-19). This is consistent with the dimensions of the wreck.

Archival Analysis:
The U-boat KTB and survivor reports in ADM 137/1516 both state the vessel was stuck aft, survivors stating the torpedo hit No4 hold. The wreck is damaged aft. The ship was enroute to Glasgow from Spain and would therefore have been steaming in a northerly route. The wreck points to the north.

Phase 2 Conclusion:
NORMANDIET (MBES/Archive). The U-boat position, dimensions of the wreck, and damage of the wreck all point to this wreck being consistent with NORMANDIET.

Phase 1: UKHO Analysis:

Original UKHO Name:
7033 UNKNOWN

Archival details:
N/A

MBES Analysis:
Cargo ship with holds fore and aft 80m long and 10.5m wide.

Additional Evidence:
Located in 1945 (UKHO).

Phase 1 Conclusion:
UNKNOWN

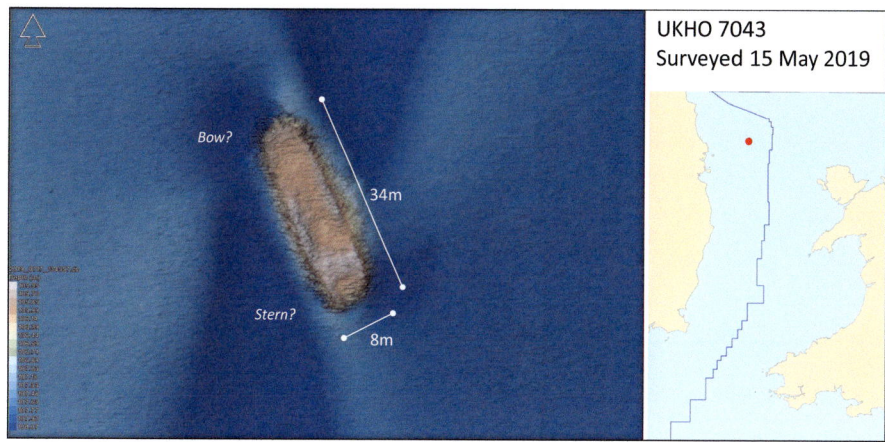

7043 PEACEFUL STAR (MBES/Archive)

Original UKHO: 7043 THE MARQUIS (POSSIBLY)	**Phase 1 Analysis:** UNKNOWN
Position WGS84: 53 36.044 N, 5 30.455 W	**Water Depth:** 108m
Type of Vessel: MFV	**Surveyed Length:** 34m
Date of Loss: 14 Mar 1941	**Circumstances:** Bombed

Phase 2: Method of wreck identification:

Positional Analysis:
The historic sinking position recorded by UKHO plots 3.8nm to the south of this wreck. The *War Losses* (1989) position plots 13.3nm to the south of this wreck.

Dimensional Analysis:
PEACEFUL STAR was under 100t (94grt), therefore its dimensions are not recorded at *Lloyd's Register*.

Archival Analysis:
No survivor report found. *War Losses* recorded that the wreck was on passage from Fleetwood to *"fishing grounds"*. All crew were picked up. The Admiralty daily war diary recorded that it was sunk while fishing off Louth (presumably the Irish county, not the Lincs town).

Phase 2 Conclusion:
PEACEFUL STAR (MBES/Archive). By the slightest of evidence, this wreck is considered to likely be PEACEFUL STAR. The wreck is small enough to be 94grt, it is in broadly the correct area and close to the position recorded by UKHO.

Phase 1: UKHO Analysis:

Original UKHO Name:
7043 THE MARQUIS (POSSIBLY)

Archival details:
THE MARQUIS was a three-mast schooner with single screw of 43.3m x 7.7m (*Lloyd's Register* 1915-16). It was sunk 11.1nm from this wreck (*War Losses* 1989).

MBES Analysis:
Barge or SV of 34m x 8m. Probably upright, possibly pointing to the north.

Additional Evidence:
Located in 1983 (UKHO).

Phase 1 Conclusion:
UNKNOWN. Dimensionally the wreck is too small to be THE MARQUIS.

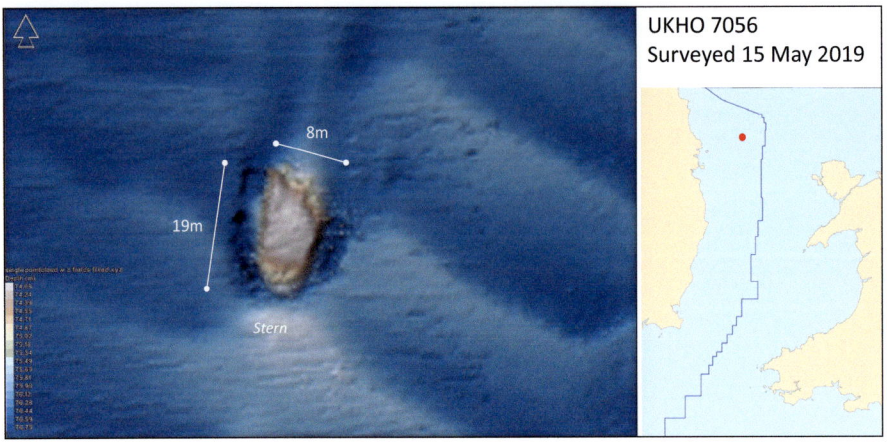

8m

19m

Stern

7056 HMS MANNERS (STERN SECTION) (MBES/Archive)

Original UKHO: 7056 GLENFORD (POSSIBLY)	**Phase 1 Analysis:** UNKNOWN
Position WGS84: 53 35.86 N, 5 21.422 W	**Water Depth:** 78m
Type of Vessel: Captain-class Frigate	**Surveyed Length:** 19m
Date of Loss: 26 Jan 1945	**Circumstances:** Torpedoed by U1051

Phase 2: Method of wreck identification:

Positional Analysis:
HMS MANNERS' stern was blown off at a position reported in the Admiralty daily diary which plots 7nm SE of this wreck.

Dimensional Analysis:
HMS MANNERS was 88.2m x 10.7m (wrecksite.eu). The wreck appears to be 8m wide.

Archival Analysis:
According to the book written by a Sub-Lt. on board, around 60ft of the ship's stern was blown off in the torpedo hit (Gibson 1993, p80). U1051 was sunk in the counterattack that followed.

Phase 2 Conclusion:
HMS MANNERS (STERN SECTION) (MBES/Archive). Positionally, dimensionally, and archivally, this portion of wreckage is consistent with being the stern of HMS MANNERS.

Phase 1: UKHO Analysis:

Original UKHO Name:
7056 GLENFORD (POSSIBLY)

Archival details:
GLENFORD was a SS of 50.29m x 7.95m (*Lloyd's Register* 1917-18). It was sunk in ballast in 1918 (*War Losses* 1989).

MBES Analysis:
Part of a ship, part of a ship's cargo or an outcrop 19m x 8m.

Additional Evidence:
Located in 1984 (UKHO).

Phase 1 Conclusion:
UNKNOWN. No evidence to suggest this is the wreck of GLENFORD. It is far too small.

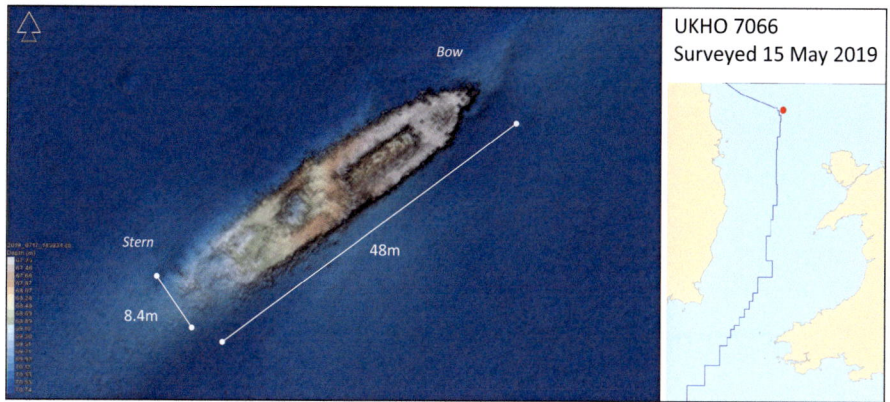

7066 GLENFORD (MBES/Archive)

Original UKHO: 7066 UNKNOWN	**Phase 1 Analysis:** UNKNOWN
Position WGS84: 53 44.2 N, 5 15.12 W	**Water Depth:** 71m
Type of Vessel: SS Cargo	**Surveyed Length:** 48m
Date of Loss: 20 Mar 1918	**Circumstances:** Gunfire from U101

Phase 2: Method of wreck identification:

Positional Analysis:
The wreck lies 5nm NE of the position given in *War Losses* (1990) and in Irish Sea (TNA ADM 137/1515). It plots 13nm north of the position given in U101's KTB.

Dimensional Analysis:
GLENFORD was 50.3m x 8m (*Lloyd's Register* 1917-18). The wreck is of similar dimensions.

Archival Analysis:
The ship was sailing from Dundalk to Dimnor, North Wales in ballast. The holds appear to be empty.

Phase 2 Conclusion:
GLENFORD (MBES/Archive) by position, dimensions, and archive this wreck is considered to be GLENFORD.

Phase 1: UKHO Analysis:

Original UKHO Name:
7066 UNKNOWN

Archival details:
N/A

MBES Analysis:
Upright SS/MV cargo vessel with holds fore and aft. 48m x 8.4m.

Additional Evidence:
Located in 1945 (UKHO).

Phase 1 Conclusion:
UNKNOWN

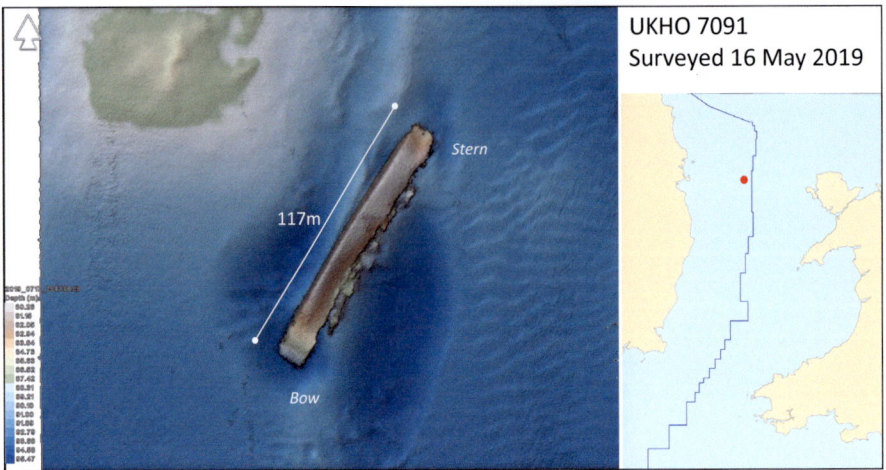

UKHO 7091
Surveyed 16 May 2019

7091 ANTEROS (MBES/Archive)

Original UKHO: 7091 UNKNOWN	**Phase 1 Analysis:** UNKNOWN
Position WGS84: 53 22.651 N, 5 25.443 W	**Water Depth:** 95m
Type of Vessel: SS Cargo	**Surveyed Length:** 117m
Date of Loss: 24 Mar 1918	**Circumstances:** Torpedoed by UB103

Phase 2: Method of wreck identification:

Positional Analysis:
UB103's KTB reports sinking the ship in a position only recorded as between the Kish light and the Skerries light. This wreck lies between the two. The master was too injured to file a report, but survivors stated the ship sank at position 11nm SE of this wreck (TNA ADM 137/1515).

Dimensional Analysis:
ANTEROS was 118.8m x 15.8m (*Lloyd's Register* 1917-18). This is consistent with the dimensions of this wreck. The drawing of the ship in the project archive seems to show features similar to those seen on the DTM.

Archival Analysis:
Both UB103's KTB and the survivors' reports state the ship was hit in the stern and sank rapidly. It was on a voyage to Port Talbot from Manchester and lies in line with its approximate course.

Phase 2 Conclusion:
ANTEROS (MBES/Archive)

Phase 1: UKHO Analysis:

Original UKHO Name:
7091 UNKNOWN

Archival details:
N/A

MBES Analysis:
Large SS or MV cargo or tanker lying on its port side. Hull partially collapsed. Length is 117m. Height from keel to upper deck is approx. 12.5m and its width as measured from the pointcloud is 14.5-15m.

Additional Evidence:
Located in 1982 (UKHO).

Phase 1 Conclusion:
UNKNOWN

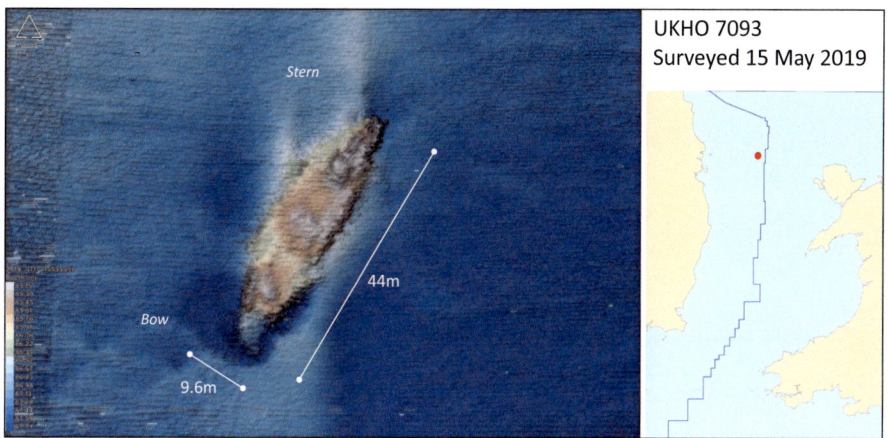

7093 THE MARQUIS (MBES/Archive)

Original UKHO: 7093 UNKNOWN	**Phase 1 Analysis:** UNKNOWN
Position WGS84: 53 28.369 N, 5 24.683 W	**Water Depth:** 89m
Type of Vessel: SS Cargo	**Surveyed Length:** 44m
Date of Loss: 8 Nov 1917	**Circumstances:** Gunfire UC75

Phase 2: Method of wreck identification:

Positional Analysis:
The UC75 KTB gives the sinking as taking place in Quadrat 027d. The wreck borders this square slightly to the west. The position in *War Losses* (1990) plots 6.4nm to the west of this wreck.

Dimensional Analysis:
THE MARQUIS was 43.3m x 7.7m (*Lloyd's Register* 1915-16). The wreck is consistent in length but looks wider on the DTM and is difficult to measure accurately. This may be due to cargo spilling out of the wreck.

Archival Analysis:
The ship was carrying a cargo of limestone (TNA ADM 137/1515). According to *Lloyd's Register* the ship has its machinery fitted aft. This can be seen in the DTM.

Phase 2 Conclusion:
THE MARQUIS (MBES/Archive). Positionally, dimensionally, and archivally the wreck is considered to most likely be THE MARQUIS.

Phase 1: UKHO Analysis:

Original UKHO Name:
7093 UNKNOWN

Archival details:
N/A

MBES Analysis:
Collapsed trawler or small coaster with possibly a motor showing at the stern. 44m x 9.6m.

Additional Evidence:
First located in 1983 (UKHO).

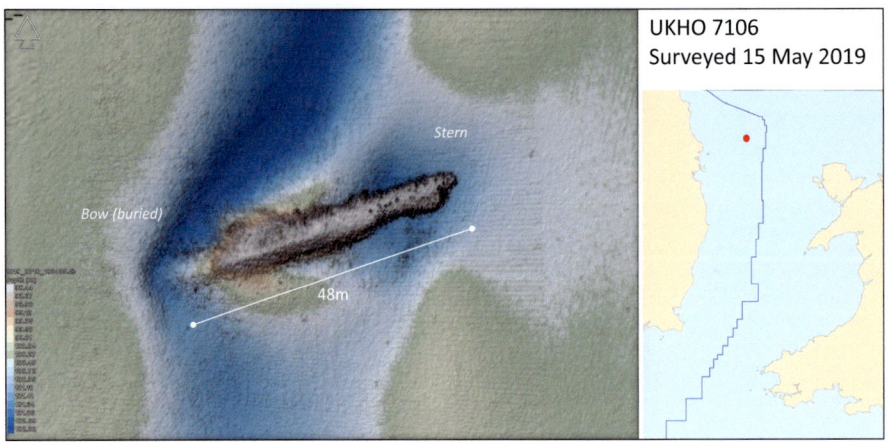

UKHO 7106
Surveyed 15 May 2019

Stern

Bow (buried)

48m

7106 U1051 (MBES/Archive)

Original UKHO: 7106 NORMANDIET (POSSIBLY)	**Phase 1 Analysis:** UNKNOWN
Position WGS84: 53 35.709 N, 5 33.705 W	**Water Depth:** 102m
Type of Vessel: Submarine	**Surveyed Length:** 48m (extant)
Date of Loss: 26 Jan 1945	**Circumstances:** Combined attack

Phase 2: Method of wreck identification:

Positional Analysis:
U1051 was recorded as being destroyed without doubt at a position which plots 7.1nm NE of this wreck (TNA ADM 199/1789).

Dimensional Analysis:
The submarine appears to have made a near vertical dive as it sank and around a quarter of its length is buried in the seabed. There is no doubt this is the wreck of a submarine.

Archival Analysis:
The U-boat came to the surface after being attacked and was rammed on the port side forward by HMS ALYMER, subsequently sinking by the bows. The U-boat was identified by its conning tower emblem which was photographed during the surface engagement (Gibson 1993, p65-66). U1051 had earlier torpedoed HMS MANNERS. That ship's stern is UKHO 7056, 5.4nm to the east.

Phase 2 Conclusion:
U1051 (MBES/Archive) By position, type of wreck, and archive there is little doubt this is the wreck of U1051.

Phase 1: UKHO Analysis:

Original UKHO Name:
7106 NORMANDIET (POSSIBLY)

Archival details:
NORMANDIET was a steamship of 81.10m x 11.30m (*Lloyd's Register* 1918-19).

MBES Analysis:
Wreck is a submarine on its starboard side, buried up to the conning tower. The pointcloud shows that the stern section aft of the conning tower is clear of the seabed.

Additional Evidence:
Located in 1983 (UKHO).

Phase 1 Conclusion:
UNKNOWN. This wreck is not NORMANDIET and is clearly a submarine.

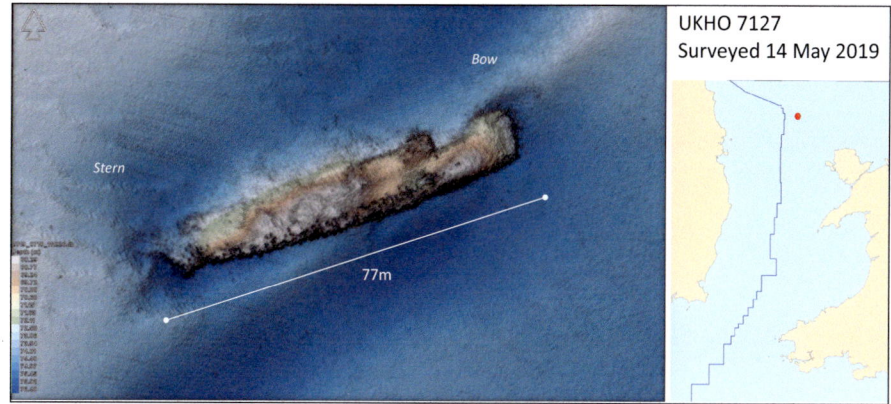

7127 CRESSADO (MBES/Archive)

Original UKHO: 7127 UNKNOWN	**Phase 1 Analysis:** UNKNOWN
Position WGS84: 53 39.729 N, 5 0.75 W	**Water Depth:** 76m
Type of Vessel: SS Cargo	**Surveyed Length:** 77m
Date of Loss: 8 May 1942	**Circumstances:** Collision with HMS POZARICA

Phase 2: Method of wreck identification:

Positional Analysis:
Positional data for the loss of this ship was very challenging to find. *Wreck Returns* (1942) states it was sunk in collision 11nm from the Skerries. The Admiralty diary states it was part of Convoy HG82, which was inbound for Liverpool. Duncan Haws (1989, p51) states it collided with HMS POZARICA. The wreck is actually 20nm NW of the Skerries.

Dimensional Analysis:
CRESSADO was 73.4m x 11m (*Lloyd's Register* 1940-41). The wreck is of broadly consistent dimensions. Photographs of the ship and the drawing in Haws (1989) show it had a distinctive well deck forward. Similar can be seen on the DTM of the wreck.

Archival Analysis:
As per positional analysis. No survivor or master's reports have been located.

Phase 2 Conclusion:
CRESSADO (MBES/Archive). The position is in the right area, the dimensions are similar, and the wreck appears to show similarities with images of the ship. None of this is conclusive, but there is enough evidence to suggest that this is CRESSADO. No other nearby wreck is a better candidate.

Phase 1: UKHO Analysis:

Original UKHO Name:
7127 UNKNOWN

Archival details:
N/A

MBES Analysis:
SS or MV lying on port side. Wreck is collapsed along its length of 77m. The width in the pointcloud is measured as 10m.

Additional Evidence:
Wreck first located in 1945 (UKHO).

Phase 1 Conclusion:
UNKNOWN

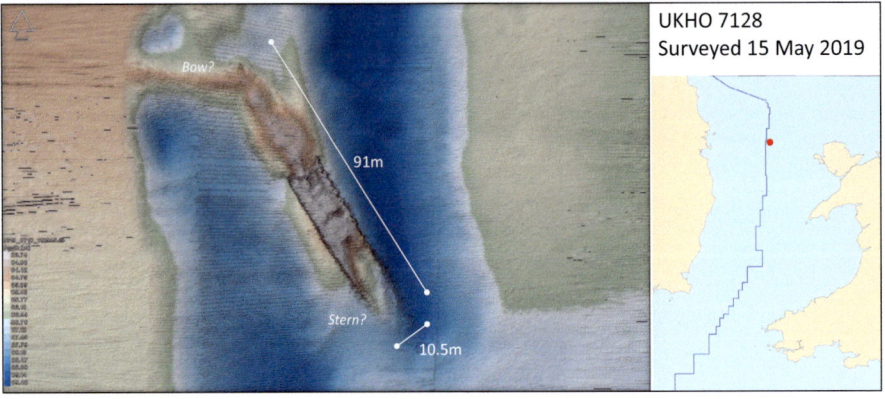

7128 HOLYHEAD (MBES/Archive)

Original UKHO: 7128 UNKNOWN		**Phase 1 Analysis:** UNKNOWN	
Position WGS84: 53 25.883 N, 5 17.367 W		**Water Depth:** 89m	
Type of Vessel: SS Passenger/Cargo		**Surveyed Length:** 91m (approx)	
Date of Loss: 30 Oct 1883		**Circumstances:** Collision with SV ALHAMBRA	

Phase 2: Method of wreck identification:

Positional Analysis:
The position given in *Shipping Casualties* (BOT 1883-84, p130) for the site of the collision plots 8nm NE of this wreck.

Dimensional Analysis:
HOLYHEAD was a custom-built ferry with an unusual, narrow design. It was 91.4m x 10.1m and was also twin screw (*Lloyd's Register* 1883-84). The wreck is of similar dimensions. Moreover, there appears to possibly be evidence of twin screws in the pointcloud data, although this is not certain.

Archival Analysis:
ALHAMBRA also sank in the collision, having been cut in half. As a timber built ship, it seems unlikely much, if any of it would show up on marine survey. Indeed, there is no candidate wreck in the MBES dataset. HOLYHEAD was damaged forward and sank 20 minutes after the collision. It should be noted the fore section of this wreck appears to be the more damaged.

Phase 2 Conclusion:
HOLYHEAD (MBES/Archive). Positionally, dimensionally, and archivally, this wreck is consistent with HOLYHEAD.

Phase 1: UKHO Analysis:

Original UKHO Name:
7128 UNKNOWN

Archival details:
N/A

MBES Analysis:
Partially upright SS of 91m x 10.5m. Bow section might be upside down and is partially buried in sand wave which appears to have been created by tide working on the wreck.

Additional Evidence:
Located in 1984 (UKHO).

Phase 1 Conclusion:
UNKNOWN

UKHO 7146
Surveyed 8 May 2019

7146 PERSEUS (MBES/Archive)

Original UKHO: 7146 UNKNOWN		**Phase 1 Analysis:** UNKNOWN	
Position WGS84: 53 1.35 N, 4 58.016 W		**Water Depth:** 70m	
Type of Vessel: SS Cargo		**Surveyed Length:** 64m (approx)	
Date of Loss: 13 Mar 1941		**Circumstances:** Air launched torpedo	

Phase 2: Method of wreck identification:

Positional Analysis:
The wreck lies 7nm north of the position given in *War Losses* (1989).

Dimensional Analysis:
PERSEUS was 68.8m x 10.6m. The wreck is foreshortened by the loss of the stern. PERSEUS can be seen in photographs to have two forward holds and one well-decked hold aft.

Archival Analysis:
PERSEUS was stuck by an air launched torpedo 12nm north of Bardsey. The explosion blew off the stern, aft of No3 hold. The ship drifted for at least two hours before foundering, sinking by the stern, and leaning to starboard (TNA ADM 199/2136).

Phase 2 Conclusion:
PERSEUS by MBES/Archive, the wreck is positionally, dimensionally, and archivally consistent with PERSEUS.

Phase 1: UKHO Analysis:

Original UKHO Name:
7146 UNKNOWN

Archival details:
N/A

MBES Analysis:
Cargo vessel lying upright with lean and collapse to port side. Stern missing. Two holds forward and one aft. Dimensions approx. 64m x 10m.

Additional Evidence:
Located in 1945 (UKHO).

Phase 1 Conclusion:
UNKNOWN

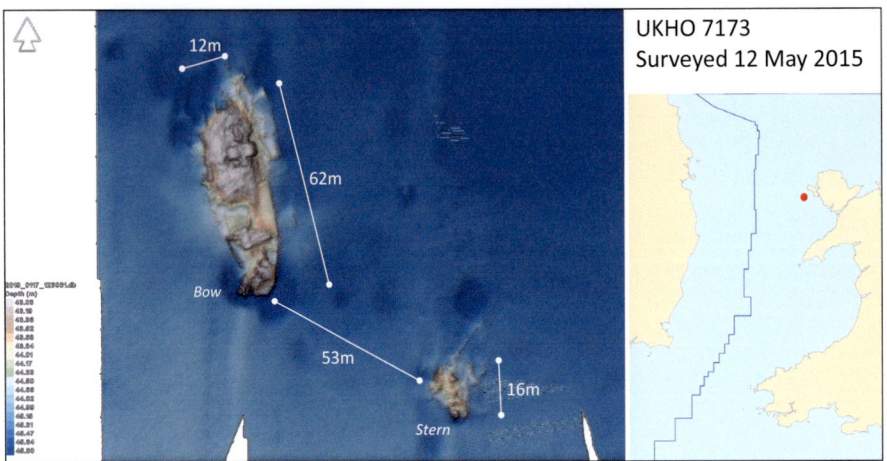

7173 NORMANDY COAST (MBES/Archive)

Original UKHO: 7173 SLIEVE BLOOM	**Phase 1 Analysis:** UNKNOWN
Position WGS84: 53 14.416 N, 4 49.134 W	**Water Depth:** 44m
Type of Vessel: SS Coaster	**Surveyed Length:** 78m (approx)
Date of Loss: 11 Jan 1945	**Circumstances:** Torpedoed by U1055

Phase 2: Method of wreck identification:

Positional Analysis:
Reported position was 15 19; 04 48W (Rohwer 1999, p188 and *War Losses* 1989, p786). The "19" may be a transcription error for 14. This would put the attack in exactly the right place. The attack also claimed ROANOKE (UKHO 7178) which has been identified as a nearby wreck to the north.

Dimensional Analysis:
Dimensionally the wreck is consistent with NORMANDY COAST, which was 79.2m x 11.6m (*Lloyd's Register* 1944-45). The structure remaining is consistent with archive photos of the ship on wrecksite.eu.

Archival Analysis:
NORMANDY COAST was sunk by U1055 on 11 Jan 1945. Survivor stated ship was hit aft and settled by the stern (TNA ADM 199/2148). It is noted the stern of this wreck is separated from the rest of the wreck. Wrecksite.eu has updated the UKHO record to show that the wreck is believed to be SS NORMANDY COAST. Means of identification by divers is not given in detail but stated as "100%" certain.

Phase 2 Conclusion:
NORMANDY COAST (MBES/Archive). The wreck is considered likely to be NORMANDY COAST by dimensions, archive, and diver identification.

Phase 1: UKHO Analysis:

Original UKHO Name:
7173 SLIEVE BLOOM

Archival details:
SLIEVE BLOOM was 91.26m x 11.32m (*Lloyd's Register* 1915-16).

MBES Analysis:
This wreck is broken into two portions. It appears as if the stern section is the smaller part lying 53m SE of the main portion of the wreck. The overall length is approximately 78m and width approx. 12m.

Additional Evidence:
Wreck located in 1945 (UKHO).

Phase 1 Conclusion:
UNKNOWN. This wreck is too short to be SLIEVE BLOOM. It is not known why UKHO considered this wreck CONFIRMED (no suffix).

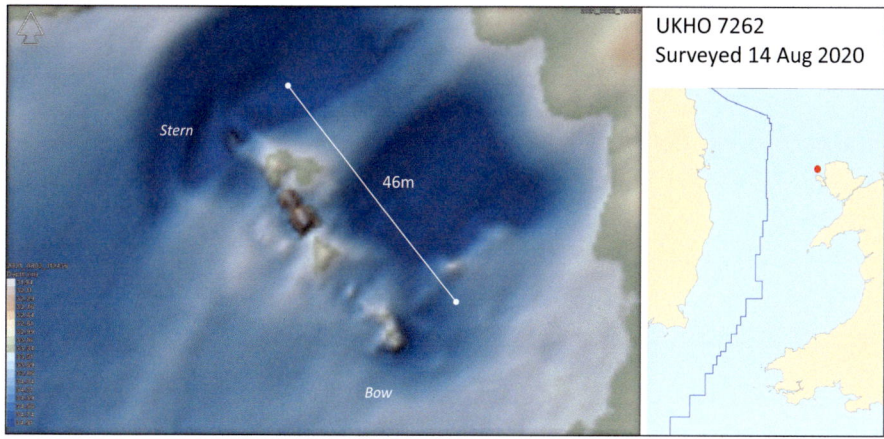

UKHO 7262
Surveyed 14 Aug 2020

7262 MARQUESS OF BUTE (MBES/Archive)

Original UKHO: 7262 RIVER HUMBER (POSSIBLY)	**Phase 1 Analysis:** UNKNOWN
Position WGS84: 53 21.465 N, 4 41.269 W	**Water Depth:** 35m
Type of Vessel: SS Cargo	**Surveyed Length:** 46m (approx)
Date of Loss: 20 Mar 1910	**Circumstances:** Collision with SS CONNEMARA

Phase 2: Method of wreck identification:

Positional Analysis:
The wreck plots 1.7nm south of the position given in Larn & Larn (2000), which was derived from *Shipping Casualties* (1910) as 3nm NWxN of Holyhead breakwater light. The wreck is misspelled in Larn & Larn as *"Marquis of Bute"*. *Wreck Returns* (1910) state the collision took place between South Stack and the Skerries, where the wreck lies.

Dimensional Analysis:
MARQUESS OF BUTE was 48.9m x 6.9m (*Lloyd's Register* 1889 as WILLIAM HINDE). This broadly conforms to the dimensions on the scan of the wreck. Single boiler and single engine.

Archival Analysis:
MARQUESS OF BUTE was on a voyage from Liverpool to Newport. *Shipping Casualties* (1909-10, p120) does not state whether the ship was inbound to Holyhead or outbound. The orientation of the wreck as seen on the UKHO scan points is as if it was entering the port. This might be due the circumstances of the collision. There is no mention of this collision in *The Times* (by online search 17 Apr 2020).

Phase 2 Conclusion:
MARQUESS OF BUTE (MBES/Archive). The positional and dimensional evidence suggests this wreck is MARQUESS OF BUTE, although the evidence is barely sufficient to make the attribution.

Phase 1: UKHO Analysis:

Original UKHO Name:
7262 RIVER HUMBER (POSSIBLY)

Archival details:
RIVER HUMBER was sunk in a collision NE of South Stack and has been identified by bell recovery as wreck No 7350. Its dimensions match the length measured of the wreck on the UKHO bathymetric map.

MBES Analysis:
Broken up wreck of a small steamer 46m long and 6m wide.

Additional Evidence:
Apparently first accurately located in 2013. No diver reports (UKHO).

Phase 1 Conclusion:
UNKNOWN. This wreck is not RIVER HUMBER, as it has been identified by the bell elsewhere.

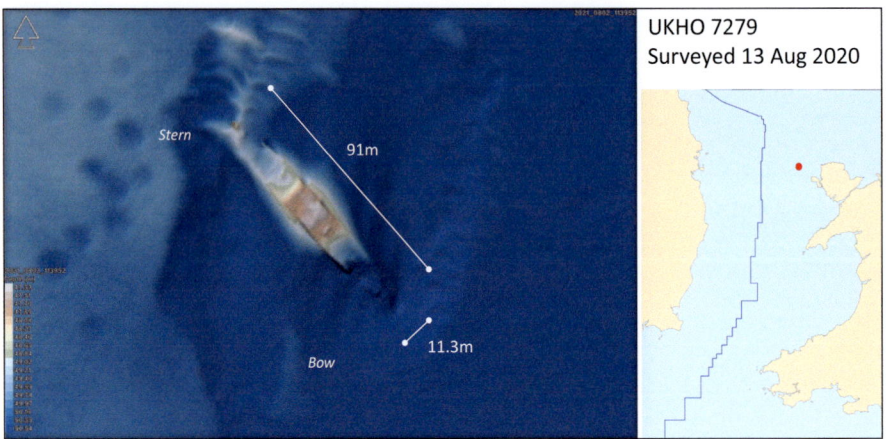

7279 SLIEVE BLOOM (MBES/Archive)

Original UKHO: 7279 UNKNOWN	**Phase 1 Analysis:** UNKNOWN
Position WGS84: 53 22.435 N, 4 52.238 W	**Water Depth:** 52m
Type of Vessel: SS Cargo/Passenger	**Surveyed Length:** 91m
Date of Loss: 30 Mar 1918	**Circumstances:** Collision with USS STOCKTON

Phase 2: Method of wreck identification:

Positional Analysis:
SLIEVE BLOOM was sunk in collision at a position which plots 3.3nm SE of this wreck (*Wreck Returns* 1918).

Dimensional Analysis:
SLIEVE BLOOM was 91.4m x 11.3m (*Lloyd's Register* 1915-16).

Archival Analysis:
The wreck was identified by the recovery of the bell by Jon Shaw (personal communication 14 Oct 2019).

Phase 2 Conclusion:
SLIEVE BLOOM (MBES/Archive). Positionally and archivally this wreck is consistent with SLIEVE BLOOM. Bell recovery by recreational divers has identified the wreck without doubt. It is for all intents and purposes CONFIRMED now.

Phase 1: UKHO Analysis:

Original UKHO Name:
7279 UNKNOWN

Archival details:
N/A

MBES Analysis:
Wreck upright with holds fore and aft collapsed. 91m x 11.3m.

Additional Evidence:
Located in 1945 (UKHO).

Phase 1 Conclusion:
UNKNOWN

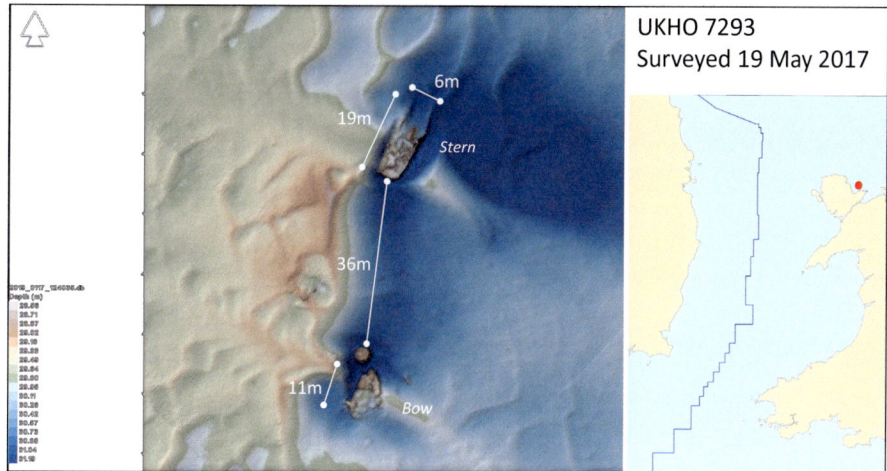

7293 KINKORTH (MBES/Archive)

Original UKHO: 7293 LADY WINDSOR (PROBABLY)		**Phase 1 Analysis:** UNKNOWN	
Position WGS84: 53 23.782 N, 4 8.422 W		**Water Depth:** 26m	
Type of Vessel: SS Drifter		**Surveyed Length:** 30m (total)	
Date of Loss: 10 Dec 1941		**Circumstances:** Mined?	

Phase 2: Method of wreck identification:

Positional Analysis:
KINKORTH sunk while heading to Fleetwood to fish. Lost with all hands. Probably mined (*War Losses* 1989, p319). Position given (82deg Point Lynas 7 miles) plots 2.6nm from this wreck site.

Dimensional Analysis:
The dimensions are correct for KINKORTH, being 30m long and 6m wide. Single boiler appears to have fallen out and lies near the bow half of the wreck. Ship in two halves 36m apart.

Archival Analysis:
Note that KINKORTH was presumed mined. The damage seen on this site is consistent with a mine hit on a small vessel.

Phase 2 Conclusion:
KINKORTH (MBES/Archive). Available evidence shows wreck is dimensionally and positionally consistent with KINKORTH. Diver evidence on wrecksite.eu is not sufficiently detailed to be definitive.

Phase 1: UKHO Analysis:

Original UKHO Name:
7293 LADY WINDSOR (PROBABLY)

Archival details:
LADY WINDSOR foundered in heavy weather in 1937. She is known to have been come ashore at Cemaes Bay and burned for her copper fittings (wrecksite.eu). This rules it out as being this wreck.

MBES Analysis:
The remains of a trawler or similar. This appears to be the remains of a metal wreck. Note that LADY WINDSOR was of timber construction (wrecksite.eu).

Additional Evidence:
Local divers now call this wreck KINKORTH, although the exact reason is not stated.

Phase 1 Conclusion:
UNKNOWN. This wreck is not likely of wood construction and shows no signs of being burned for its copper.

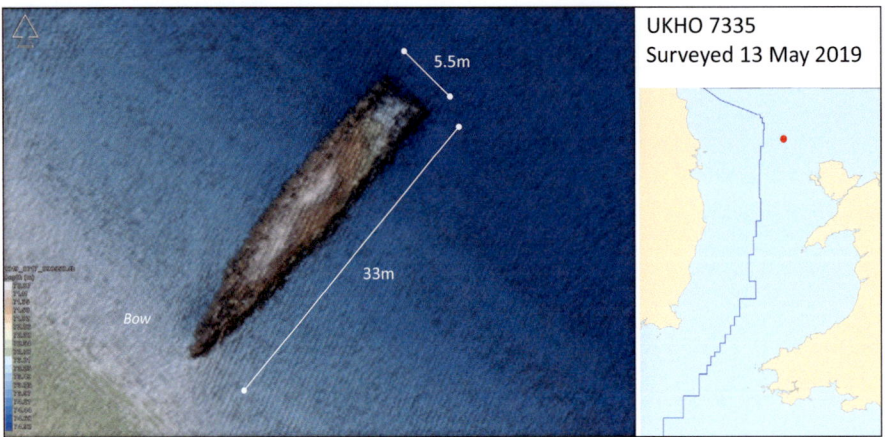

UKHO 7335
Surveyed 13 May 2019

5.5m

33m

Bow

7335 CRESSIDA (MBES/Archive)

Original UKHO: 7335 GLENCONA (PROBABLY)	**Phase 1 Analysis:** UNKNOWN
Position WGS84: 53 28.864 N,4 58.1 W	**Water Depth:** 75m
Type of Vessel: SS Cargo	**Surveyed Length:** 33m (incomplete)
Date of Loss: 17 Mar 1918	**Circumstances:** Torpedoed by U103

Phase 2: Method of wreck identification:

Positional Analysis:
The wreck lies 3.5nm NE of the position given in *War Losses* (1990) and in Irish Sea (TNA ADM 137/1515). The KTB position for this attack plots 3.5nm NW of the wreck.

Dimensional Analysis:
CRESSIDA was a converted yacht of 39.3m x 5.6m (*Lloyd's Register* 1918-19). The wreck is a match on width, although shorter in length, possibly due to the circumstances of its sinking.

Archival Analysis:
The survivors state that the ship was hit in the stern and sunk within one minute (TNA ADM 137/1515). A similar description is given in U103's KTB. It appears from the MBES scan that the stern was completely blown off.
In a notably rare mistake, the official German historian Arno Spindler attributed this attack to the loss of SS SEA GULL. This becomes evident when studying the KTB with his handwritten annotations.

Phase 2 Conclusion:
CRESSIDA (MBES/Archive). By position, archive, and dimensions it is likely this is the wreck of SS CRESSIDA.

Phase 1: UKHO Analysis:

Original UKHO Name:
7335 GLENCONA (PROBABLY)

Archival details:
This wreck cannot be GLENCONA because the ship drifted ashore in Morecambe Bay after catching fire 15m NWxW of the Skerries (approx. where this wreck lies). GLENCONA was identified in 1926 in Morecambe Bay (Larn & Larn 2000). Sinking date in UKHO is also inconsistent with Larn & Larn.

MBES Analysis:
MBES scan reveals small wreck of approx. 33m long. Pointcloud shows is upright with stern to the north. The stern appears to be missing and what may be a boiler can be discerned in the centre of the ship.

Additional Evidence:
Wreck located in 1976 (UKHO).

Phase 1 Conclusion:
UNKNOWN. GLENCONA is ruled out as the identity of this wreck.

7347 BRITISH VISCOUNT (MBES/Archive)

Original UKHO: 7347 MANCHESTER	**Phase 1 Analysis:** UNKNOWN
Position WGS84: 53 30.388 N, 4 50.027 W	**Water Depth:** 69m
Type of Vessel: SS Petroleum Tanker	**Surveyed Length:** 96m (approx)
Date of Loss: 23 Feb 1918	**Circumstances:** Torpedoed by U91

Phase 2: Method of wreck identification:

Positional Analysis:
The wreck lies 3.6nm south of the position in *War Losses* (1990), which is almost identical to the master's position (TNA ADM 137/1514). The U-boat KTB position plots 10.8nm NW of the wreck.

Dimensional Analysis:
The ship was 95.09m x 12.29m (*Lloyd's Register* 1916-17), which is consistent with the dimensions on the MBES scan.

Archival Analysis:
The ship was hit on the starboard side by No4 tank and immediately bent into a "V" shape. It was abandoned, but not seen to sink. It was later reported to have sunk 20 minutes after being hit. The vessel was known to buckle in heavy weather (TNA ADM 137/1514).

Phase 2 Conclusion:
BRITISH VISCOUNT (MBES/Archive). The wreck is consistent with being a tanker, showing what appears to be cylindrical tanks in its structure. It also shows features as described in the text of the sinking (split amidships and damage on starboard side) and in its dimensions.

Phase 1: UKHO Analysis:

Original UKHO Name:
7347 MANCHESTER

Archival details:
MANCHESTER (ex-GLANWERN 1882) sank in a collision in 1905 at a position which plots 22nm north of this wreck (BOT 1905, p134). The ship was 64.64m x 9.14m (*Lloyd's Register* 1883).

MBES Analysis:
Steamship, upright with hole amidships and damaged bow. Approximately 96m long and 12.2m wide.

Additional Evidence:
Wreck located in 1976. Bell supposedly recovered from this wreck in 2002 (UKHO).

Phase 1 Conclusion:
UNKNOWN. The wreck cannot be MANCHESTER as it is far too large. The bell reported coming from GLANWERN clearly was not recovered from this wreck site.

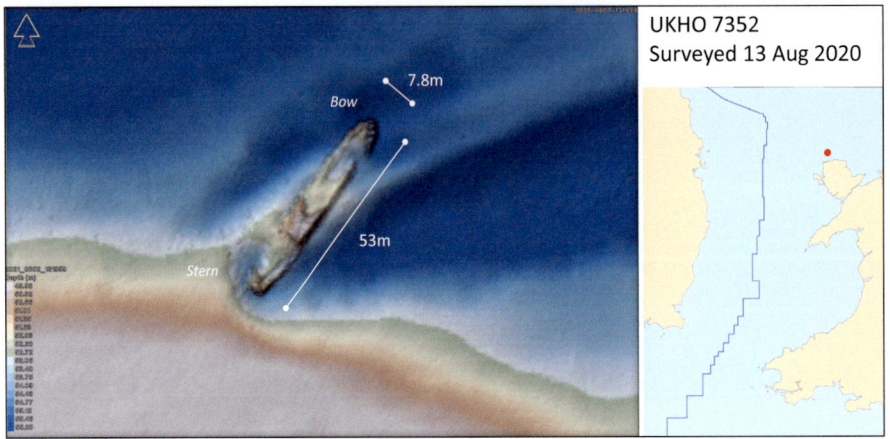

7352 COGNAC (MBES/Archive)

Original UKHO: 7352 UNKNOWN	**Phase 1 Analysis:** UNKNOWN
Position WGS84: 53 31.248 N, 4 32.279 W	**Water Depth:** 54m
Type of Vessel: SS Cargo	**Surveyed Length:** 53m
Date of Loss: 11 Nov 1898	**Circumstances:** Collision with SS VOLTAIC

Phase 2: Method of wreck identification:

Positional Analysis:
This wreck lies at a position 7nm east of the collision point given in *Shipping Casualties* (1898-99, p154).

Dimensional Analysis:
COGNAC was 51.81m x 7.64m (*Lloyd's Register* 1889-90). This is consistent with the dimensions of the wreck.

Archival Analysis:
The wreck shows machinery just aft of midships. This is not inconsistent with *Lloyd's Register*, which also described the vessel as having three holds, consistent with the wreck.

Phase 2 Conclusion:
COGNAC (MBES/Archive). This wreck is positionally, archivally, and dimensionally consistent with COGNAC.

Phase 1: UKHO Analysis:

Original UKHO Name:
7352 UNKNOWN

Archival details:
N/A

MBES Analysis:
Wreck of a coaster 53m x 7.8m long pointing NE. Two holds forward. One hold aft. Central engine room.

Additional Evidence:
Located in 1976.

Phase 1 Conclusion:
UNKNOWN

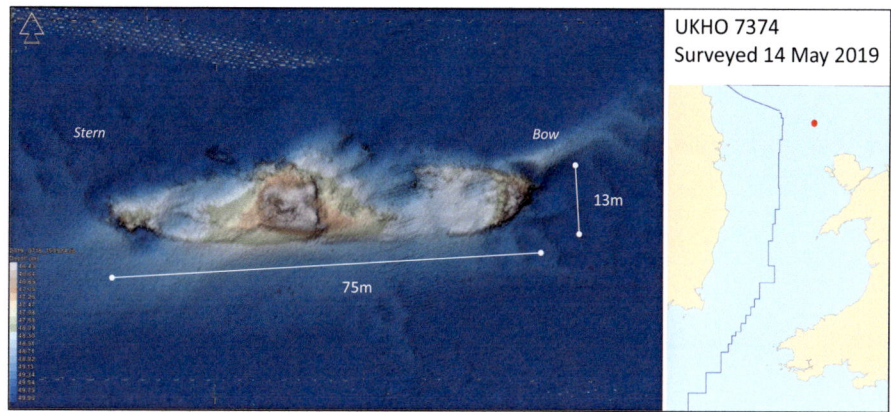

Stern

Bow

13m

75m

7374 SARPFOS (MBES/Archive)

Original UKHO: 7374 U242 (PROBABLY)	**Phase 1 Analysis:** UNKNOWN
Position WGS84: 53 42.362 N, 4 56.967 W	**Water Depth:** 51m
Type of Vessel: SS Cargo	**Surveyed Length:** 75m
Date of Loss: 24 Feb 1918	**Circumstances:** Torpedoed by U105

Phase 2: Method of wreck identification:

Positional Analysis:
The wreck lies 5.5nm to the SE of the position given in *War Losses* (1990) and in the KTB of U105. The wreck is known to have drifted for over an hour after initially being hit (KTB). Wind was from the NW, so the wreck would be expected to have drifted SE.

Dimensional Analysis:
SARPFOS was 77.54m x 11.93m (*Lloyd's Register* 1916-17). This is consistent with the dimensions of this wreck site.

Archival Analysis:
U105's KTB states the ship was hit twice, with a gap of 40 minutes between shots. It drifted for 60 minutes before sinking. The master's account shows the ship was initially hit port side by engine room and the ship was abandoned. The crew escaped into an SV which made off. It was not seen to sink. Cargo was coal (TNA ADM 137/1514).

Phase 2 Conclusion:
SARPFOS (MBES/Archive). Positionally, dimensionally, and archivally this wreck is most likely to be SARPFOS.

Phase 1: UKHO Analysis:

Original UKHO Name:
7374 U242 (PROBABLY)

Archival details:
N/A

MBES Analysis:
SS coaster of 75m length and 13m width, with twin boilers and a single engine. Bows to the east.

Additional Evidence:
Located in 1945 (UKHO).

Phase 1 Conclusion:
UNKNOWN. The wreck is not a U-boat.

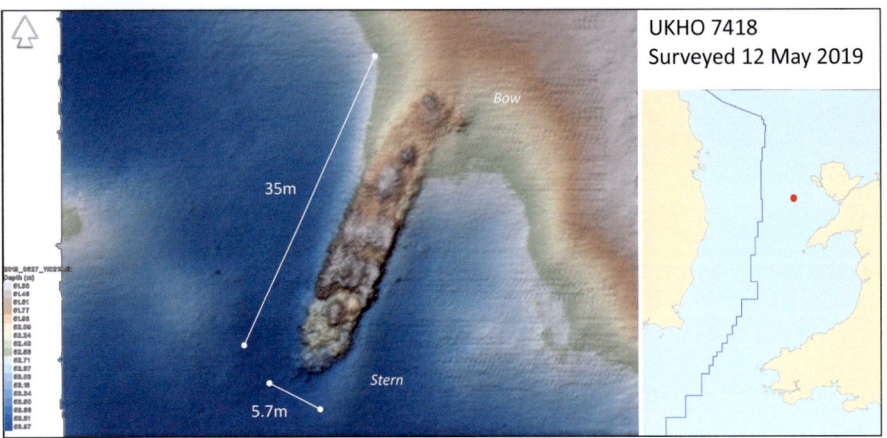

UKHO 7418
Surveyed 12 May 2019

7418 FOSTIMORE (MBES/Archive)

Original UKHO: 7418 UNKNOWN	**Phase 1 Analysis:** UNKNOWN
Position WGS84: 53 7.216 N, 4 58.183 W	**Water Depth:** 53m
Type of Vessel: SS Cargo	**Surveyed Length:** 35m
Date of Loss: 7 Sep 1898	**Circumstances:** Collision with SS DOTTEREL

Phase 2: Method of wreck identification:

Positional Analysis:
The wreck plots 0.5nm south of the position of the collision recorded in *Shipping Casualties* (BOT 1898-99, p152).

Dimensional Analysis:
FOSTIMORE was 39.14m x 6.49m (*Lloyd's Register* 1898-99).

Archival Analysis:
Sank in collision with SS DOTTEREL 5nm WNW of the Carnarvon Bay LV. No lives lost (BOT 1898-99, p152)

Phase 2 Conclusion:
FOSTIMORE (MBES/Archive). Positionally, archivally, and dimensionally, this wreck is most probably FOSTIMORE.

Phase 1: UKHO Analysis:

Original UKHO Name:
7418 UNKNOWN

Archival details:
N/A

MBES Analysis:
Upright coaster with engine aft. Dimensions of 35m x 5.7m.

Additional Evidence:
Located in 1980 (UKHO).

Phase 1 Conclusion:
UNKNOWN

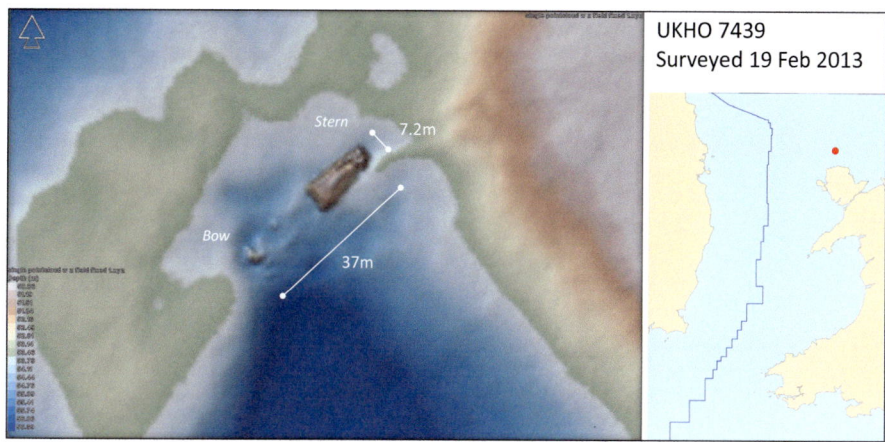

7439 HILDENA (MBES/Archive)

Original UKHO: 7439 UNKNOWN	**Phase 1 Analysis:** UNKNOWN
Position WGS84: 53 33.397 N, 4 28.67 W	**Water Depth:** 54m
Type of Vessel: SS Coaster	**Surveyed Length:** 37m (approx)
Date of Loss: 14 Jan 1918	**Circumstances:** Collision with SS HIRANO MARU

Phase 2: Method of wreck identification:

Positional Analysis:
This wreck lies 6.7nm NE of the collision position given in *Shipping Casualties* (1921) p37.

Dimensional Analysis:
HILDENA was 36.65m x 7.16m (*Lloyd's Register* 1917-18). This is consistent with the dimensions of the wreck. Photo of the ship shows it had machinery aft, which is similar to this wreck.

Archival Analysis:
HILDENA was heading SE at the time of the collision. It was hit amidships and sank immediately (BOT *Wreck Report*). It is noted the wreck points SE and is damaged from bow to aft of midships.

Phase 2 Conclusion:
This wreck is positionally, dimensionally, and archivally consistent with HILDENA.

Phase 1: UKHO Analysis:

Original UKHO Name:
7439 UNKNOWN

Archival details:
N/A

MBES Analysis:
Small coaster/FV with prominent rounded stern and forward hold. Upright and points SW. Dimensions of 37x7.2m.

Additional Evidence:
Located in 1988 (UKHO).

Phase 1 Conclusion:
UNKNOWN

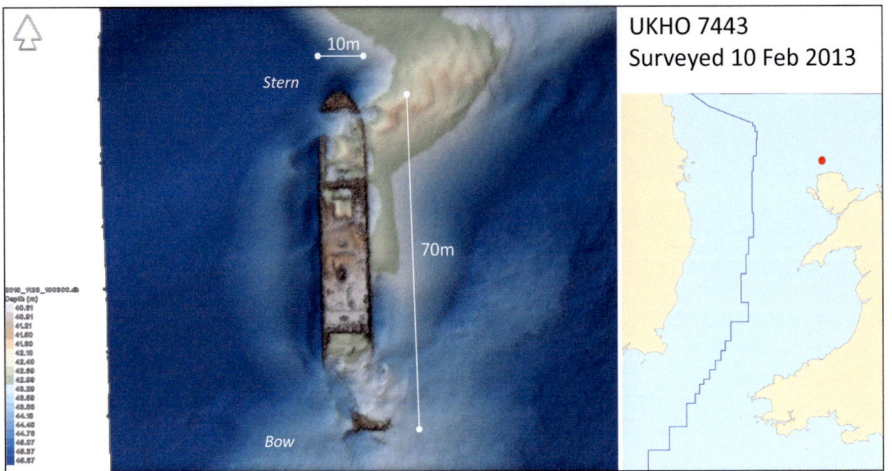

7443 BENHOLM (MBES/Archive)

Original UKHO: 7443 BRITISH VISCOUNT (POSSIBLY)	**Phase 1 Analysis:** UNKNOWN
Position WGS84: 53 30.714 N, 4 29.387 W	**Water Depth:** 46m
Type of Vessel: SS Cargo	**Surveyed Length:** 70m (approx)
Date of Loss: 14 May 1898	**Circumstances:** Collision with SS KLONDYKE

Phase 2: Method of wreck identification:

Positional Analysis:
This wreck plots 3.1nm north of the collision point recorded in *Shipping Casualties* (BOT 1897-98 p151).

Dimensional Analysis:
BENHOLM was 71.93m x 10.21m (*Lloyd's Register* 1889). This is consistent with the dimensions of the wreck.

Archival Analysis:
The ship was in ballast when sunk. The DTM of the wreck shows the holds to be empty.

Phase 2 Conclusion:
BENHOLM (MBES/Archive). Positionally, dimensionally, and archivally this wreck is consistent with BENHOLM.

Phase 1: UKHO Analysis:

Original UKHO Name:
7443 BRITISH VISCOUNT (POSSIBLY)

Archival details:
BRITISH VISCOUNT was a tanker 95.9m long. U-boat (KTB) position is 20nm from wreck. The *War Losses* (1990) position is 13m from wreck.

MBES Analysis:
Wreck resembles a steamship with holds forward and aft. Wreck is approx. 70m long with fore section collapsed. Beam is measured at approx. 10m.

Additional Evidence:
Located in 1987 (UKHO).

Phase 1 Conclusion:
UNKNOWN. BRITISH VISCOUNT is eliminated as candidate identity due size of wreck.

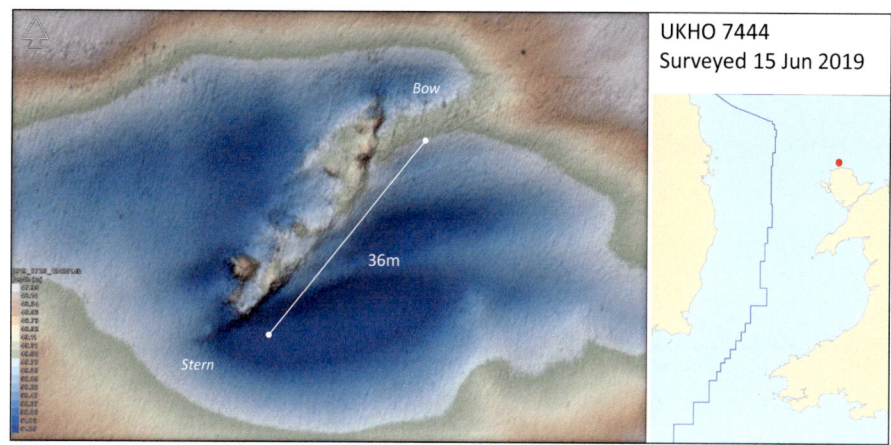

7444 HERBERT (MBES/Archive)

Original UKHO: 7444 LEEDS (POSSIBLY)	**Phase 1 Analysis:** UNKNOWN
Position WGS84: 53 29.147 N, 4 28.037 W	**Water Depth:** 49m
Type of Vessel: SS Cargo	**Surveyed Length:** 39m
Date of Loss: 19 Nov 1895	**Circumstances:** Foundered

Phase 2: Method of wreck identification:

Positional Analysis:
The wreck is recorded in *Shipping Casualties* as sinking 2.5nm east of this wreck (BOT 1895-6, p110).

Dimensional Analysis:
HERBERT was 35.15m x 5.84m (*Lloyd's Register* 1883). This is consistent with this wreck site.

Archival Analysis:
HERBERT foundered with a cargo of clay. No loss of life (BOT 1895-6, p110). No discernible cargo on DTM.

Phase 2 Conclusion:
HERBERT (MBES/Archive). Positionally and dimensionally this wreck is likely to be HERBERT.

Phase 1: UKHO Analysis:

Original UKHO Name:
7444 LEEDS (POSSIBLY)

Archival details:
LEEDS was a PSS which sank in a gale in Jan 1852 (Michael 2008, p94). The position was reported to be in the Liverpool Bay roadstead, outside of the study area. It was 42.98m x 7.62m (https://www.liverpool.ac.uk/~cmi/books/earlySS/earlySS.html).

MBES Analysis:
Steamship with boiler at the aft end. Very collapsed with a length of 36m. Width difficult to discern with accuracy.

Additional Evidence:
Located in 1987.

Phase 1 Conclusion:
UNKNOWN. Diver suggestion that this might be LEEDS (UKHO) cannot be substantiated from available archival records. Moreover, the MBES scan seems to suggest that the wreck is not a PSS.

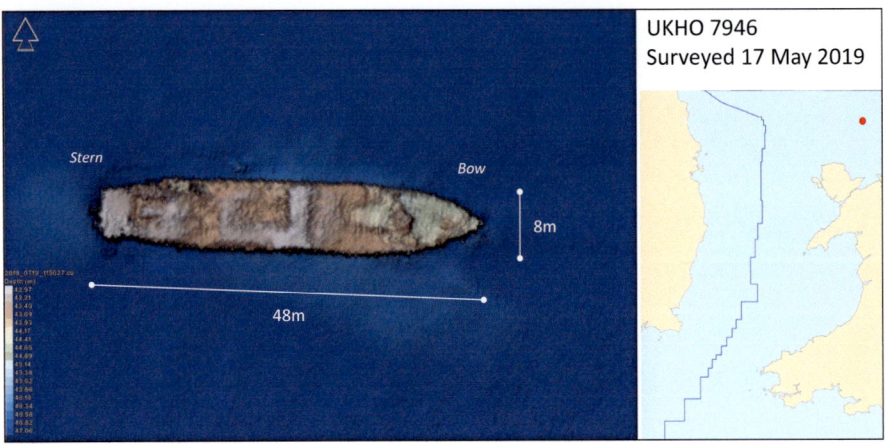

Stern   Bow

8m

48m

7946 CLYTIE (MBES/Archive)

Original UKHO: 7946 SEA GULL (POSSIBLY)	**Phase 1 Analysis:** UNKNOWN
Position WGS84: 53 34.93 N, 3 59.107 W	**Water Depth:** 47m
Type of Vessel: SS Cargo	**Surveyed Length:** 48m
Date of Loss: 6 Dec 1905	**Circumstances:** Collision

Phase 2: Method of wreck identification:

Positional Analysis:
The wreck lies at a position which plots 6.4nm NE of the position recorded in *Shipping Casualties* (BOT 1905-06 p135).

Dimensional Analysis:
CLYTIE was 49.42m x 7.77m with machinery placed aft (*Lloyd's Register* 1904-05).

Archival Analysis:
CLYTIE was in collision with CORRWG (BOT 1905-06, p135). It appears to have been holed in the engine room and rapidly sank while attempting to tow CORRWG (wrecksite.eu). Machinery situated aft.

Phase 2 Conclusion:
CLYTIE (MBES/Archive). The wreck is positionally, dimensionally, and archivally consistent with CLYTIE.

Phase 1: UKHO Analysis:

Original UKHO Name:
7946 SEA GULL (POSSIBLY)

Archival details:
SEA GULL was 68.58m x 10.05m (*Lloyd's Register* 1918-19).

MBES Analysis:
Upright wreck of 48m x 8m, with bows to east.

Additional Evidence:
Located in 1945.

Phase 1 Conclusion:
UNKNOWN. The wreck is too small to be SEA GULL. The actual SEA GULL has been identified as wreck No 7370.

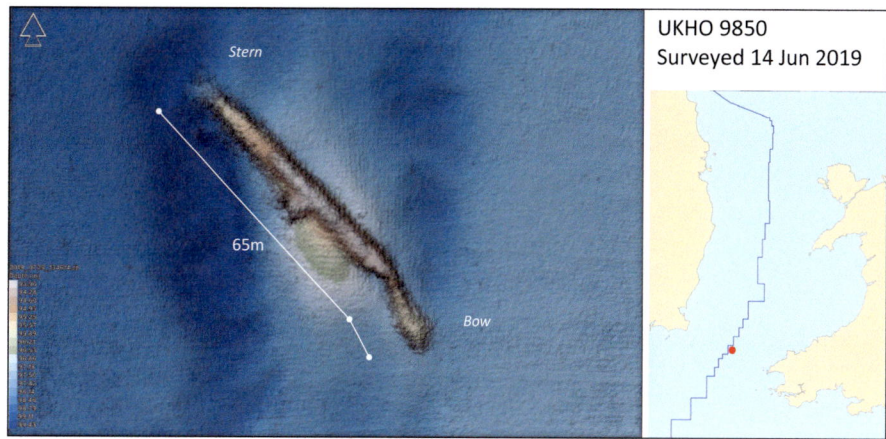

Stern

65m

Bow

9850 U1172 (MBES/Archive)

Original UKHO: 9850 ETHEL	**Phase 1 Analysis:** UNKNOWN
Position WGS84: 52 1.14 N, 5 48.74 W	**Water Depth:** 101m
Type of Vessel: Submarine	**Surveyed Length:** 65m
Date of Loss: After 23 Jan and prior to 5 May 1945	**Circumstances:** Mined

Phase 2: Method of wreck identification:

Positional Analysis:
The wreck of this U-boat lies in the "E1" minefield, laid as part of Operation "CH" on 23 Jan 1945 (MoD BR 1736 (56) (1) p213). There is little doubt it was mined. It appears to possibly have originally been detected on 5 May 1945 by EG14 and attacked (TNA ADM 199/1786 sec 4 report No 124). In this report it is stated that EG14 was of the view there possibly were two U-boats present. This is in fact the case, (see UKHO 9852), which lies 0.5nm NE of this wreck.

Dimensional Analysis:
The wreck is dimensionally consistent with the Type VII U-boat of 1945 era.

Archival Analysis:
All the U-boats recorded as destroyed in the study area have been accounted for during this study, except for U1172. This U-boat was listed post war as being destroyed off the Arklow Bank on 27 Jan 1945. The actual attack report was assessed at "B" (probably sunk) and therefore, not a confirmed kill. Problematically the attack developed after two ships had been hit by torpedoes (TNA ADM 199/1786 sec 3 report No 118). The loss of these ships has been attributed to U825, which survived the patrol and reported being heavily attacked after attacking the ships (NHB FDS D/NHB/22/2(849)). Furthermore, and equally problematic, is the absence of a U-boat wreck in the area where this took place. The assumption now must be that 1) U1172 has not been found, or 2) more likely it was not present, and it was U825 which was attacked, but survived. It is known U1172 operated in the Irish Sea and did not return. Therefore, it seems probable that it could be either this U-boat, or UKHO 9852. It is currently thought more likely to be this one, because it is pointing outbound from the area where U1172 is credited with probably sinking VIGSNES (see UKHO 7361).

Phase 2 Conclusion:
U1172 (MBES/Archive). Evidentially, and only on the balance of probability, this wreck could be U1172, but this is not certain by any means. If a U-boat is ever found off the Arklow Bank, this would need revising.

Phase 1: UKHO Analysis:

Original UKHO Name:
9850 ETHEL

Archival details:
ETHEL was a 100-ton SV sunk 26 Apr 1918 by gunfire.

MBES Analysis:
Wreck of a submarine lying on starboard side beam ends. Damaged forward. Most likely to be a WW2-era U-boat.

Additional Evidence:
Located in 1977.

Phase 1 Conclusion:
UNKNOWN. This wreck is not a steamship.

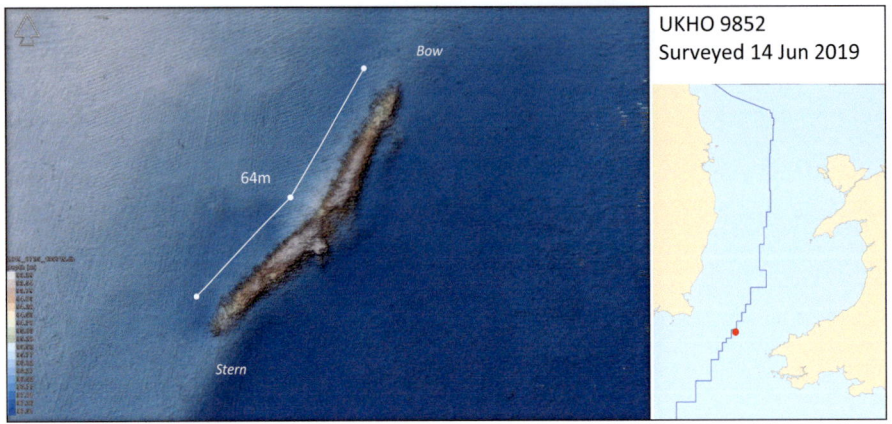

UKHO 9852
Surveyed 14 Jun 2019

9852 U242 (MBES/Archive)

Original UKHO: 9852 UNKNOWN		**Phase 1 Analysis:** UNKNOWN	
Position WGS84: 52 1.562 N, 5 48.168 W		**Water Depth:** 98m	
Type of Vessel: Submarine		**Surveyed Length:** 64m	
Date of Loss: 5 Apr 1945		**Circumstances:** Mined	

Phase 2: Method of wreck identification:

Positional Analysis:
The wreck of this U-boat lies 0.9nm from the position of a U-boat which was known to have been mined (the event was witnessed) on 5 Apr 1945 (ADM 199/1786 sec 4 report No 124).

Dimensional Analysis:
The wreck is dimensionally consistent with the Type VII U-boat of 1945 era.

Archival Analysis:
The U-boat's identity was established after the war as U242, because only two U-boats were in the Irish Sea at that time. The other was U1024 (see UKHO 7377), which was also sunk (NHB FDS 476/97). The wreck lies in the "E1" minefield, laid as part of Operation "CH" on 23 Jan 1945 (MOD BR 1736 (56) (1), p213). Problematically another U-boat wreck lies very nearby (see UKHO 9850) and is not archivally recorded and therefore is a mystery site. It could also be U242, but since this wreck points inbound and U242 is not recorded as having any successes in the area, it is thought marginally more likely to be this wreck.

Phase 2 Conclusion:
U242 (MBES/Archive). This wreck is probably the U-boat recorded as mined on 5 Apr 1945, later established to be U242. This is not definitive (see above and UKHO 9850) as it could be U1172, or potentially another loss previously assessed to lie elsewhere.

Phase 1: UKHO Analysis:

Original UKHO Name:
9852 UNKNOWN

Archival details:
TBC during Phase 2. This U-boat has been mined.

MBES Analysis:
Wreck of a submarine lying on its starboard side, broken almost in half amidships. Dimensionally consistent with a WW2-era submarine.

Additional Evidence:
Located in 1972 (UKHO).

Phase 1 Conclusion:
UNKNOWN. This wreck is a submarine.

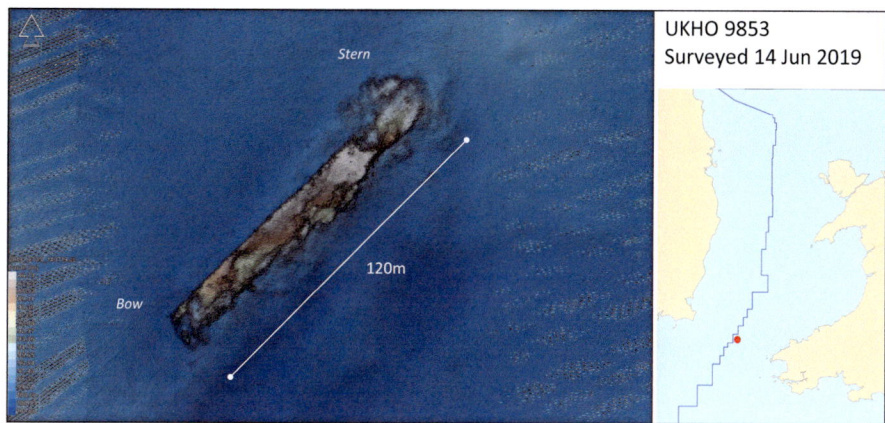

9853 AMIRAL ZEDE (MBES/Archive)

Original UKHO: 9853 U1169 (POSSIBLY)	**Phase 1 Analysis:** UNKNOWN
Position WGS84: 52 1.746 N, 5 46.537 W	**Water Depth:** 103m
Type of Vessel: SS Cargo	**Surveyed Length:** 120m
Date of Loss: 19 Nov 1917	**Circumstances:** Torpedoed by UC77

Phase 2: Method of wreck identification:

Positional Analysis:
This wreck plots 2.9nm NE of the position given in *War Losses* (1990) and 5.7nm north of the position given in UC77's KTB.

Dimensional Analysis:
AMIRAL ZEDE was 123.77m x 15.92m (*Lloyd's Register* 1915-16). This is consistent with this wreck.

Archival Analysis:
The ship was sailing from Newport to Rosslare to join a convoy bound to Buenos Aires. Upon a submarine being spotted the ship turned to port, prior to being hit abaft the engine room on the port side (TNA ADM 137/1358).

Phase 2 Conclusion:
AMIRAL ZEDE (MBES/Archive). The wreck is dimensionally, positionally, and archivally consistent with AMIRAL ZEDE. It lies on the route from Newport to Rosslare.

Phase 1: UKHO Analysis:

Original UKHO Name:
9853 U1169 (POSSIBLY)

Archival details:
In 1946 U1169 was listed as mined in a position 4.2nm west of this wreck site (TNA ADM 199/1789).

MBES Analysis:
Wreck of a SS/MV cargo vessel lying on its port side 120m long. The wreck is on its side and measures approx. 15m wide.

Additional Evidence:
Originally located in 1945 (UKHO).

Phase 1 Conclusion:
UNKNOWN. This wreck is not a submarine.

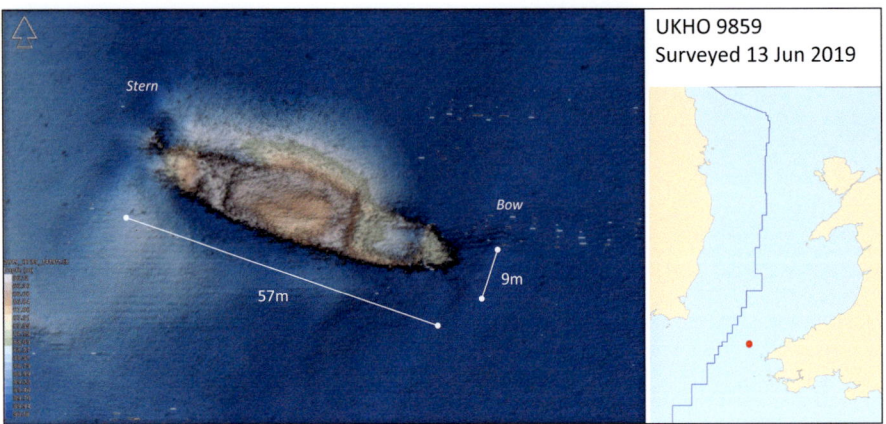

Stern

Bow

57m

9m

9859 SAPPHIRE (MBES/Archive)

Original UKHO: 9859 PENTWYN (POSSIBLY)	**Phase 1 Analysis:** UNKNOWN
Position WGS84: 52 1.074 N, 5 32.879 W	**Water Depth:** 89m
Type of Vessel: MV Coaster	**Surveyed Length:** 57m
Date of Loss: 12 Apr 1939	**Circumstances:** Collision

Phase 2: Method of wreck identification:

Positional Analysis:
The wreck plots 4.2nm north of the position reported to UKHO (UKHO 12124).

Dimensional Analysis:
SAPPHIRE was 61.01m x 9.90m with machinery aft (*Lloyd's Register* 1938-39). The dimensions are consistent with this wreck.

Archival Analysis:
SAPPHIRE was on a voyage from Llandulas to Dagenham with a cargo of stone when it collided with CLAN CAMERON, 15nm off St. David's Head. It was taken in tow by the German ship NEPTUNIA but foundered at the position reported in UKHO 12124.

Phase 2 Conclusion:
SAPPHIRE (MBES/Archive). This wreck is positionally, dimensionally, and archivally consistent with SAPPHIRE. Of note is the aft machinery and laden cargo as seen in the MBES scan.

Phase 1: UKHO Analysis:

Original UKHO Name:
9859 PENTWYN (POSSIBLY)

Archival details:
PENTWYN was sunk by torpedo on 16 Oct 1918 (*War Losses* 1990). It was 108.51m x 15.39m (*Lloyd's Register* 1915-16).

MBES Analysis:
Small coaster, upright of 57m x 9m. Bows to east. The wreck appears to be laden with a bulk cargo.

Additional Evidence:
Located in 1982 (UKHO).

Phase 1 Conclusion:
UNKNOWN. The wreck is much too small to be PENTWYN.

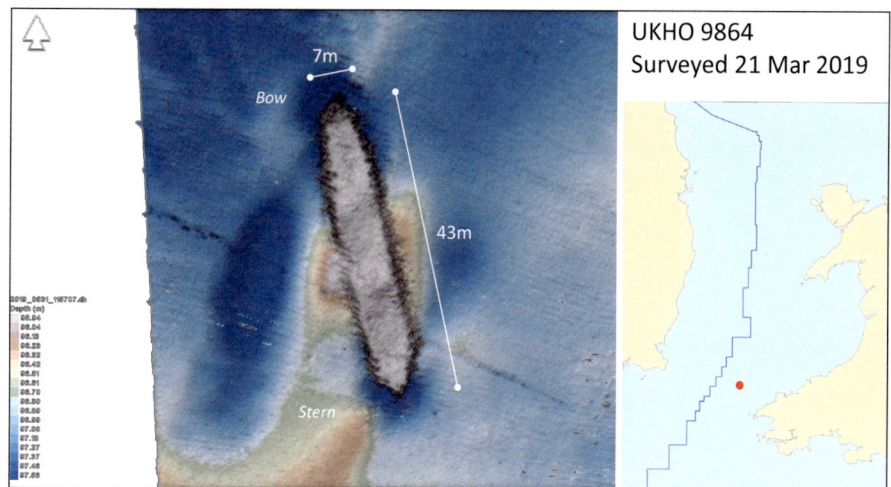

UKHO 9864
Surveyed 21 Mar 2019

9864 HEENVLIET (MBES/Archive)

Original UKHO: 9864 ST PATRICK (POSSIBLY)	**Phase 1 Analysis:** UNKNOWN
Position WGS84: 52 7.528 N, 5 27.591 W	**Water Depth:** 95m
Type of Vessel: SS Cargo	**Surveyed Length:** 43m
Date of Loss: 17 Oct 1941	**Circumstances:** Collision

Phase 2: Method of wreck identification:

Positional Analysis:
The wreck plots 0.6nm from the position recorded in *Wreck Returns* (1941 Q4 sec f Collisions p9).

Dimensional Analysis:
HEENVLIET was 42.90m x 7.49m with machinery aft (*Lloyd's Register* 1940-41).

Archival Analysis:
HEENVLIET was in collision with SS VESTLAND which was part of convoy BB89 heading to Barry. The crew were picked up by HMS POZARICA and taken to Barry Docks. HEENVLIET was sailing independently with a cargo of coal bound to Londonderry from Barry (*Wreck Returns* 1941 Q4 sec f Collisions p9).

Phase 2 Conclusion:
HEENVLIET (MBES/Archive). This wreck is positionally, dimensionally, and archivally consistent for HEENVLIET.

Phase 1: UKHO Analysis:

Original UKHO Name:
9864 ST PATRICK (POSSIBLY)

Archival details:
ST PATRICK was bombed in 1941 (*War Losses* 1989). It was 85.3m x 12.5m (*Lloyd's Register* 1940-41).

MBES Analysis:
Small coaster/FV possibly upside-down. 43m x 7m, bows to the south. Area of damage midships on starboard side.

Additional Evidence:
Located in 1980 (UKHO).

Phase 1 Conclusion:
UNKNOWN. The wreck is far too small to be ST PATRICK.

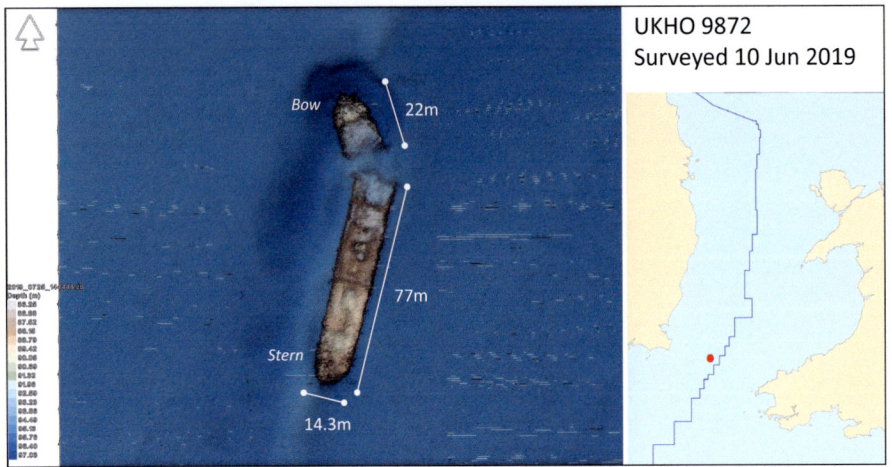

UKHO 9872
Surveyed 10 Jun 2019

9872 HIGHCLIFFE (MBES/Archive)

Original UKHO: 9872 UNKNOWN	**Phase 1 Analysis:** UNKNOWN
Position WGS84: 52 6.332 N, 5 51.721 W	**Water Depth:** 94m
Type of Vessel: SS Cargo	**Surveyed Length:** 99m (approx)
Date of Loss: 3 Sep 1918	**Circumstances:** Torpedoed UB87

Phase 2: Method of wreck identification:

Positional Analysis:
The wreck plots 10.4nm north of UB87's reported position (KTB). It plots 1.45nm from the position given in the survivor's report (TNA ADM 137/1502).

Dimensional Analysis:
HIGHCLIFFE was 102.18m x 14.17m (*Lloyd's Register* 1917-1918).

Archival Analysis:
HIGHCLIFFE was enroute from Glasgow to Milford Haven to join a convoy. The ship was torpedoed on the starboard side next to the foremast. The U-boat was not sighted, and the crew abandoned ship. One casualty. Ship sank by the head (TNA ADM 137/1502).

Phase 2 Conclusion:
HIGHCLIFFE (MBES/Archive). Positionally, dimensionally, and archivally this wreck is consistent with HIGHCLIFFE. The master's position is very close, and the damage on the wreck is consistent with a torpedo strike as described.

Phase 1: UKHO Analysis:

Original UKHO Name:
9872 UNKNOWN

Archival details:
N/A

MBES Analysis:
Upright cargo vessel with bows pointing north and broken off. Approx 99m x 14.3m.

Additional Evidence:
Located in 1977 (UKHO).

Phase 1 Conclusion:
UNKNOWN

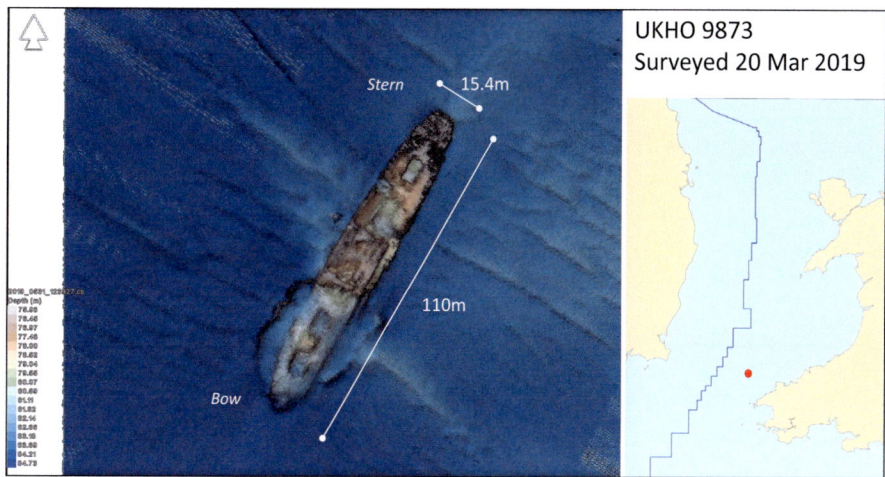

9873 EMPIRE GUNNER (MBES/Archive)

Original UKHO: 9873 PORT TOWNSVILLE (PROBABLY)	**Phase 1 Analysis:** UNKNOWN
Position WGS84: 52 6.938 N, 5 23.943 W	**Water Depth:** 81m
Type of Vessel: SS Cargo	**Surveyed Length:** 110m
Date of Loss: 6 Sep 1941	**Circumstances:** Bombed

Phase 2: Method of wreck identification:

Positional Analysis:
The wreck plots 5.3nm west of the position given in *War Losses* (1989).

Dimensional Analysis:
EMPIRE GUNNER was 112.77m x15.84m (*Lloyd's Register* 1940-41).

Archival Analysis:
The master's account states the ship was hit by a bomb in the after well deck, which caused the ship to split. It sank by the head after being abandoned (TNA ADM 199/2138). It was carrying a cargo of iron ore.

Phase 2 Conclusion:
EMPIRE GUNNER (MBES/Archive). This wreck is dimensionally, positionally, and archivally consistent with being EMPIRE GUNNER. A split in the hull aft of the bridge can be seen in the MBES pointcloud.

Phase 1: UKHO Analysis:

Original UKHO Name:
9873 PORT TOWNSVILLE (PROBABLY)

Archival details:
PORT TOWNSVILLE was bombed in Mar 1941 (*War Losses* 1989). It was 151.28m x 19.86m (*Lloyd's Register* 1941-42).

MBES Analysis:
Upright cargo ship with two holds fore and aft. Bows point southwest. Approx 110m x 15.4m. The fore section has collapsed.

Additional Evidence:
Located in 1945 (UKHO).

Phase 1 Conclusion:
UNKNOWN. This is wreck is far too small to be PORT TOWNSVILLE.

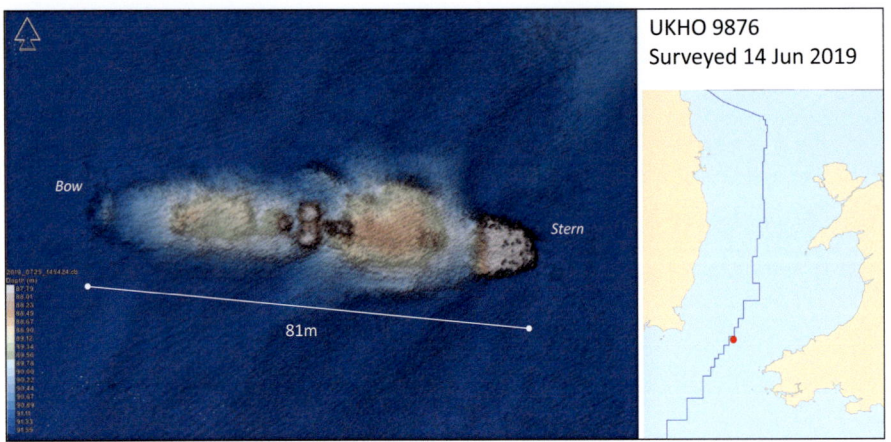

Bow

Stern

81m

9876 ISIDORO (MBES/Archive)

Original UKHO: 9876 UNKNOWN	**Phase 1 Analysis:** UNKNOWN
Position WGS84: 52 6.704 N, 5 43.047 W	**Water Depth:** 95m
Type of Vessel: SS Cargo	**Surveyed Length:** 81m
Date of Loss: 17 Aug 1915	**Circumstances:** Gunfire from U38

Phase 2: Method of wreck identification:

Positional Analysis:
This wreck plots 2.5nm NE of the position given in *War Losses* (1989). The U38 KTB position cannot be discerned.

Dimensional Analysis:
ISIDORO was 82.6m x 12.4m (*Lloyd's Register* 1915-16).

Archival Analysis:
ISIDORO was sunk by gunfire after the crew had abandoned the vessel (TNA ADM 137/1131). *Lloyd's Register* states it had two boilers. These can be seen on the DTM. The raised poop deck seen on the DTM is similar to that seen in the photo of the ship in the project archive.

Phase 2 Conclusion:
ISIDORO (MBES/Archive). This wreck is dimensionally, positionally, and archivally consistent with ISIDORO.

Phase 1: UKHO Analysis:

Original UKHO Name:
9876 UNKNOWN

Archival details:
N/A

MBES Analysis:
Wreck of collapsed SS of 81m length. Stern to the east is the most intact area. Two main boilers showing clearly on scan.

Additional Evidence:
Located in 1977 (UKHO).

Phase 1 Conclusion:
UNKNOWN

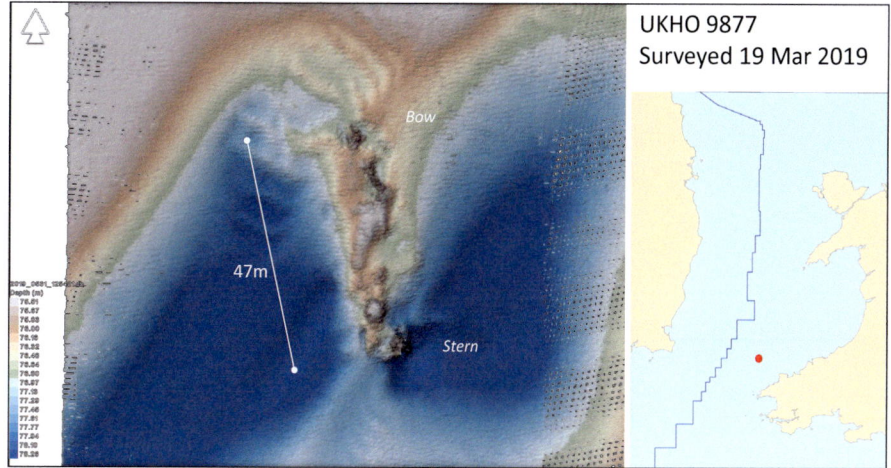

9877 THORNFIELD (MBES/Archive)

Original UKHO: 9877 EMPIRE GUNNER (POSSIBLY)	**Phase 1 Analysis:** UNKNOWN
Position WGS84: 52 8.506 N, 5 19.78 W	**Water Depth:** 74m
Type of Vessel: SS Cargo	**Surveyed Length:** 47m (approx)
Date of Loss: 17 Aug 1915	**Circumstances:** Gunfire from U38

Phase 2: Method of wreck identification:

Positional Analysis:
This wreck is located 4nm NE of the position given in *War Losses* (1989). The U38 KTB position cannot be discerned.

Dimensional Analysis:
THORNFIELD was 48.76m x 7.95m, well-decked with machinery aft (*Lloyd's Register* 1915-16).

Archival Analysis:
THORNFIELD (Admiralty Transport No 1608) was sunk by gunfire after the crew abandoned ship. It was seen to sink stern first (TNA ADM 137/1131). Machinery is seen aft on the wreck.

Phase 2 Conclusion:
THORNFIELD (MBES/Archive). This wreck is dimensionally, positionally, and archivally consistent with THORNFIELD.

Phase 1: UKHO Analysis:

Original UKHO Name:
9877 EMPIRE GUNNER (POSSIBLY)

Archival details:
EMPIRE GUNNER was bombed in 1941 (*War Losses* 1989). It was 112.77m x 15.84m (*Lloyd's Register* 1941-42).

MBES Analysis:
Broken up wreck with possible boiler showing at the southern end. Approx 47m long with boiler at the stern end of the wreck.

Additional Evidence:
Located in 1945 (UKHO).

Phase 1 Conclusion:
UNKNOWN. This wreck is much too small to be EMPIRE GUNNER.

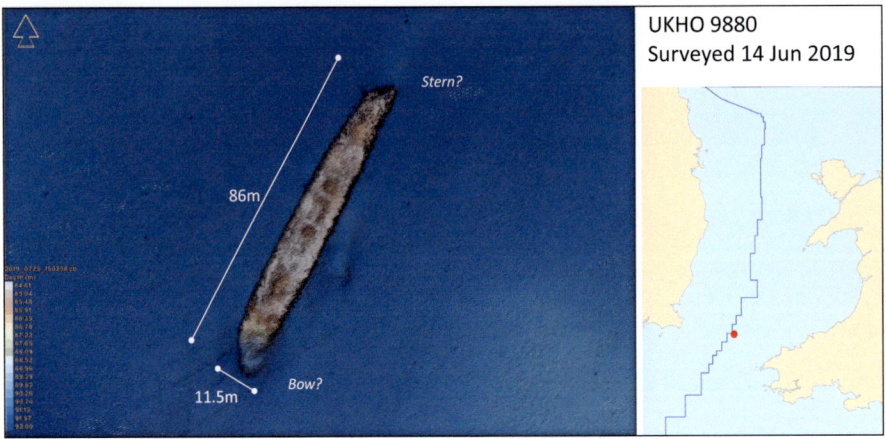

Stern?

86m

Bow?

11.5m

9880 GLENBY (MBES/Archive)

Original UKHO: 9880 UNKNOWN	**Phase 1 Analysis:** UNKNOWN
Position WGS84: 52 7.682 N, 5 42.556 W	**Water Depth:** 93m
Type of Vessel: SS Cargo	**Surveyed Length:** 86m
Date of Loss: 17 Aug 1915	**Circumstances:** Gunfire from U38

Phase 2: Method of wreck identification:

Positional Analysis:
The wreck plots 5.6nm south of the position reported by the master as 30nm north of The Smalls (TNA ADM 137/2959 No 232). The position in U38's KTB cannot be discerned.

Dimensional Analysis:
GLENBY was 84.63m x 12.01m (*Lloyd's Register* 1915-16).

Archival Analysis:
GLENBY (Admiralty Transport No 590) was sunk by gunfire, while the crew were lowering the boat. Two were killed. The U-boat approached the survivors to ask the identity of the vessel and then headed east to attack another (TNA ADM 137/1131). The vessel sank stern first (TNA ADM 137/2959). Bow and stern show damage. Heading of the wreck is difficult to discern.

Phase 2 Conclusion:
GLENBY (MBES/Archive). This wreck is positionally and dimensionally consistent with GLENBY.

Phase 1: UKHO Analysis:

Original UKHO Name:
9880 UNKNOWN

Archival details:
N/A

MBES Analysis:
Upright cargo ship with visible holds. Bows to SW have collapsed. Dimensions 86m x 11.5m.

Additional Evidence:
Located in 1977 (UKHO).

Phase 1 Conclusion:
UNKNOWN

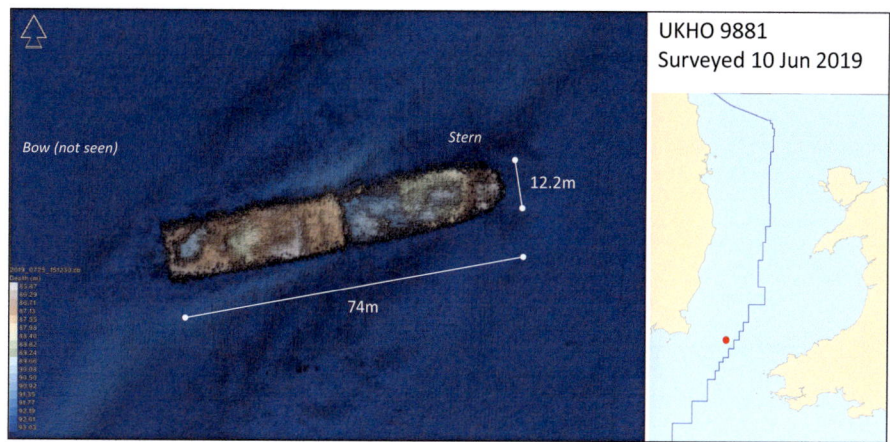

9881 ALU MENDI (MBES/Archive)

Original UKHO: 9881 UNKNOWN		**Phase 1 Analysis:** UNKNOWN	
Position WGS84: 52 7.708 N, 5 50.799 W		**Water Depth:** 93m	
Type of Vessel: SS Cargo		**Surveyed Length:** 74m (incomplete)	
Date of Loss: 28 Apr 1917		**Circumstances:** Explosives by UC65	

Phase 2: Method of wreck identification:

Positional Analysis:
The wreck lies 6.2 nm NE of the position given in *War Losses* (1990) and 6.4nm south of the U-boat position (KTB).

Dimensional Analysis:
ALU MENDI was 83.82m x 12.70m and well-decked (*Lloyd's Register* 1917-18).

Archival Analysis:
ALU MENDI was stopped by UC65 which used the ship's boat to plant four bombs on board the ship, which then rapidly sank (TNA ADM 137/1311).

Phase 2 Conclusion:
ALU MENDI (MBES/Archive). Positionally, dimensionally (on width), and archivally this is consistent with ALU MENDI. The absence of the fore part of the ship is notable and unexplained, although it should be noted one bomb was placed in No2 hold.

Phase 1: UKHO Analysis:

Original UKHO Name:
9881 UNKNOWN

Archival details:
N/A

MBES Analysis:
Upright cargo ship with bows missing. 74m x 12.2m. Stern to east.

Additional Evidence:
Located in 1977 (UKHO).

Phase 1 Conclusion:
UNKNOWN

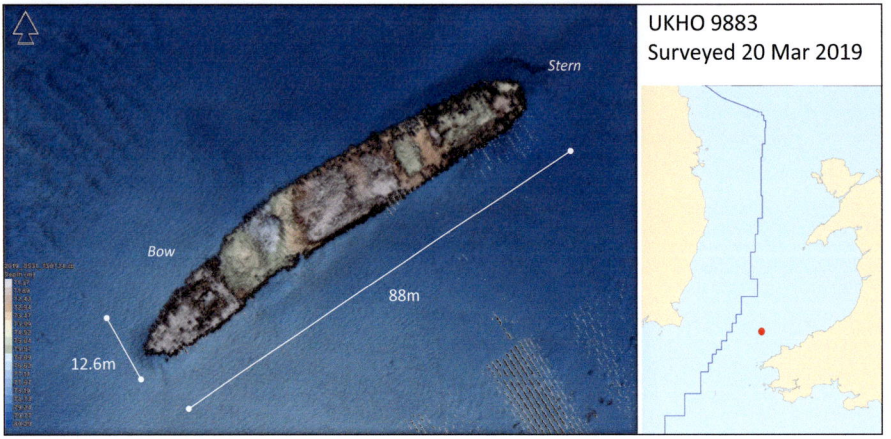

UKHO 9883
Surveyed 20 Mar 2019

9883 ST PATRICK (MBES/Archive)

Original UKHO: 9883 UNKNOWN	**Phase 1 Analysis:** UNKNOWN
Position WGS84: 52 7.977 N, 5 22.667 W	**Water Depth:** 77m
Type of Vessel: SS Cargo	**Surveyed Length:** 88m (approx)
Date of Loss: 13 Jun 1941	**Circumstances:** Bombed

Phase 2: Method of wreck identification:

Positional Analysis:
This wreck lies 4.2nm north of the position recorded by *War Losses* (1989).

Dimensional Analysis:
ST PATRICK was 85.34m x 12.50m (*Lloyd's Register* 1940-41).

Archival Analysis:
The survivor account (Chief Engineer) stated the ship was hit by four bombs around the bridge area and rapidly folded up and sank. 29 of the crew and passengers were not saved (TNA ADM 199/2137).

Phase 2 Conclusion:
ST PATRICK (MBES/Archive). The wreck is dimensionally, positionally, and archivally consistent with ST PATRICK. Of note is the damage to the wreck, consistent with a vessel which folded as it sank.

Phase 1: UKHO Analysis:

Original UKHO Name:
9883 UNKNOWN

Archival details:
N/A

MBES Analysis:
Upright cargo vessel with holds fore and aft. Bow to SW. Approx. 88m x 12.6m.

Additional Evidence:
Located in 1945 (UKHO).

Phase 1 Conclusion:
UNKNOWN

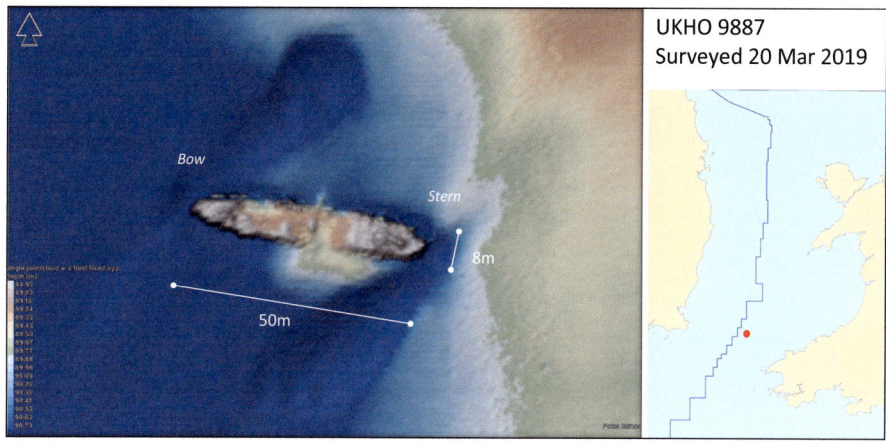

UKHO 9887
Surveyed 20 Mar 2019

Bow

Stern

8m

50m

9887 THE QUEEN (MBES/Archive)

Original UKHO: 9887 UNKNOWN	**Phase 1 Analysis:** UNKNOWN
Position WGS84: 52 9.261 N, 5 35.741 W	**Water Depth:** 92m
Type of Vessel: SS Cargo	**Surveyed Length:** 50m
Date of Loss: 17 Aug 1915	**Circumstances:** Gunfire from U38

Phase 2: Method of wreck identification:

Positional Analysis:
This wreck plots 15.7nm SW of the position given in *War Losses* (1990). The U-boat position cannot be discerned from the KTB, but according to U38's KTB, GLENBY (see 9880), THE QUEEN, and ISIDORO (see 9876) were all sunk within one hour. GLENBY's survivors stated that U38 passed eastward of them to sink another ship, which must have beeen THE QUEEN, conversely the master of THE QUEEN reported that U38 then passed to the westward to sink another ship in view of them, which must have been ISIDORO. It is not clear why the master's reported position for the loss of THE QUEEN is so inaccurate. It is given as a range/bearing from The Smalls 40nm NNE (TNA ADM 137/1131).

Dimensional Analysis:
THE QUEEN was 53.34m x 8.22m, well-decked with machinery aft (*Lloyd's Register* 1915-16).

Archival Analysis:
THE QUEEN was sunk by gunfire and was seen to sink stern first.

Phase 2 Conclusion:
THE QUEEN (MBES/Archive). Dimensionally and archivally this wreck is consistent with THE QUEEN. The cluster of three wrecks conforming to GLENBY, THE QUEEN, and ISISDORO place the wreck where it is, explaining the abnormally long distance to the master's position.

Phase 1: UKHO Analysis:

Original UKHO Name:
9887 UNKNOWN

Archival details:
N/A

MBES Analysis:
Small wreck, upright 50m x 8m with machinery in the stern.

Additional Evidence:
Located in 1945 (UKHO).

Phase 1 Conclusion:
UNKNOWN

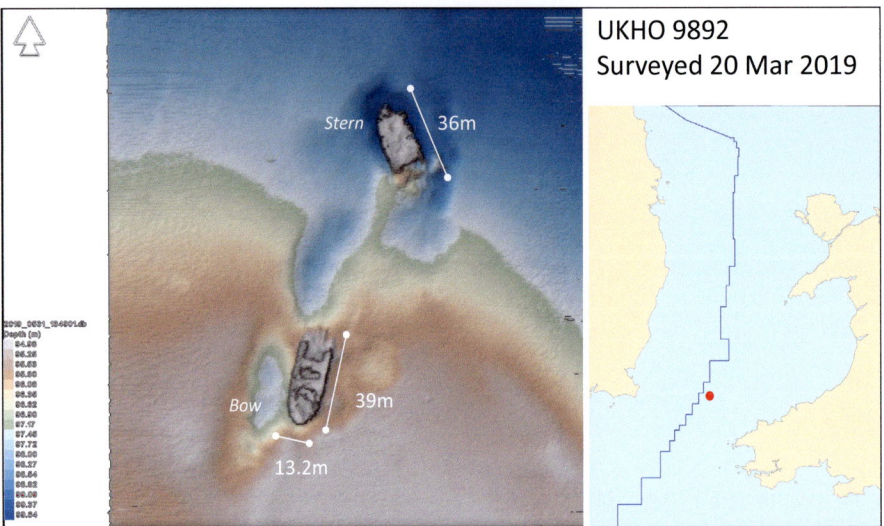

9892 SORELDOC (MBES/Archive)

Original UKHO: 9892 EMPIRE PANTHER (PROBABLY)	**Phase 1 Analysis:** UNKNOWN
Position WGS84: 52 10.811 N, 5 34.675 W	**Water Depth:** 94m
Type of Vessel: SS Cargo	**Surveyed Length:** 65m (approx)
Date of Loss: 28 Feb 1945	**Circumstances:** Torpedoed by U775

Phase 2: Method of wreck identification:

Positional Analysis:
This wreck lies at a position 4.2nm south of the position given in *War Losses* (1989).

Dimensional Analysis:
SORELDOC was 77.10m x 13.10m, bridge on the forecastle and machinery aft (*Lloyd's Register* 1944-45).

Archival Analysis:
SORELDOC was torpedoed and sunk by U775 (Rohwer 1999, p191). The ship reportedly broke in half and sank in four minutes, killing 15 of the crew (Hocking 1969, p655). The 21 surviving crew were safely landed at Milford Haven the same day (TNA ADM 199/1443).

Phase 2 Conclusion:
SORELDOC (MBES/Archive). Positionally, archivally, and dimensionally this is the wreck of SORELDOC. Of particular note are the dimensions of the vessel, which is wide for its length. The ship was built to operate in US/Canadian waterways

Phase 1: UKHO Analysis:

Original UKHO Name:
9892 EMPIRE PANTHER (PROBABLY)

Archival details:
EMPIRE PANTHER was reported mined in 1943 (*War Losses* 1989). It was 125.0m x 16.6m (*Lloyd's Register* 1943-44).

MBES Analysis:
Cargo vessel or barge in two distinct halves which are 80m apart on the seabed. Ship's dimensions were approx. 75m x 13.2m.

Additional Evidence:
First located in 1980 (UKHO).

Phase 1 Conclusion:
UNKNOWN. This wreck is far too small to be EMPIRE PANTHER.

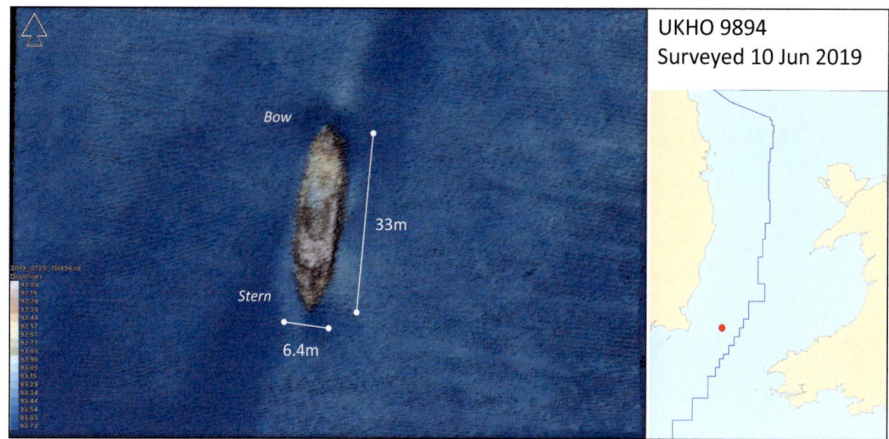

9894 WHEATFLOWER (MBES/Archive)

Original UKHO: 9894 TWEED (POSSIBLY)	**Phase 1 Analysis:** UNKNOWN
Position WGS84: 52 10.583 N, 5 50.421 W	**Water Depth:** 94m
Type of Vessel: SS Cargo	**Surveyed Length:** 33m
Date of Loss: 19 Feb 1918	**Circumstances:** Gunfire from U86

Phase 2: Method of wreck identification:

Positional Analysis:
This wreck plots 3.6nm west of the position given by U86 (KTB). The position in *War Losses* (1990) plots on the Irish coast and must be considered a mistake (10nm NWxN Tuskar). It appears to have been later corrected by the Admiralty (see TNA ADM 137/1489) to a position which plots 6.2nm SW of this wreck.

Dimensional Analysis:
WHEATFLOWER was 33.57m x 6.70m, well-decked with machinery aft (*Lloyd's Register* 1917-18).

Archival Analysis:
The ship was bound for Dublin from Newport carrying coal. The cook was killed in the sinking. The rest of the crew were saved (TNA ADM 137/1489). The wreck is on the Newport to Dublin route.

Phase 2 Conclusion:
WHEATFLOWER (MBES/Archive). This wreck is positionally, dimensionally, and archivally consistent with WHEATFLOWER.

Phase 1: UKHO Analysis:

Original UKHO Name:
9894 TWEED (POSSIBLY)

Archival details:
TWEED was torpedoed in Mar 1918. It was 81.68m x 11.48m (*Lloyd's Register* 1918-19).

MBES Analysis:
Small SS/MV/FV/Coaster of 33m x 6.4m. Probably upright with bows pointing north.

Additional Evidence:
Located in 1979 (UKHO).

Phase 1 Conclusion:
UNKNOWN. This wreck is far too small to be TWEED.

9896 PORT TOWNSVILLE (MBES/Archive)

Original UKHO: 9896 STRATHNAIRN (POSSIBLY)	**Phase 1 Analysis:** UNKNOWN
Position WGS84: 52 11.571 N, 5 27.153 W	**Water Depth:** 96m
Type of Vessel: MV Cargo	**Surveyed Length:** 142m
Date of Loss: 4 Mar 1941	**Circumstances:** Bombed

Phase 2: Method of wreck identification:

Positional Analysis:
This wreck lies 6.1nm north of the position given in *War Losses* (1990).

Dimensional Analysis:
PORT TOWNSVILLE was 151.28m x 19.86m (*Lloyd's Register* 1940-41). It is the largest ship to have been sunk in the entire study area.

Archival Analysis:
The master's account of the loss of this ship states the fatal bomb strike hit in the area of No3 hold. The ship was abandoned and last seen sinking by the head (TNA ADM 199/2136).

Phase 2 Conclusion:
PORT TOWNSVILLE (MBES/Archive). Positionally and archivally this wreck is consistent with PORT TOWNSVILLE. The wreck measures slightly short, but this is explained by the fact that it sank by the head. It can be noted the ship is longer than the sea is deep at this location. The MBES scan shows that the bow seems to have bent upwards on impact. Photos of the ship show it had two islands, both of which can be discerned in the MBES data.

Phase 1: UKHO Analysis:

Original UKHO Name:
9896 STRATHNAIRN (POSSIBLY)

Archival details:
STRATHNAIRN was 112.7m long, with single screw (Larn & Larn 2000). U22's position (KTB) for the sinking of STRATHNAIRN plots 31nm from this wreck.

MBES Analysis:
Large steamship bows pointing SW. Length approx. 142m. Wreck lies on port side. The height of the wreck is approx. 16.5m.

Additional Evidence:
Known wreck since 1945 (UKHO).

Phase 1 Conclusion:
UNKNOWN. This cannot be STRATHNAIRN based on length.

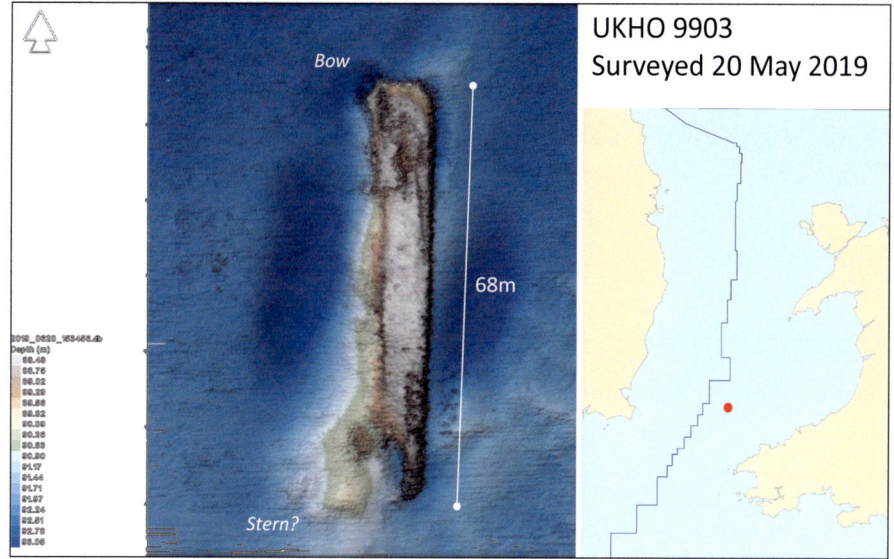

9903 BOSCASTLE (MBES/Archive)

Original UKHO: 9903 GRESHAM (POSSIBLY)	**Phase 1 Analysis:** UNKNOWN
Position WGS84: 52 14.722 N, 5 22.28 W	**Water Depth:** 91m
Type of Vessel: SS Cargo	**Surveyed Length:** 68m
Date of Loss: 7 Apr 1918	**Circumstances:** Torpedoed by U111

Phase 2: Method of wreck identification:

Positional Analysis:
This wreck lies 5.6nm west of the position given in the master's report (TNA ADM 137/2964) and 7.9nm north of the position given in U111's KTB.

Dimensional Analysis:
BOSCASTLE was 90.80m x 13.60m (*Lloyd's Register* 1917-18).

Archival Analysis:
U111's KTB states the ship was hit between the funnel and the after mast and that it broke in two, with the stern sinking immediately (KTB). The survivors' report states the ship sank in four minutes after being hit on the port side. It rolled over to port as it sank. It was steaming north when it was hit (TNA ADM 137/1493)

Phase 2 Conclusion:
BOSCASTLE (MBES/Archive). This wreck is positionally consistent with BOSCASTLE. It seems from the archival data that the ship broke in half. This MBES scan appears to only represent the fore part of the ship, with the after part not being seen, but the break being exactly as described by U111. It should be noted that the shape of the bow of this vessel is similar in shape to that seen in the drawing of BOSCASTLE in the project archive. The evidence is not conclusive, but this wreck appears to be the best fit. It is just possible that the unidentifiable wreck 9898 could be associated with this one.

Phase 1: UKHO Analysis:

Original UKHO Name:
9903 GRESHAM (POSSIBLY)

Archival details:
GRESHAM was 105.5m x 15.1m (*Lloyd's Register* 1915-16). Position of wreck is 1.8nm from *War Losses* (1990) position. This is probably why the wreck is associated with GRESHAM. U-boat position (KTB) is 5nm distant. Torpedo hit No4 hold and took out steering. Subsequently reported by survivors as being hit again and sunk within three hours (ADM 137/2964). UKHO card states ship was towed.

MBES Analysis:
Wreck 68m long lying N/S with bow north.

Additional Evidence:
Located in 1976.

Phase 1 Conclusion:
UNKNOWN. Too short for GRESHAM.

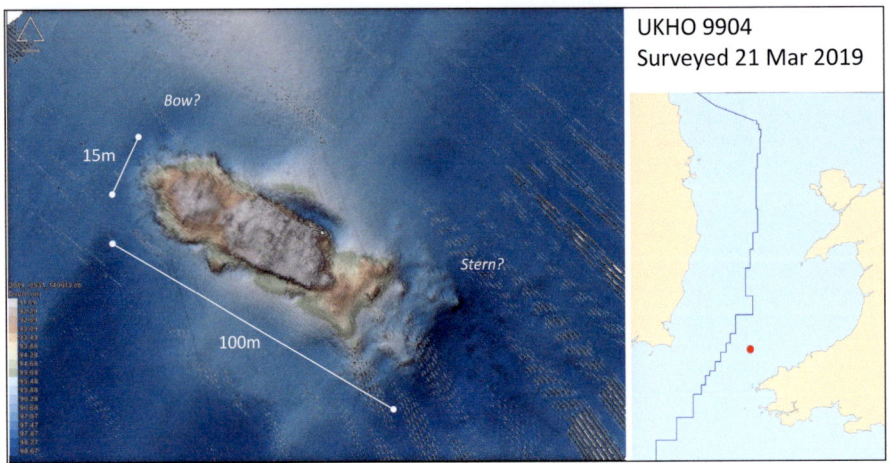

UKHO 9904
Surveyed 21 Mar 2019

Bow?

15m

100m

Stern?

9904 GRESHAM (MBES/Archive)

Original UKHO: 9904 UNKNOWN	**Phase 1 Analysis:** UNKNOWN
Position WGS84: 52 9.445 N, 5 24.588 W	**Water Depth:** 95m
Type of Vessel: SS Cargo	**Surveyed Length:** 100m (approx)
Date of Loss: 26 Apr 1918	**Circumstances:** Torpedoed by U91

Phase 2: Method of wreck identification:

Positional Analysis:
This wreck plots 4.5nm south of the position given in *War Losses* (1990) and 8nm south of the position given by U91 (KTB).

Dimensional Analysis:
GRESHAM was 105.50m x 15.10m (*Lloyd's Register* 1917-18).

Archival Analysis:
U91's KTB states the ship was hit under the mainmast by the first torpedo. A 2nd failed and the 3rd hit amidships, and the ship then sank (KTB). The survivors' account states the ship was hit up to three times. The first in No4 hold, the 2nd and 3rd (possibly the boilers exploding) were amidships (TNA ADM 137/1494). Interestingly the 2nd officer witnessed the 2nd torpedo pass under the ship. This is consistent with U91's KTB of the 2nd torpedo not hitting. He also noted the ship sank by the stern. The vessel, being longer than the sea is deep, it may have stood on end, collapsing the stern.

Phase 2 Conclusion:
GRESHAM (MBES/Archive). This wreck is positionally and archivally consistent with GRESHAM. The length of the wreck is difficult to discern with accuracy but is at least 100m long. The width is correct for GRESHAM.

Phase 1: UKHO Analysis:

Original UKHO Name:
9904 UNKNOWN

Archival details:
N/A

MBES Analysis:
Collapsed SS of approx. 100m x 15m if it is all present. Stern (possible) has disintegrated into a scour.

Additional Evidence:
Located in 1945 (UKHO).

Phase 1 Conclusion:
UNKNOWN

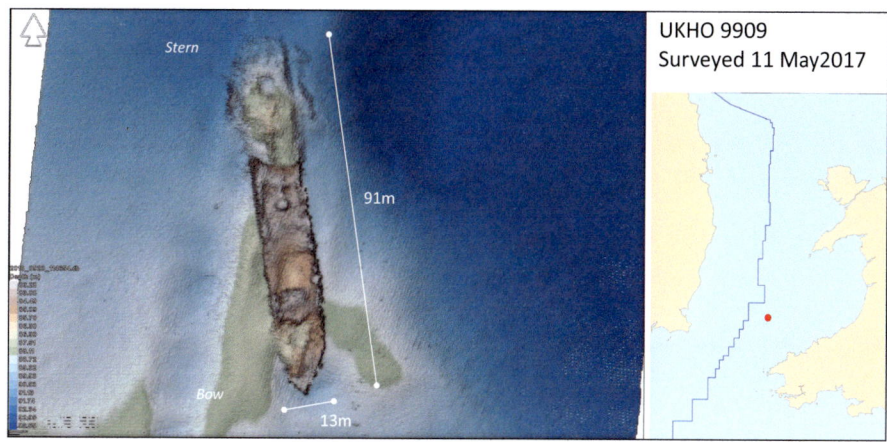

Stern

91m

Bow

13m

9909 LYCIA (MBES/Archive)

Original UKHO: 9909 BOLTONHALL (POSSIBLY)	**Phase 1 Analysis:** UNKNOWN
Position WGS84: 52 17.122 N, 5 20.197 W	**Water Depth:** 90m
Type of Vessel: SS Cargo	**Surveyed Length:** 91m (approx)
Date of Loss: 11 Feb 1917	**Circumstances:** Explosives by UC65

Phase 2: Method of wreck identification:

Positional Analysis:
The wreck lies 8.4nm NE of the position (given as R/B from S Bishop Light) given in the master's report (TNA ADM 137/1304). The U-boat KTB chart does not give accurate positions.

Dimensional Analysis:
LYCIA was 93.87m x 13.18m (*Lloyd's Register* 1916-17).

Archival Analysis:
LYCIA was travelling north from Swansea to Liverpool when the U-boat was sighted on the starboard beam. The ship turned away and fired with its stern gun. LYCIA was abandoned 45 minutes later, and the U-boat closed and sunk the ship with bombs. It sank stern first going vertical before disappearing (TNA ADM 137/1304).

Phase 2 Conclusion:
LYCIA (MBES/Archive). The wreck is archivally, positionally, and dimensionally consistent with LYCIA. Of note is the collapsed stern, which relates to the manner in which the ship sank. The heading south matches the course during the stern chase.

Phase 1: UKHO Analysis:

Original UKHO Name:
9909 BOLTONHALL (POSSIBLY)

Archival details:
BOLTONHALL was 104m x 14.7m (Larn & Larn 2000). Hit in after end of engine room and sank rapidly (ADM 137/2964). The *War Losses* (1990) position is 11 miles from wreck. U-boat position is 8.6nm (KTB).

MBES Analysis:
Upright cargo ship with four holds. Stern section has collapsed to seabed. Base of probable funnel is prominent. Hole in starboard side of No1 hold. Wreck points south. Wreck is 91m x 13m.

Additional Evidence:
Located in 1945.

Phase 1 Conclusion:
UNKNOWN. Too short for BOLTONHALL.

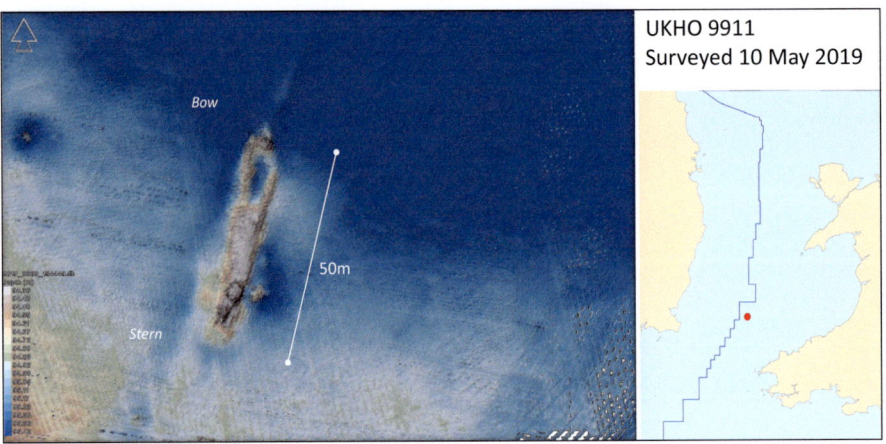

Bow

50m

Stern

9911 U1302 (MBES/Archive)

Original UKHO: 9911 SORELDOC (POSSIBLY)	**Phase 1 Analysis:** UNKNOWN	
Position WGS84: 52 15.856 N, 5 30.479W	**Water Depth:** 93m	
Type of Vessel: Submarine	**Surveyed Length:** 50m (approx)	
Date of Loss: 7 Mar 1945	**Circumstances:** Depth Charges	

Phase 2: Method of wreck identification:

Positional Analysis:
This wreck plots 5.6nm SW of the position given for the destruction of a U-boat on 7 Mar 1945 (TNA ADM 199/1786 v3 p39).

Dimensional Analysis:
U1302 was 67.20m x 6.18m Gröner (1991).

Archival Analysis:
The evidence of destruction included human tissue and the attack was graded "A Known Sunk" in 1945. The attacks on the wreckage were persistent and went on for some time. The wreckage on this site is flattened and this is consistent with heavy depth charge attacks, as seen in the case of U678 (see McCartney 2014, p153-356). It should be noted too that this wreck was located in May 1945 when it gave off oil and wreckage (presumably as a result of being depth-charged previously) (UKHO).

Phase 2 Conclusion:
U1302 (MBES/Archive). Dimensionally, archivally, and positionally this could be the wreck of U1302. Certainly, it is the best candidate near to the destruction position, although the details in the MBES scan are not conclusive. There is a lesser possibly this U-boat could be at 9898, a wreck which is not currently identified (Dec 2019).

Phase 1: UKHO Analysis:

Original UKHO Name:
9911 SORELDOC (POSSIBLY)

Archival details:
SORELDOC was torpedoed in Feb 1945 (*War Losses* 1989). It was 77.1m x 13.10m (*Lloyd's Register* 1944-45).

MBES Analysis:
Small broken down wreck, prominent high point a at southerly end. Approx. 50m long.

Additional Evidence:
Located in 1945 when it gave off oil and wreckage (presumably after being depth-charged) (UKHO).

Phase 1 Conclusion:
UNKNOWN. The wreck is too short to be SORELDOC.

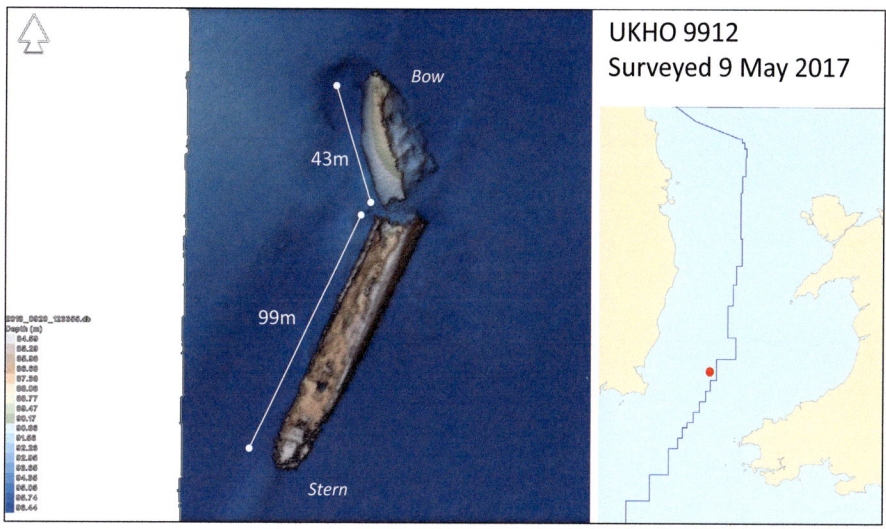

UKHO 9912
Surveyed 9 May 2017

9912 MESABA (MBES/Archive)

Original UKHO: 9912 CITY OF GLASGOW (PROBABLY)	**Phase 1 Analysis:** UNKNOWN
Position WGS84: 52 16.789 N, 5 39.19 W	**Water Depth:** 96m
Type of Vessel: SS Liner	**Surveyed Length:** 142m
Date of Loss: 1 Sep 1918	**Circumstances:** Torpedoed by UB118

Phase 2: Method of wreck identification:

Positional Analysis:
This wreck plots 3.7nm SW of the position reported by UB118 (KTB) and 0.7nm east of the position recorded in *War Losses* (1990).

Dimensional Analysis:
MESABA was 146.94m x 15.90m (*Lloyd's Register* 1918-19). The second largest vessel in the study area.

Archival Analysis:
Both MESABA and CITY OF GLASGOW (see9913) were part of combined convoys OE21 and OL32. The course of the convoy was 200deg true and both were hit about the same time on their starboard sides. MESABA (OE21 Liverpool for Montreal, in ballast (500 tons manganese ore)) was hit forward and sank in nine minutes by the bow, making a swift vertical plunge with propellers still revolving. CITY OF GLASGOW (OL32 Manchester for Montreal in ballast) was hit amidships, breaking its back, and taking an hour to sink (TNA ADM 137/1502). UB118 fired only one torpedo at each ship (KTB).

Phase 2 Conclusion:
MESABA (MBES/Archive). Positionally, dimensionally, and archivally, this is likely the wreck of MESABA.

Phase 1: UKHO Analysis:

Original UKHO Name:
9912 CITY OF GLASGOW (PROBABLY)

Archival details:
CITY OF GLASGOW torpedoed in Sep 1918. At a position 0.7nm from this wreck (*War Losses* 1990). U-boat position (KTB) is 3.7 miles from wreck. It was 135m x 16.2m (Larn & Larn 2002).

MBES Analysis:
Large wreck in two portions of 43m and 99m with an overall length of approximately 142m. Wreck is approx. 15.2m wide. The break is forward. Small deck house on stern discernible in the DTM and pointcloud. Decks are collapsing down, and the wreck is slumping to the port side.

Additional Evidence:
Wreck was located in January 1945 (UKHO).

Phase 1 Conclusion:
UNKNOWN. Wreck is too long to be CITY OF GLASGOW. Distribution of wreckage does match photo of CITY OF GLASGOW sinking in the project archive and it can be eliminated as a candidate for this wreck.

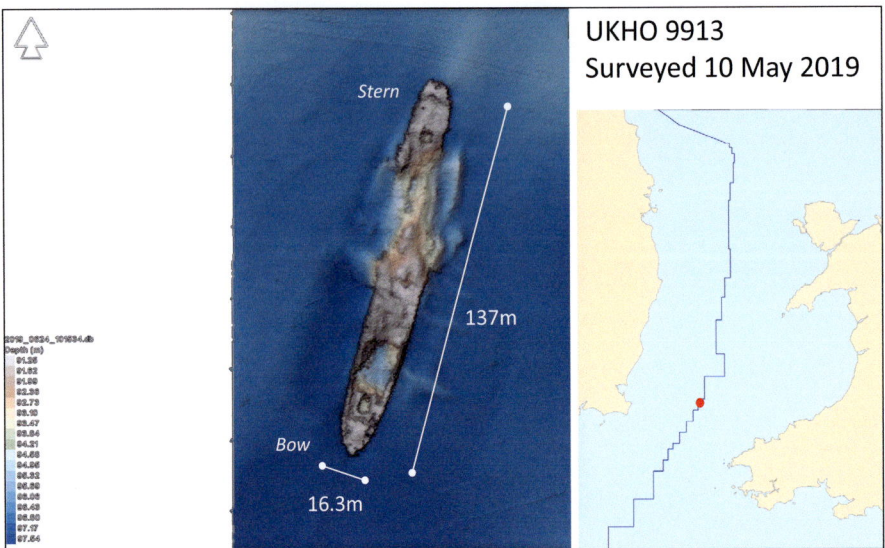

UKHO 9913
Surveyed 10 May 2019

Stern

137m

Bow

16.3m

9913 CITY OF GLASGOW (MBES/Archive)

Original UKHO: 9913 MESABA (PROBABLY)	**Phase 1 Analysis:** UNKNOWN
Position WGS84: 52 15.246 N, 5 38.506 W	**Water Depth:** 97m
Type of Vessel: SS liner	**Surveyed Length:** 137m
Date of Loss: 1 Sep 1918	**Circumstances:** Torpedoed by UB118

Phase 2: Method of wreck identification:

Positional Analysis:
This wreck lies 5nm south of the position reported by UB118 (KTB) and 2.75 nm south of the position recorded in *War Losses* (1990).

Dimensional Analysis:
CITY OF GLASGOW was 135.02m x 16.30m (*Lloyd's Register* 1917-18)

Archival Analysis:
Both MESABA (see 9912) and CITY OF GLASGOW were part of combined convoys OE21 and OL32. The course of convoy was 200deg true and both were hit about the same time on their starboard sides. MESABA (OE21 Liverpool for Montreal, in ballast (500 tons manganese ore)) was hit forward sank in nine minutes by the bow, making a swift vertical plunge. CITY OF GLASGOW (OL32 Manchester for Montreal in ballast) was hit amidships, breaking its back, and taking an hour to sink (TNA ADM 137/1502). UB118 fired only one torpedo at each ship (KTB).

Phase 2 Conclusion:
CITY OF GLASGOW (MBES/Archive). Positionally, dimensionally, and archivally, this is CITY OF GLASGOW.

Phase 1: UKHO Analysis:

Original UKHO Name:
9913 MESABA (PROBABLY)

Archival details:
MESABA was torpedoed in Sep 1918 (*War Losses* 1990). It was 146.94m x 15.9m (*Lloyd's Register* 1918-19).

MBES Analysis:
Large upright SS cargo of 137m x 16.3m. Appears to have broken aft of the bridge area.

Additional Evidence:
Located in 1980 (UKHO).

Phase 1 Conclusion:
UNKNOWN. This wreck is too short to be MESABA.

9914 KYANITE (MBES/Archive)

Original UKHO: 9914 U1302 (POSSIBLY)	**Phase 1 Analysis:** UNKNOWN
Position WGS84: 52 19.772 N, 5 17.731 W	**Water Depth:** 83m
Type of Vessel: SS Cargo	**Surveyed Length:** 50m
Date of Loss: 15 Feb 1917	**Circumstances:** Explosives UC65

Phase 2: Method of wreck identification:

Positional Analysis:
This wreck plots 7.8nm west of the position given in *War Losses* (1990). The U-boat position cannot be accurately plotted from the KTB chart.

Dimensional Analysis:
KYANITE was 53.39m x 8.55m, well-decked with machinery aft (*Lloyd's Register* 1915-16).

Archival Analysis:
KYANITE was stopped by warning shot, abandoned, and then sunk by two bombs. One in No2 hold and one in the engine room (TNA ADM 137/1305).

Phase 2 Conclusion:
KYANITE (MBES/Archive) This wreck is positionally, dimensionally and archivally consistent with KYANITE. Note: there a slight possibility this could be THE QUEEN, based on the master's position, but this was refuted by other evidence (see 9887).

Phase 1: UKHO Analysis:

Original UKHO Name:
9914 U1302 (POSSIBLY)

Archival details:
U1302 was graded "A Known Sunk" in 1945 at a position 5.4nm from this wreck. Its destruction has never fallen under suspicion because surface evidence contained human remains and items of German manufacture. The position was buoyed after the attack. The buoyed position was probably 52 19 N; 05 23 W as this is the position given in the assessed ASW report (TNA ADM 199/1786). The U-boat therefore is likely to be nearby (see UKHO 9911).

MBES Analysis:
Small FV/Coaster with forward hold and bridge structure aft.

Additional Evidence:
Located in Apr 1945 and at the time considered to be a U-boat wreck (UKHO).

Phase 1 Conclusion:
UNKNOWN. This wreck is not a submarine.

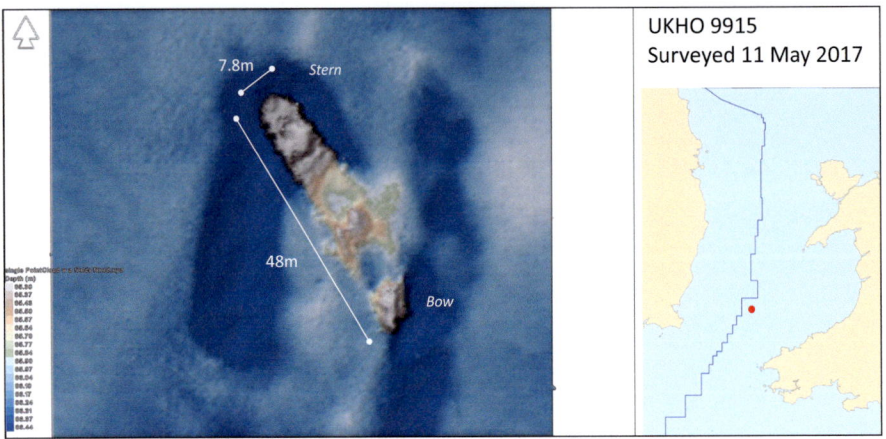

7.8m Stern

48m

Bow

9915 VOLTAIRE (MBES/Archive)

Original UKHO: 9915 ROTULA (POSSIBLY)	**Phase 1 Analysis:** UNKNOWN
Position WGS84: 52 18.6 N, 5 28.417 W	**Water Depth:** 89m
Type of Vessel: SS Cargo	**Surveyed Length:** 50m
Date of Loss: 11 Feb 1917	**Circumstances:** Explosives by UC65

Phase 2: Method of wreck identification:

Positional Analysis:
The wreck plots 17.4nm NW of the position given in *War Losses* (1990).

Dimensional Analysis:
VOLTAIRE was 45.72m x 7.31m with refrigeration machinery (*Lloyd's Register* 1915-16).

Archival Analysis:
According to UC65's KTB VOLTAIRE sank at 0900, 90 minutes later LYCIA sank. So, the wrecks should be within a few miles of each other. LYCIA (UKHO 9909) plots 5.2nm west of this wreck.

Phase 2 Conclusion:
VOLTAIRE (MBES/Archive). This wreck is a good fit for VOLTAIRE by its dimensions and its archival connection with LYCIA. The positional distance from *War Losses* notwithstanding.

Phase 1: UKHO Analysis:

Original UKHO Name:
9915 ROTULA (POSSIBLY)

Archival details:
ROTULA was sunk in 1941 (*War Losses* 1989). It was 146.61m x 18m (*Lloyd's Register* 1940-41).

MBES Analysis:
Small coaster 50m long and very collapsed except for raised stern area.

Additional Evidence:
Located in 1972.

Phase 1 Conclusion:
UNKNOWN. This wreck is far too small to be ROTULA.

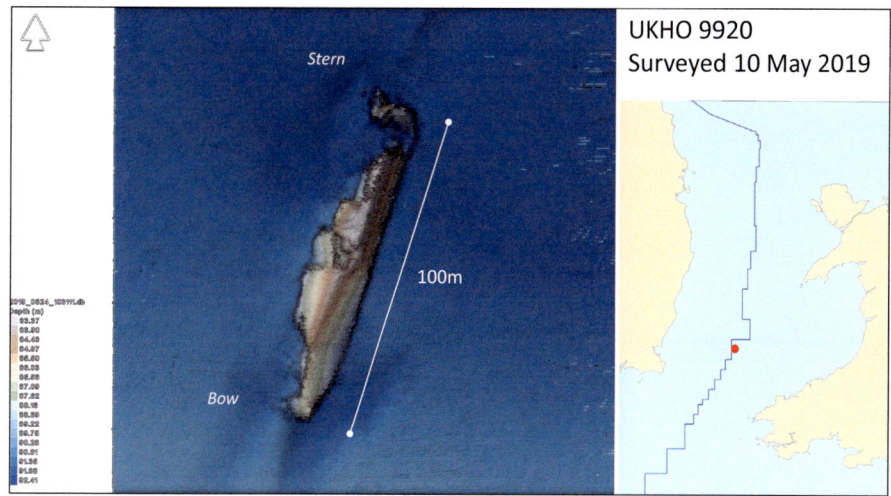

9920 BOLTONHALL (MBES/Archive)

Original UKHO: 9920 PRINS FREDERIK HENDRIK (POSSIBLY)	**Phase 1 Analysis:** UNKNOWN
Position WGS84: 52 20.195N; 5 34.132 W	**Water Depth:** 92m
Type of Vessel: SS Cargo	**Surveyed Length:** 100m
Date of Loss: 20 Aug 1918	**Circumstances:** Torpedoed by UB92

Phase 2: Method of wreck identification:

Positional Analysis:
This wreck plots 4.17nm west of the position given in *War Losses* (1990) and 5.05nm west of the position given by UB92 (KTB).

Dimensional Analysis:
BOLTONHALL was 103.98m x 14.68m (*Lloyd's Register* 1917-18).

Archival Analysis:
UB92's KTB states that the ship was hit by one torpedo and sank in five minutes. It was bound from Partington to Milford Haven and was hit in the aft end of the engine room, port side, blowing a hole on the port side. It listed to port and sank by the stern. Five were killed. Ship abandoned 29nm north of the N Bishop. (TNA ADM 137/1501).

Phase 2 Conclusion:
BOLTONHALL (MBES/Archive). Positionally, dimensionally, and archivally this wreck is consistent with BOLTONHALL.

Phase 1: UKHO Analysis:

Original UKHO Name:
9920 PRINS FREDERIK HENDRIK (POSSIBLY)

Archival details:
PRINS FREDERIK HENDRIK was bombed on 8 Mar 1941. It was 76.3m x 12.9m (*Lloyd's Register* 1939-40).

MBES Analysis:
Large wreck lying on is starboard side and collapsing to an upside-down orientation. 100m long.

Additional Evidence:
Located in 1945 (UKHO).

Phase 1 Conclusion:
UNKNOWN. This wreck is too long to be PRINS FREDERIK HENDRIK.

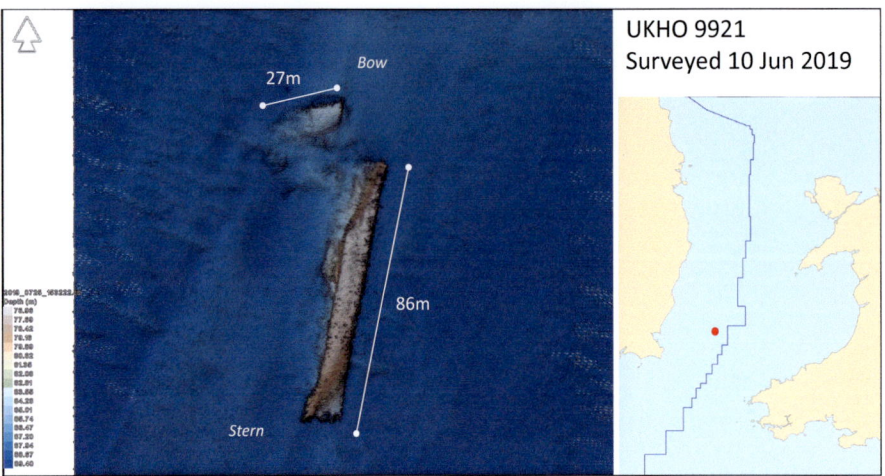

9921 CITY OF DUNDEE (MBES/Archive)

Original UKHO: 9921 ARCADIAN (POSSIBLY)	**Phase 1 Analysis:** UNKNOWN
Position WGS84: 52 21.549 N, 5 47.461 W	**Water Depth:** 89m
Type of Vessel: SS Cargo	**Surveyed Length:** 113m (approx)
Date of Loss: 4 Oct 1908	**Circumstances:** Collision with SS MATINA

Phase 2: Method of wreck identification:

Positional Analysis:
This wreck plots 4.4nm SE of the position given for the collision in the Board of Trade Inquiry (BOT Wreck Report No 7216).

Dimensional Analysis:
CITY OF DUNDEE was 110.21m x 12.97m (*Lloyd's Register* 1907-08).

Archival Analysis:
CITY OF DUNDEE was travelling SW and was struck by MATINA, travelling NE on the port bow in No1 hatch, almost completely severing the bows. The ships remained locked together as MATINA held into the stricken ship and tried to rescue its survivors. It sunk 14 minutes later, rolling to port. MATINA's excessive speed in thick fog was considered the main causal factor in the disaster (BOT Inquiry No 7216).

Phase 2 Conclusion:
CITY OF DUNDEE (MBES/Archive). Positionally and dimensionally this wreck is consistent with CITY OF DUNDEE. Archivally the damage to the wreck is identical to that described. However, it should be noted the vessel is pointing north, meaning that it must have rotated while locked with MATINA post-collision before sinking.

Phase 1: UKHO Analysis:

Original UKHO Name:
9921 ARCADIAN (POSSIBLY)

Archival details:
ARCADIAN was sunk in a collision 21.3nm north of this wreck (BOT 1910). It was struck on the starboard side abaft of the bridge and sank very rapidly (Hocking 1969 v1, p42). ARCADIAN was 97.53m x 12.80m (*Lloyd's' Register* 1909-10).

MBES Analysis:
Large steamship wreck lying on port side with bows broken off to the north. Overall length approximately 113m.

Additional Evidence:
Located in 1980 (UKHO).

Phase 1 Conclusion:
UNKNOWN. This wreck appears too short to be ARCADIAN, the damage seen is not consistent with that reported at the time of the collision. Also, the position of the wreck is over 20nm from the reported collision point.

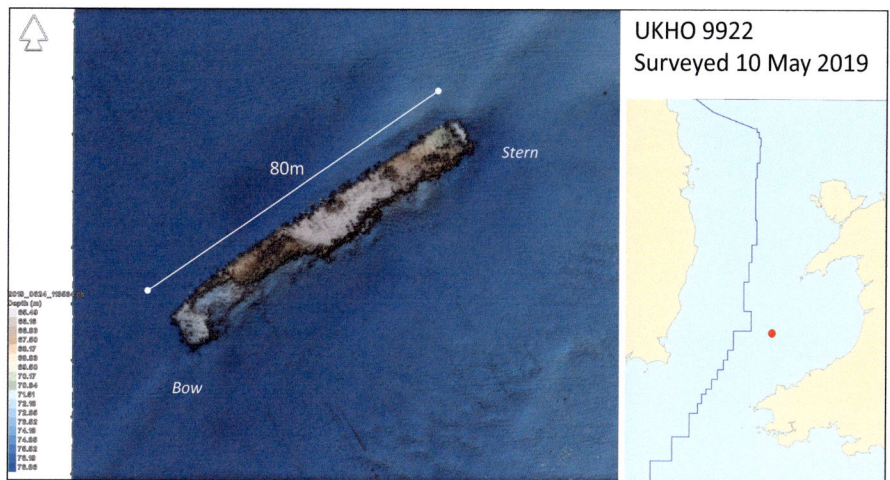

UKHO 9922
Surveyed 10 May 2019

Stern

80m

Bow

9922 GREENLAND (MBES/Archive)

Original UKHO: 9922 UNKNOWN	**Phase 1 Analysis:** UNKNOWN
Position WGS84: 52 22.661 N, 5 9.414 W	**Water Depth:** 74m
Type of Vessel: SS Cargo	**Surveyed Length:** 80m
Date of Loss: 14 Feb 1917	**Circumstances:** Explosives by UC65

Phase 2: Method of wreck identification:

Positional Analysis:
This wreck plots 7.6nm south of the position given in *War Losses* (1990). UC65's reported position cannot be accurately discerned from its KTB, but the ship was sunk on the same day as FERGA and MARGARITA.

Dimensional Analysis:
GREENLAND was 79.24m x 10.97 (*Lloyd's Register* 1915-16).

Archival Analysis:
GREENLAND was stopped by UC65 which used GREENLAND'S boat to place bombs and sink the ship. It then headed away towards MARGARITA and sank it next (TNA 137/2961). The ship sank stern first. The submarine also used its gun to finish it off (TNA ADM 137/1305).

Phase 2 Conclusion:
GREENLAND (MBES/Archive) Positionally, dimensionally, and archivally, this wreck is consistent with GREENLAND. Due to the vague positional recording by both U-boat and ship, it should be noted that 82347 (currently thought to be DUNSLEY) and 9929 (SERULA) cannot be ruled out as alternative candidates.

Phase 1: UKHO Analysis:

Original UKHO Name:
9922 UNKNOWN

Archival details:
N/A

MBES Analysis:
Large wreck lying on port side with noticeable damage in the holds forward of the bridge. 80m long.

Additional Evidence:
Located in 1945 (UKHO).

Phase 1 Conclusion:
UNKNOWN

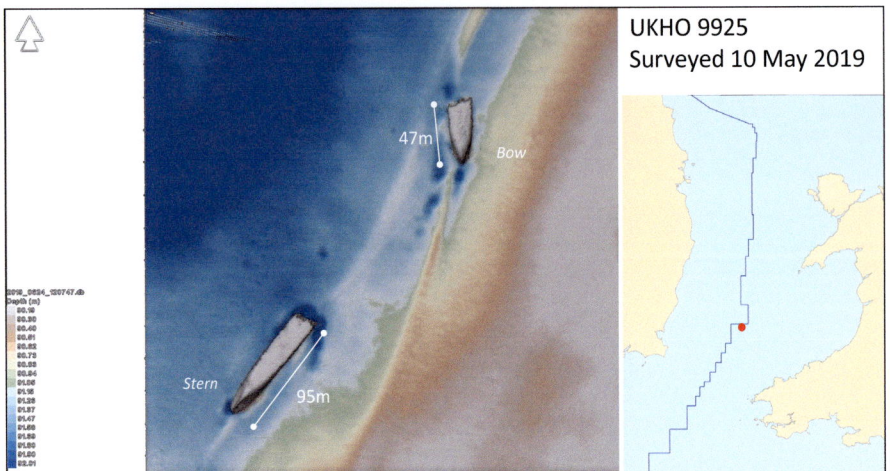

9925 ROTULA (MBES/Archive)

Original UKHO: 9925 LANDONIA (PROBABLY)	**Phase 1 Analysis:** UNKNOWN
Position WGS84: 52 22.8 N, 5 28.45 W	**Water Depth:** 90m
Type of Vessel: MV Tanker	**Surveyed Length:** 142m
Date of Loss: 1 Mar 1941	**Circumstances:** Bombed and later sunk by Allied gunfire

Phase 2: Method of wreck identification:

Positional Analysis:
The wreck plots 2.9nm SW of the position given in *War Losses* (1989) and 0.36nm from the position reported by HMT ARMANA where it sank the derelict by gunfire (TNA ADM 199/2225).

Dimensional Analysis:
ROTULA was 146.61m x 18.10m (*Lloyd's Register* 1940-41). The 3rd largest wreck in the study area.

Archival Analysis:
ROTULA was inbound from Halifax with a cargo of aviation spirit. It was bombed, catching fire aft and was abandoned. 16 crew killed (of 49) (TNA ADM 199/2136). The wreck sank stern first and anchored itself to the bottom while the fore section still floated. On 3 Mar HMT ARMANA sank the wreck with gunfire, the sea still blazing fiercely (TNA ADM 199/2225).

Phase 2 Conclusion:
ROTULA (MBES/Archive) Positionally, dimensionally, and archivally this wreck cannot be any other than ROTULA.

Phase 1: UKHO Analysis:

Original UKHO Name:
9925 LANDONIA (PROBABLY)

Archival details:
LANDONIA was torpedoed in Apr 1918 and sank at once (TNA 137/2964). The *War Losses* (1990) position plots 11.4nm SW of this wreck. The U-boat position (KTB) plots 13.4nm north of the wreck. These two archival positions plot 24nm miles apart. To add to the archival muddle, the master's position plots 13nm to the east of this wreck (TNA 137/2964). LANDONIA was 91.70m x 13.59m (*Lloyd's Register* 1918-19).

MBES Analysis:
Upside-down shipwreck in two halves which are 175m apart. The fore section lies to the north. The overall length of the wreck is measured at approx. 142m. The width is 18.3m.

Additional Evidence:
Located during WW2 (UKHO).

Phase 1 Conclusion:
UNKNOWN. This wreck is too large to be LANDONIA.

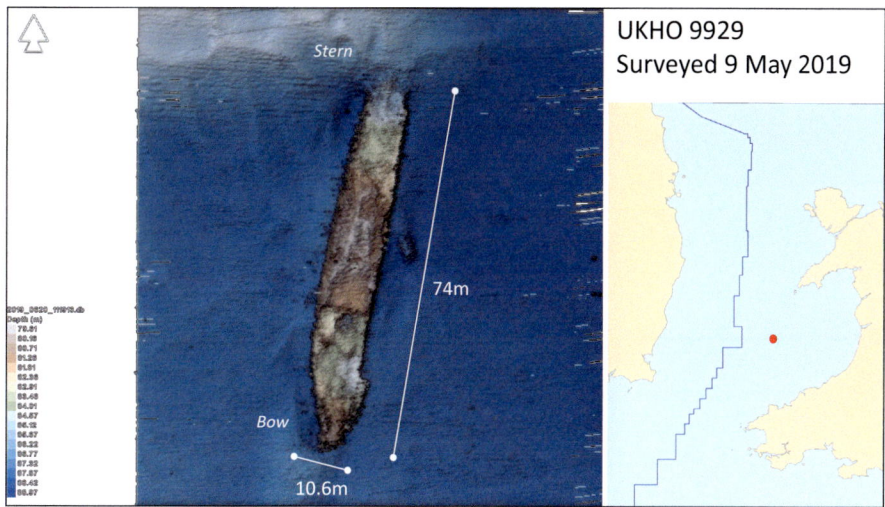

9929 SERULA (MBES/Archive)

Original UKHO: 9929 UNKNOWN	**Phase 1 Analysis:** UNKNOWN
Position WGS84: 52 27.136 N, 5 7.672 W	**Water Depth:** 81m
Type of Vessel: SS Cargo	**Surveyed Length:** 74m (approx)
Date of Loss: 16 Sep 1918	**Circumstances:** Torpedoed by UB64

Phase 2: Method of wreck identification:

Positional Analysis:
The wreck plots 10.1 nm north of the position reported by the 2nd mate, the senior surviving crewmember (TNA ADM 137/1503) and plots 9.1nm SW of the position reported in UB64's KTB.

Dimensional Analysis:
SERULA was 79.20m x 10.70m, well-decked (*Lloyd's Register* 1917-18).

Archival Analysis:
Manchester for Rouen. The 2nd mate reported that it was hit on the starboard quarter abreast the stern hatch. The ship immediately sank, stern first. He and eight other survivors clung to driftwood. The U-boat surfaced, picked them up and deposited them on the ship's raft before asking the usual questions and heading off (TNA ADM 137/1503).

Phase 2 Conclusion:
SERULA (MBES/Archive). Positionally, dimensionally and archivally, this wreck is consistent with SERULA. Due to the vague positional recording by both U-boat and ship, it should be noted that 82347 (DUNSLEY) and 9922 (GREENLAND) cannot be entirely ruled out as alternative candidates. In this instance, the missing stern and the heading supports the view that it is probably SERULA. Layout of holds appears to conform with the photo of the vessel in the project archive.

Phase 1: UKHO Analysis:

Original UKHO Name:
9929 UNKNOWN

Archival details:
N/A

MBES Analysis:
Upright SS with holds fore and aft. Dimensions of 74m x 10.6m. Stern possibly truncated.

Additional Evidence:
Located in 1945 (UKHO).

Phase 1 Conclusion:
UNKNOWN

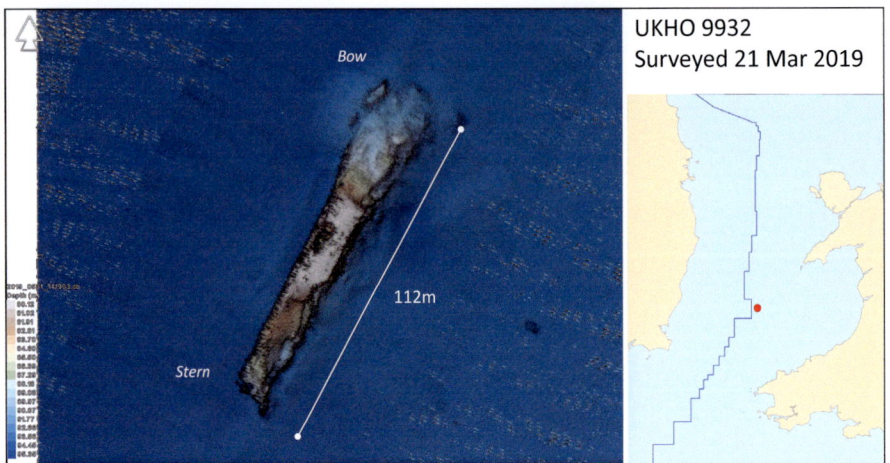

UKHO 9932
Surveyed 21 Mar 2019

9932 STRATHNAIRN (MBES/Archive)

Original UKHO: 9932 UNKNOWN	**Phase 1 Analysis:** UNKNOWN
Position WGS84: 52 28.154 N, 5 14.946 W	**Water Depth:** 91m
Type of Vessel: SS Cargo	**Surveyed Length:** 112m (approx)
Date of Loss: 15 Jun 1915	**Circumstances:** Torpedoed by U22

Phase 2: Method of wreck identification:

Positional Analysis:
The wreck plots 1.8nm south of the position reported by U22 (KTB) and 14.3 nm north of the position reported in *War Losses* (1990). The senior surviving crewmember was the 2nd officer.

Dimensional Analysis:
STRATHNAIRN was 112.77m x 15.90m (*Lloyd's Register* 1915-16).

Archival Analysis:
Cardiff for Archangel with 7,000 tons of coal. Hit at 2130. Listed to port. One boat lost during launch, including the master (TNA ADM 137/2959). Torpedo hit just aft of the funnel (KTB). The vessel was still seen afloat at 2300 (TNA ADM 137/1061).

Phase 2 Conclusion:
STRATHNAIRN (MBES/Archive). Dimensionally and archivally and by the KTB position, this wreck is consistent with STRATHNAIRN.

Phase 1: UKHO Analysis:

Original UKHO Name:
9932 UNKNOWN

Archival details:
N/A

MBES Analysis:
Large wreck lying on starboard side, broken up forward. Length of approx. 112m and approx. 13m high.

Additional Evidence:
Located in 1980 (UKHO).

Phase 1 Conclusion:
UNKNOWN

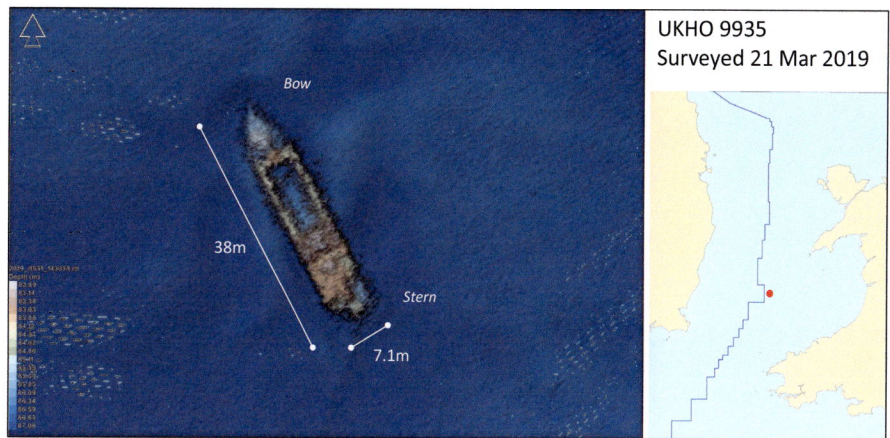

9935 SALLAGH (MBES/Archive)

Original UKHO: 9935 UNKNOWN	**Phase 1 Analysis:** UNKNOWN
Position WGS84: 52 28.075 N, 5 15.804 W	**Water Depth:** 85m
Type of Vessel: SS Cargo	**Surveyed Length:** 38m
Date of Loss: 10 Feb 1917	**Circumstances:** Explosives by UC65

Phase 2: Method of wreck identification:

Positional Analysis:
The wreck lies 6.9nm SW of the position reported by GREENLAND when it picked up the survivors at 1630, some eight hours after the sinking (TNA ADM 137/1304). The position cannot be accurately ascertained from UC65's KTB.

Dimensional Analysis:
SALLAGH was 40.10m x 7.03m (*Lloyd's Register* 1916-17).

Archival Analysis:
Stopped and evacuated. Continued shellfire killing the chief engineer. The U-boat used the ship's boat to plant bombs in the ship, departing the scene on the surface (TNA ADM 137/2961). The abandonment followed the U-boat flying the abandon ship signal. The boat was being readied when the fatality occurred. Five men were placed on the sub while the bombs were being planted. Bandages were supplied to the injured. Two bombs were placed on board. One in the peak and one in the engine room. Two more shells were fired before the ship exploded and sank. The U-boat was last seen 5nm south with a rigged mast (TNA ADM 137/1304).

Phase 2 Conclusion:
SALLAGH (MBES/Archive). This wreck is positionally, dimensionally, and archivally consistent with SALLAGH.

Phase 1: UKHO Analysis:

Original UKHO Name:
9935 UNKNOWN

Archival details:
N/A

MBES Analysis:
Small wreck of a MV/SS coaster or FV. Upright and 38m x 7.1m. Bows to NE.

Additional Evidence:
Located in 1945 (UKHO).

Phase 1 Conclusion:
UNKNOWN

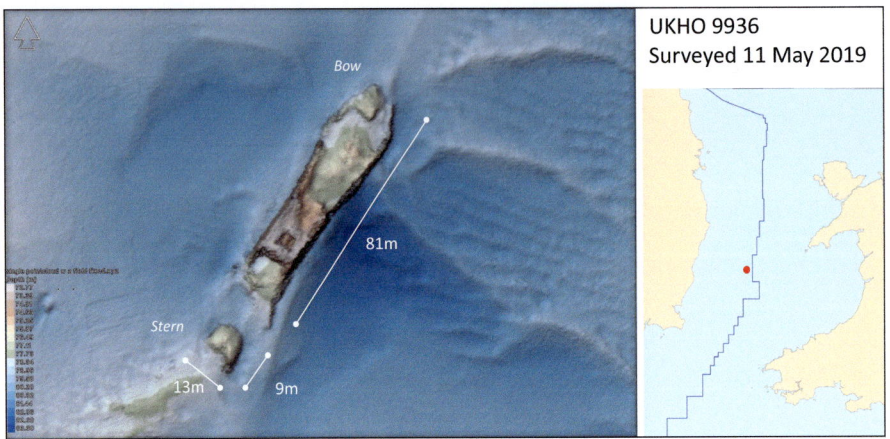

Bow

81m

Stern

13m   9m

9936 LANDONIA (MBES/Archive)

Original UKHO: 9936 YTURRI BIDE (POSSIBLY)	**Phase 1 Analysis:** UNKNOWN
Position WGS84: 52 36.286 N, 5 34.207 W	**Water Depth:** 85m
Type of Vessel: SS Cargo	**Surveyed Length:** 90m (approx)
Date of Loss: 21 Apr 1918	**Circumstances:** Torpedoed by U91

Phase 2: Method of wreck identification:

Positional Analysis:
This wreck plots 6.2nm west of the position given by U91 (KTB). Due to only five survivors, the British reports are vague. The survivors' report gives a position 13.3nm south of this wreck (TNA ADM 137/1494).

Dimensional Analysis:
LANDONIA was 91.70m x 13.59m (*Lloyd's Register* 1918-19).

Archival Analysis:
U91's KTB states the ship was hit in the stern and sank immediately by the stern. The torpedo hit the starboard side of No3 hold and the ship immediately sank. 21 killed and five survivors (one later died) pulled from the sea after three hours immersed. The U-boat surfaced and took the chief gunner prisoner, as reported by the U-boat's radio (survivors said it was the master who was taken prisoner). Bilbao for Glasgow carrying iron ore, possibly explaining its rapid sinking. (TNA ADM 137/2961).

Phase 2 Conclusion:
LANDONIA (MBES/Archive). This wreck is archivally, positionally, and dimensionally consistent with LANDONIA. Of note is the shape of the stern, as seen on the MBES DTM and in the ship plans in the project archive.

Phase 1: UKHO Analysis:

Original UKHO Name:
9936 YTURRI BIDE (POSSIBLY)

Archival details:
YTURRI BIDE was sunk in 1918 (*War Losses* 1990). It was 57.61m x 7.92m (*Lloyd's Register* 1918-19).

MBES Analysis:
Upright wreck of a SS with holds fore and aft. It appears if the stern has been reversed, having been blown off the ship. Ship is approx. 90m long and is 13m wide.

Additional Evidence:
Located in 1980 (UKHO).

Phase 1 Conclusion:
UNKNOWN. The wreck is much too wide to be YTURRI BIDE.

10.8m Stern

74m Bow

9937 NOR (MBES/Archive)

Original UKHO: 9937 ELSENA (POSSIBLY)	**Phase 1 Analysis:** UNKNOWN
Position WGS84: 52 33.294 N, 5 37.222 W	**Water Depth:** 83m
Type of Vessel: SS Cargo	**Surveyed Length:** 74m (approx)
Date of Loss: 14 Dec 1917	**Circumstances:** Torpedoed by UB65

Phase 2: Method of wreck identification:

Positional Analysis:
This wreck plots 5.5nm south of the position given in UB65's KTB and 12.2nm south of the position given in *War Losses* (1990). The survivor report in Irish Sea (TNA ADM 137/1362) only gives a position as *"off Arklow Bank"* and it is not known where the *War Losses* position originated.

Dimensional Analysis:
NOR was 74.70m x 11.02m (*Lloyd's Register* 1915-16).

Archival Analysis:
Caen to Glasgow in ballast. Hit in the after hold, mainmast falling away. The ship sank in 30 minutes. U-boat surfaced and interrogated the crew in the usual manner. One boat capsized killing one. The rest got into the other boat and made for Wales with the wind, one later dies of exposure. An hour later they were picked up (TNA ADM 137/1362).

Phase 2 Conclusion:
NOR (MBES/Archive). Positionally, dimensionally, and archivally this wreck is consistent with NOR.

Phase 1: UKHO Analysis:

Original UKHO Name:
9937 ELSENA (POSSIBLY)

Archival details:
ELSENA was sunk in 1917 (*War Losses* 1990). It was 43.28m x 7.34m (*Lloyd's Register* 1917-18).

MBES Analysis:
Upright SS cargo vessel with holds fore and aft. Approximately 74m long and 10.8m wide.

Additional Evidence:
Located in 1980 (UKHO).

Phase 1 Conclusion:
UNKNOWN. The wreck is far too large to be ELSENA.

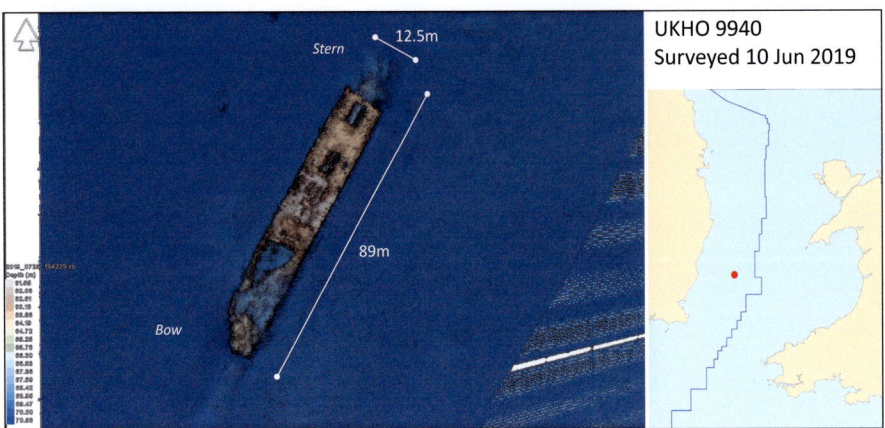

UKHO 9940
Surveyed 10 Jun 2019

Stern
12.5m

89m

Bow

9940 GLORIA (MBES/Archive)

Original UKHO: 9940 ENNISTOWN (POSSIBLY)	**Phase 1 Analysis:** UNKNOWN
Position WGS84: 52 31.532 N, 5 44.598 W	**Water Depth:** 69m
Type of Vessel: SS Cargo	**Surveyed Length:** 89m (approx)
Date of Loss: 2 Dec 1933	**Circumstances:** Fire

Phase 2: Method of wreck identification:

Positional Analysis:
This wreck plots 10.4nm WSW from the position the wreck was abandoned in (*The Times* 4 Dec 1933).

Dimensional Analysis:
GLORIA was 89.91m x 12.80m (*Lloyd's Register* 1917-18).

Archival Analysis:
Larne to Port Talbot in ballast (*Wreck Returns* 1933 Qtr.4 p5). The ship was abandoned on fire with the stern holds awash. Ship on fire. Heading SW and sinking rapidly (*The Times* 4 Dec 1933).

Phase 2 Conclusion:
GLORIA (MBES/Archive). Positionally, dimensionally, and archivally this wreck is consistent with GLORIA.

Phase 1: UKHO Analysis:

Original UKHO Name:
9940 ENNISTOWN (POSSIBLY)

Archival details:
ENNISTOWN (ex-EIMSTAD) was sunk in 1917 in ballast (*War Losses* 1990). It was 59.43m x 9.14m (*Lloyd's Register* 1915-16).

MBES Analysis:
Large steamer of 89m x 12.5m. Upright with stern to NE. Bow area has collapsed. Stern may have had prop salvage, or damaged as ship sunk.

Additional Evidence:
Located in 1980 (UKHO).

Phase 1 Conclusion:
UNKNOWN. This wreck is far too large to be ENNISTOWN.

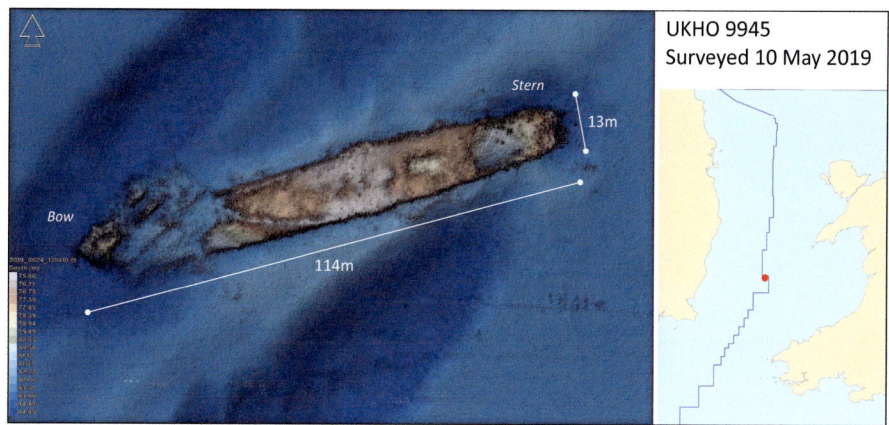

9945 BURUTU (MBES/Archive)

Original UKHO: 9945 SPENSER (POSSIBLY)		**Phase 1 Analysis:** UNKNOWN	
Position WGS84: 52 31.0 N, 5 25.1 W		**Water Depth:** 84m	
Type of Vessel: SS Cargo Liner		**Surveyed Length:** 114m	
Date of Loss: 3 Oct 1918		**Circumstances:** Collision with SS CITY OF CALCUTTA	

Phase 2: Method of wreck identification:

Positional Analysis:
This wreck plots 1.4nm SW of the position given in the OL37 record (TNA ADM 137/2618) and 5.3nm south of the HL50 record (TNA ADM 137/2657).

Dimensional Analysis:
BURUTU was 109.72m x 13.46m (*Lloyd's Register* 1917-18). Drawings show the engine room was aft of the centre of the ship.

Archival Analysis:
BURUTU was part of inbound convoy HL50 when outbound convoy OL37 passed through it. CITY OF CALCUTTA colliding with and sinking BURUTU with heavy loss of life (TNA ADM 137/2657). 150 people were killed (BOT 1921 Pt2 Table A p39 (3)). BURUTU was struck on the port side (Larn & Larn 2000).

Phase 2 Conclusion:
BURUTU (MBES/Archive). Positionally, dimensionally, and archivally, the wreck is consistent with BURUTU.

Phase 1: UKHO Analysis:

Original UKHO Name:
9945 SPENSER (POSSIBLY)

Archival details:
SPENSER was torpedoed by U61 at a position reported by the U-boat (KTB) which plots 8.2nm north of this wreck. The *War Losses* (1990) position is also nearby, plotting 7.3nm NW. SPENSER was a four-masted steamship of 117.40m x 15.54m (*Lloyd's Register* 1917-18). The ship's certification papers in the project archive show that the ship had five cargo hold hatches. The master's report (TNA ADM 137/2963) states the ship was hit in hold No5 and sank within the hour.

MBES Analysis:
Upright steamship cargo of 114m length and 13m width. Collapsed forward into a scour and partially collapsed aft. Ship appears to have had a maximum of four hatches.

Additional Evidence:
Located in 1980 (UKHO).

Phase 1 Conclusion:
UNKNOWN. After detailed measurement it is concluded it is too narrow to be SPENSER.

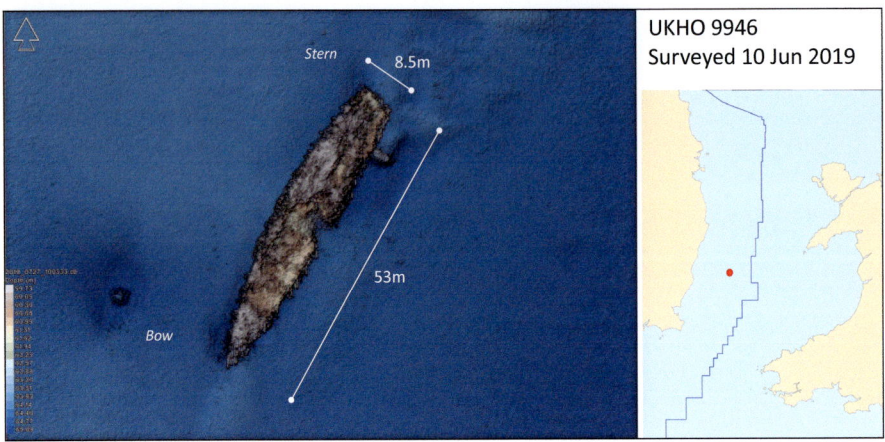

UKHO 9946
Surveyed 10 Jun 2019

Stern 8.5m

53m

Bow

9946 HMY KETHAILES (MBES/Archive)

Original UKHO: 9946 ARDGLASS (POSSIBLY)	**Phase 1 Analysis:** UNKNOWN
Position WGS84: 52 35.702 N, 5 49.274 W	**Water Depth:** 63m
Type of Vessel: HM Yacht	**Surveyed Length:** 53m (approx)
Date of Loss: 11 Oct 1917	**Circumstances:** Collision with SS LEICESTERSHIRE

Phase 2: Method of wreck identification:

Positional Analysis:
This wreck plots 5.4nm NE of the position recorded by the Admiralty as "*11nm NE of the Blackwater light vessel*" (Hepper 2006, p106).

Dimensional Analysis:
HMY KETHAILES was 55.6m x 8.3m (Hepper 2006, p106).

Archival Analysis:
Collision killed 17 of the crew with only 11 survivors (Hepper 2006, p106).

Phase 2 Conclusion:
HMY KETHAILES (MBES/Archive). Positionally, dimensionally, and archivally this wreck is consistent with KETHAILES.

Phase 1: UKHO Analysis:

Original UKHO Name:
9946 ARDGLASS (POSSIBLY)

Archival details:
ARDGLASS was stopped and blown up by UC65 4nm SSW of the S Arklow LV (*War Losses* 1990) at a position which plots 8nm west of this wreck. It was 60.96m x 8.86m (*Lloyd's Register* 1917-18).

MBES Analysis:
Upright coaster of approx. 53 length with a width of 8.5m. The wreck might not be complete, as the stern looks slightly cut off. Hole in port side amidships.

Additional Evidence:
Located in 1980 (UKHO).

Phase 1 Conclusion:
UNKNOWN. The wreck is too short to be ARDGLASS, even allowing for a portion of missing stern.

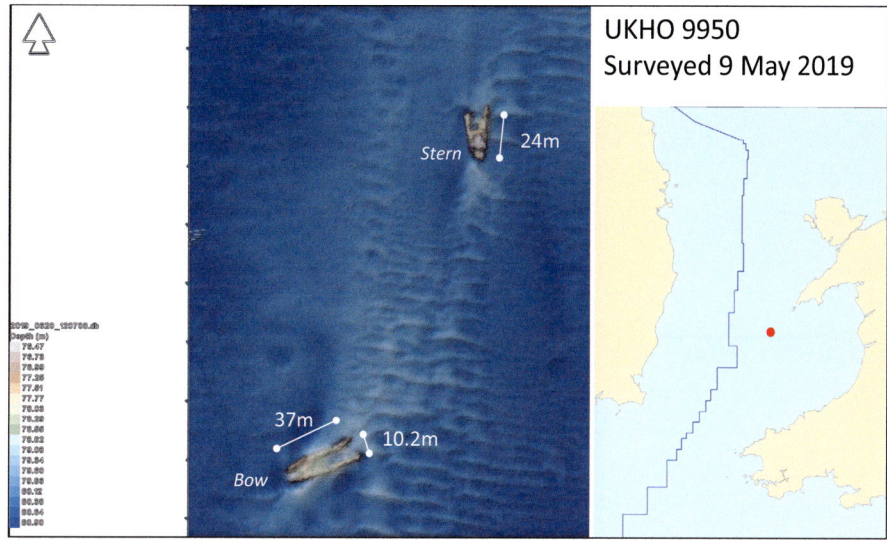

9950 LCT 326 (MBES/ARCHIVE)

Original UKHO: 9950 FOYLE (POSSIBLY)	**Phase 1 Analysis:** UNKNOWN
Position WGS84: 52 36.659 N, 5 3.082 W	**Water Depth:** 78m
Type of Vessel: Landing Craft Tank (LCT)	**Surveyed Length:** 61m (approx)
Date of Loss: 2 Feb 1943	**Circumstances:** Foundered

Phase 2: Method of wreck identification:

Positional Analysis:
LCT 326 is listed as having been lost to accident or mine off the Isle of Man (Admiralty 1947, p53). However, research for this project found that it was in convoy from Troon to Appledore and was last observed at 53 00; 04 58W, which is 25nm north of this wreck, NW of Bardsey Island. The weather was Force 6 from the SW (TNA ADM 217/13).

Dimensional Analysis:
The dimensions and appearance of this wreck on the DTM point to it most likely being LCT 326. LCT 326 was 58.2m x 9.4m.

Archival Analysis:
The proceedings of convoy show that LCT 326 would have passed over the area where this wreck lies after it was last sighted (TNA ADM 217/13).

Phase 2 Conclusion:
LCT 326 (MBES/Archive). This vessel appears to have foundered in heavy seas sometime after it was last seen. The dimensions and character of the MBES scan leave little doubt that this wreck is LCT 326. The vessel appears to have broken in half forward of the bridge with both halves staying afloat long enough to have become separated by 130m.

Phase 1: UKHO Analysis:

Original UKHO Name:
9950 FOYLE (POSSIBLY)

Archival details:
FOYLE foundered with a cargo of coal on 9 Aug 1897 (BOT 1897) in a position approx. 7nm SW of this wreck. FOYLE was a SS of 48.80m x 6.4m (*Lloyd's Register* 1897).

MBES Analysis:
Shipwreck in two parts which are 133m apart. The overall length is approx. 61m with a width of 10.2m. The stern portion is to the north, and it appears the vessel had its machinery and bridge aft.

Additional Evidence:
Located in 1945 (UKHO).

Phase 1 Conclusion:
UNKNOWN. This wreck is too long to be FOYLE.

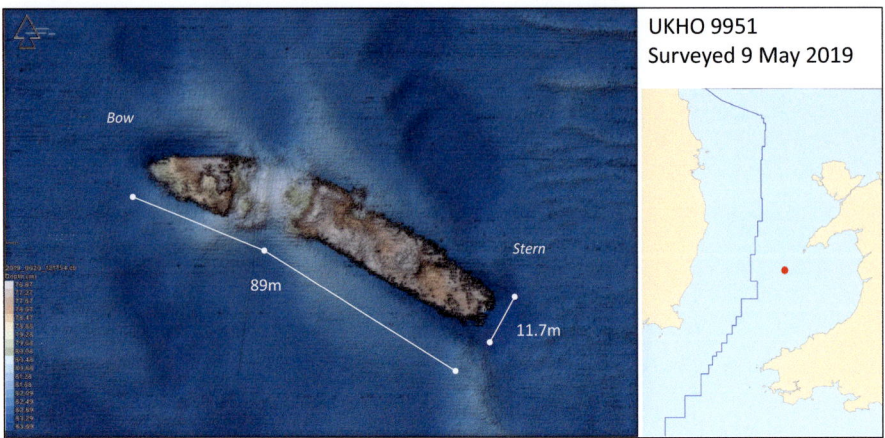

9951 ROBERT EGGLETON (MBES/ARCHIVE)

Original UKHO: 9951 RFA VITOL (POSSIBLY)	**Phase 1 Analysis:** UNKNOWN
Position WGS84: 52 37.902 N, 5 4.169 W	**Water Depth:** 80m
Type of Vessel: SS Cargo	**Surveyed Length:** 89m (approx)
Date of Loss: 28 Dec 1917	**Circumstances:** Torpedoed by U91

Phase 2: Method of wreck identification:

Positional Analysis:
This wreck plots 2.5nm west of the position reported in *War Losses* (1990). The U-boat's KTB position simply states "*bei Bardsey*" and therefore must be somewhere in the local area.

Dimensional Analysis:
ROBERT EGGLETON was 88.45m x 11.92m (*Lloyd's Register* 1915-16).

Archival Analysis:
Ship was hit by torpedoes on starboard side, firstly by No2 hold forward and then amidships and sank by the head (TNA ADM 137/1360) and U91 (KTB).

Phase 2 Conclusion:
ROBERT EGGLETON (MBES/Archive). This wreck is positionally, dimensionally, and archivally consistent with ROBERT EGGLETON. Of note is the damage to No2 hold and to the mid ship area.

Phase 1: UKHO Analysis:

Original UKHO Name:
9951 RFA VITOL (POSSIBLY)

Archival details:
RFA VITOL was sunk by U110 on 7 Mar 1918. The reported position (Hepper 2006, p123) plots 3.6nm west of this wreck. Published sources state the ship was 104.39m x 12.65m (Puddefoot 2010, p172). In fact, it was shorter. The plans retained at the National Maritime Museum reveal its length was actually 97.5m. This loss is absent from Larn & Larn (2000), *War Losses* (1990) and Admiralty (1947). Survivors stated the ship was initially torpedoed in the starboard side between engine room and boilers. After the vessel was abandoned a second explosion was heard from the lifeboats, but not witnessed (TNA ADM 137/1491). U110 was destroyed on its patrol, so there is no evidence from the German side.

MBES Analysis:
Upright SS of approximately 89m x 11.7m. The wreck is broken in half forward of the bridge.

Additional Evidence:
This area was known for a large oil patch in 1918 and became associated with RFA VITOL at that time. It was located by survey in 1981. The position where the oil was reported coming from plots 0.7nm south of this wreck (TNA ADM 137/1515).

Phase 1 Conclusion:
UNKNOWN. This wreck is at least 10m too short to be RFA VITOL.

9953 BOSCAWEN (MBES/Archive)

Original UKHO: 9953 UNKNOWN	**Phase 1 Analysis:** UNKNOWN
Position WGS84: 52 38.011 N, 5 1.174 W	**Water Depth:** 76m
Type of Vessel: SS Cargo	**Surveyed Length:** 82m
Date of Loss: 21 Aug 1918	**Circumstances:** Torpedoed by UB92

Phase 2: Method of wreck identification:

Positional Analysis:
The wreck lies 10nm SE of the position given by UB92 (KTB) and 16nm SE of the position given in *War Losses* (1990).

Dimensional Analysis:
BOSCAWEN was 85.04m x 12.39m (*Lloyd's Register* 1916-17).

Archival Analysis:
The torpedo hit between the forecastle and No1 hold on the port side and the ship immediately sank by the head. It is noted by the Admiralty assessors that the ship was east of its allotted course (TNA ADM 137/1501).

Phase 2 Conclusion:
BOSCAWEN (MBES/Archive). Dimensionally and archivally the wreck is consistent with BOSCAWEN. The position is at the extreme end of what one would expect. However, in this case the damage to the wreck and its orientation (on port side) closely the match the torpedo hit as described.

Phase 1: UKHO Analysis:

Original UKHO Name:
9953 UNKNOWN

Archival details:
N/A

MBES Analysis:
Large SS lying on its port side and 82m long. Bows SW.

Additional Evidence:
Located in 1945 (UKHO).

Phase 1 Conclusion:
UNKNOWN

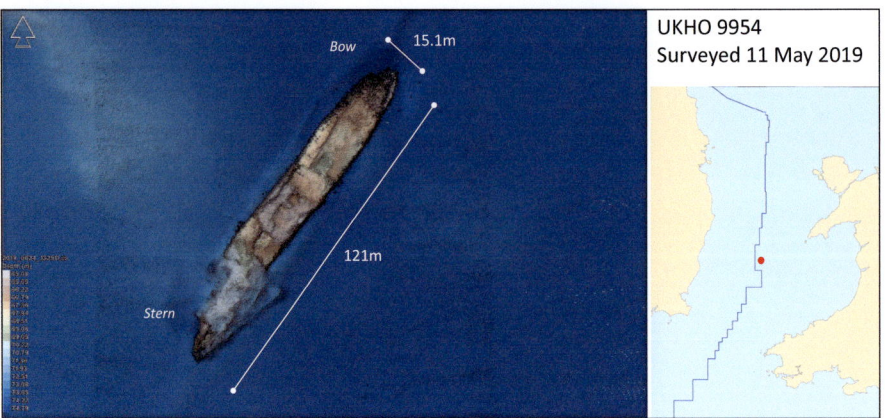

9954 SPENSER (MBES/Archive)

Original UKHO: 9954 BIRCHWOOD (POSSIBLY)	**Phase 1 Analysis:** UNKNOWN
Position WGS84: 52 36.883 N, 5 24.067 W	**Water Depth:** 74m
Type of Vessel: SS Cargo	**Surveyed Length:** 121m
Date of Loss: 6 Jan 1918	**Circumstances:** Torpedoed by U61

Phase 2: Method of wreck identification:

Positional Analysis:
This wreck plots 4.8nm east of the position given in *War Losses* (1990) and 3.2nm SE of the position given by U61 (KTB).

Dimensional Analysis:
SPENSER was 117.40m x 15.54m (*Lloyd's Register* 1917-18).

Archival Analysis:
Mount's Bay to Liverpool in convoy HD16. Hit on the port side under No5 hatch, throwing the propeller shaft out of line, rolled to port, and sank by the stern (TNA ADM 137/1488). U61 torpedoed HALBARDIER three hours later.

Phase 2 Conclusion:
SPENSER (MBES/Archive) Positionally, dimensionally, and archivally this wreck is consistent with SPENSER. Of note is the large size of the vessel and the damage which matches descriptions of the sinking. HALBARDIER (9955) lies 3nm SW of this wreck.

Phase 1: UKHO Analysis:

Original UKHO Name:
9954 BIRCHWOOD (POSSIBLY)

Archival details:
Birchwood was sunk on 3 Jan 1918 by U61. The *War Losses* (1990) position plots 18nm south of this wreck and the U-boat position plots 31nm south of this wreck. It was 96.62m x 13.91m (*Lloyd's Register* 1917-18).

MBES Analysis:
SS cargo vessel upright with marked collapse of the stern area. Bows to NE, 121m x 15.1m.

Additional Evidence:
Located in 1945 (UKHO).

Phase 1 Conclusion:
UNKNOWN. This wreck is too large to be BIRCHWOOD and plots a long way from the reported sinking locations.

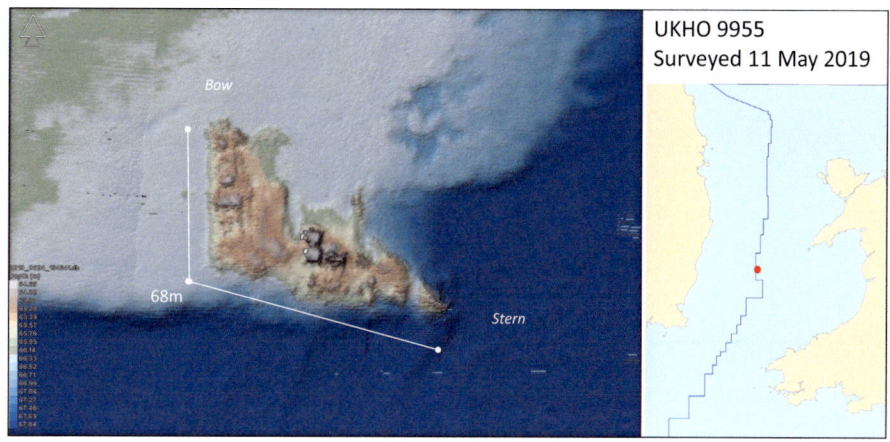

9955 HALBERDIER (MBES/Archive)

Original UKHO: 9955 FRANCOISE DE LILLE (POSSIBLY)	**Phase 1 Analysis:** UNKNOWN
Position WGS84: 52 35.619 N, 5 27.731 W	**Water Depth:** 65m
Type of Vessel: SS Cargo	**Surveyed Length:** 68m (approx)
Date of Loss: 6 Jan 1918	**Circumstances:** Torpedoed by U61

Phase 2: Method of wreck identification:

Positional Analysis:
This wreck lies 6nm south of the position reported by the master (TNA ADM 137/1488) and 4.8nm NE of the position reported by U61 (KTB).

Dimensional Analysis:
HALBERDIER was 70.10m x 10.66m (*Lloyd's Register* 1917-18).

Archival Analysis:
Manchester for London. Torpedoed in the stokehold area, under the boilers (TNA ADM 137/1488).

Phase 2 Conclusion:
HALBERDIER (MBES/Archive). Positionally, dimensionally, and archivally this wreck is consistent with HALBERDIER.

Phase 1: UKHO Analysis:

Original UKHO Name:
9955 FRANCOISE DE LILLE (POSSIBLY)

Archival details:
FRANCOISE DE LIILE was a MFV which sank in 1963 (UKHO).

MBES Analysis:
Wreck of a much collapsed SS of approx. 68m overall length. Twin boilers and engine are evident on the scan.

Additional Evidence:
Located in 1980 (UKHO).

Phase 1 Conclusion:
UNKNOWN. This is the wreck of a steamship, not an MFV.

Bow

61m

20m

68m

Stern

13.7m

9957 WAESLAND (MBES/Archive)

Original UKHO: 9957 KIRKBY (PROBABLY)	**Phase 1 Analysis:** UNKNOWN
Position WGS84: 52 40.223 N, 5 27.174 W	**Water Depth:** 71m
Type of Vessel: SS Liner	**Surveyed Length:** 129m (approx)
Date of Loss: 5 Mar 1902	**Circumstances:** Collision with SS HARMONIDES

Phase 2: Method of wreck identification:

Positional Analysis:
This wreck lies 7.6nm south of the position given in *Shipping Casualties* (BOT 1901-2 Appendix C, p145).

Dimensional Analysis:
WAESLAND was 132.61m x 12.72m (*Lloyd's Register* 1889-90).

Archival Analysis:
Built as SS RUSSIA for Cunard. Sold to the Red Star Line in 1880 and lengthened by 23.5m in the same year (*Lloyd's Register* 1880-81). Had been fitted with a triple expansion engine by 1901 (*Lloyd's Register* 1901-02). It collided with SS HARMONIDES which hit the ship amidships, rebounding into it a second time and WAESLAND was fatally damaged and quickly awash (*The Times* 8 Mar 1902) The collision was on the port side (*The Sphere* 15 Mar 1902).

Phase 2 Conclusion:
WAESLAND (MBES/Archive). Positionally, dimensionally, and archivally this wreck is consistent with WAESLAND. Of note is the vessel's size and the damage coming from the port side, matching the accounts of the accident. It is also noteworthy that the extension added in 1880 appears to be the area of the ship which gave way and is seen as collapsed on the MBES scan of the wreck.

Phase 1: UKHO Analysis:

Original UKHO Name:
9957 KIRKBY (PROBABLY)

Archival details:
U-boat KTB could not be read. 5nm from *War Losses* (1990) position. KIRKBY was 96m long and the wreck plots and 18nm from Larn & Larn (2000) position. Survivors stated the crew evacuated under gunfire and the ship was sunk by a torpedo hitting port side forward. Ship sinking by bows (TNA ADM 137/2959).

MBES Analysis:
Steamship with large engine and a pair of boilers visible in the DTM. Broken amidships with a break of 20m length. The combined lengths of the two halves is 129m. Width of the wreck is approx. 13.5m.

Additional Evidence:
Wreck was first located in 1939 (UKHO).

Phase 1 Conclusion:
UNKNOWN. This wreck is too long to be KIRKBY.

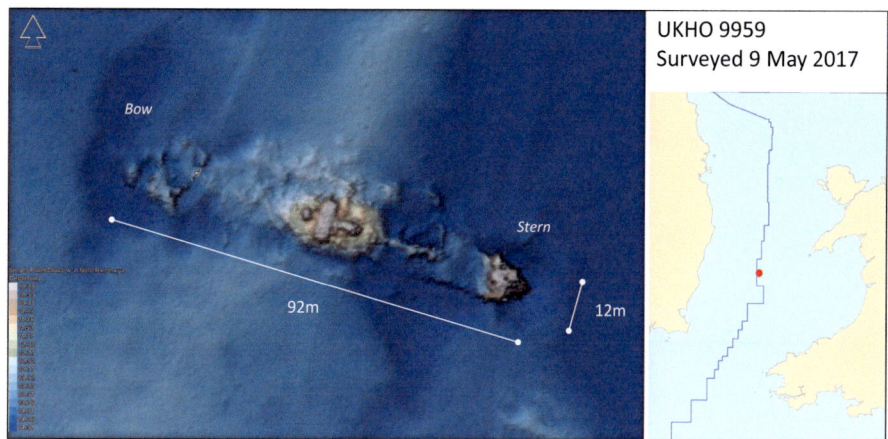

Bow

Stern

92m

12m

9959 KIRKBY (MBES/Archive)

Original UKHO: 9959 GALATEA (POSSIBLY)	**Phase 1 Analysis:** UNKNOWN
Position WGS84: 52 38.619 N, 5 26.439 W	**Water Depth:** 68m
Type of Vessel: SS Cargo	**Surveyed Length:** 92m (approx)
Date of Loss: 17 Aug 1915	**Circumstances:** Torpedoed by U38

Phase 2: Method of wreck identification:

Positional Analysis:
This wreck plots 2.8nm south of the position given in the master's account (TNA ADM 137/2959). The U-boat position cannot be discerned from its KTB.

Dimensional Analysis:
KIRKBY was 96.01m x 12.31m (*Lloyd's Register* 1915-16).

Archival Analysis:
KIRKBY was carrying coal from Barry to an undisclosed location. The U-boat was spotted on the port beam and KIRKBY turned away but could not escape. Boats were lowered and U38 then struck the ship with a torpedo on the fore part of the beam on the port side. The ship sank by the head (TNA ADM 137/2959).

Phase 2 Conclusion:
KIRKBY (MBES/Archive) Positionally, dimensionally, and archivally, this wreck is consistent with KIRKBY.

Phase 1: UKHO Analysis:

Original UKHO Name:
9959 GALATEA (POSSIBLY)

Archival details:
GALATEA was 69.6m x 10.7m (*Lloyd's Register* 1943-44).

MBES Analysis:
Steamship with twin boilers and a compound engine. Length of approx. 92m. The stern is the highest point.

Additional Evidence:
First surveyed in 1980 (UKHO).

Phase 1 Conclusion:
UNKNOWN. This wreck is too long to be GALATEA.

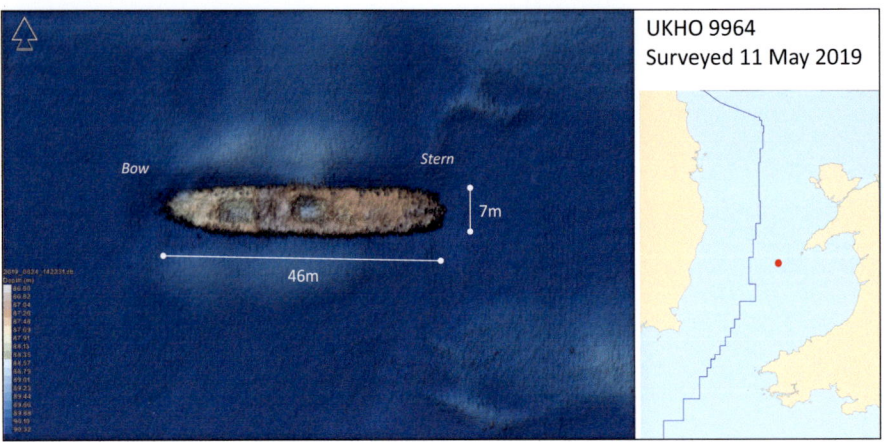

Bow

Stern

7m

46m

9964 PULTENEY (MBES/Archive)

Original UKHO: 9964 UNKNOWN	**Phase 1 Analysis:** UNKNOWN
Position WGS84: 52 41.469 N, 5 6.598 W	**Water Depth:** 89m
Type of Vessel: SS Cargo	**Surveyed Length:** 46m
Date of Loss: 18 Aug 1934	**Circumstances:** Collision with SS THELMA

Phase 2: Method of wreck identification:

Positional Analysis:
Only recorded as lost off Bardsey.

Dimensional Analysis:
PULTENEY was 44.24m x 7.16m, well-decked with machinery aft (*Lloyd's Register* 1934).

Archival Analysis:
PULTENEY sank following collision with SS THELMA in fog "*off Bardsey*" carrying a cargo of stone (*The Times* 21 Aug 1934)

Phase 2 Conclusion:
PULTENEY (MBES/Archive) This wreck is dimensionally and archivally consistent with PULTENEY. Positionally it is "*off Bardsey*" and is not (at the time of assessment, Oct 2019) easily reconcilable to any other local event. This attribution is therefore on the thinnest of evidence.

Phase 1: UKHO Analysis:

Original UKHO Name:
9964 UNKNOWN

Archival details:
N/A

MBES Analysis:
Upright coaster with two holds forward of superstructure. 46m x 7m.

Additional Evidence:
Located and attacked in 1942 then declared a non-sub contact (UKHO).

Phase 1 Conclusion:
UNKNOWN. The wreck is a small FV/Coaster sunk prior to 1942.

9970 FRANCOISE DE LILLE (MBES/Archive)

Original UKHO: 9970 UNKNOWN	**Phase 1 Analysis:** UNKNOWN
Position WGS84: 52 42.981 N, 5 25.507 W	**Water Depth:** 74m
Type of Vessel: MFV	**Surveyed Length:** 40m
Date of Loss: 20 Dec 1963	**Circumstances:** Foundered

Phase 2: Method of wreck identification:

Positional Analysis:
This wreck plots 5.2nm north of the position reported by UKHO (Wreck No 9955).

Dimensional Analysis:
FRANCOISE DE LILLE's dimensions have not been established. Diesel engine, single shaft (wrecksite.eu).

Archival Analysis:
FRANCOISE DE LILLE foundered in 1963 with eight crew rescued (UKHO).

Phase 2 Conclusion:
FRANCOISE DE LILLE (MBES/Archive). The MBES Scan reveals an intact vessel which is likely therefore to be more modern than old. This wreck is positionally close to the reported sinking position. It is of dimensions synonymous with an MFV. None of this evidence is conclusive though. An alternative identity could be PORTHMEOR (sunk 1941), coinciding with the location date of 1945 (UKHO). PORTHMEOR was 43.34m long and sunk in collision. The position of this event could not be discerned during the archival research.

Phase 1: UKHO Analysis:

Original UKHO Name:
9970 UNKNOWN

Archival details:
N/A

MBES Analysis:
Small wreck pointing south. 41m long with a distinctive tubular feature running along the central area.

Additional Evidence:
Located in May 1945 (UKHO).

Phase 1 Conclusion:
UNKNOWN

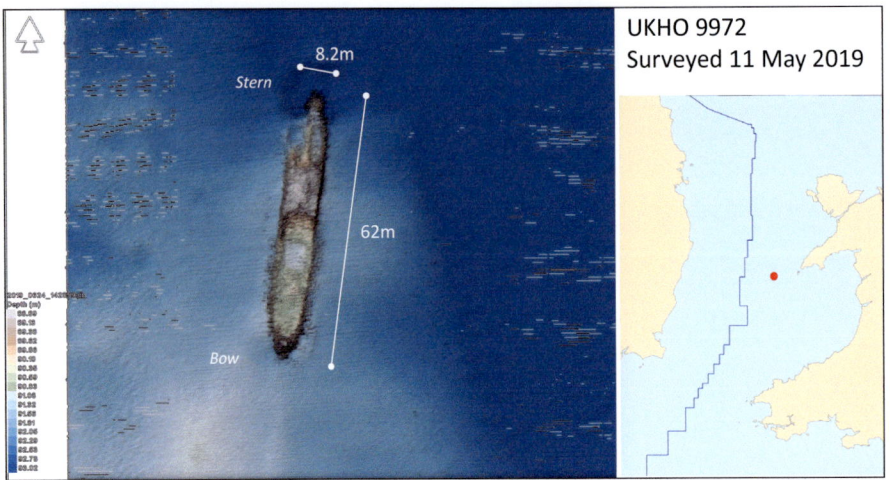

9972 KORSNAES (MBES/Archive)

Original UKHO: 9972 UNKNOWN	**Phase 1 Analysis:** UNKNOWN
Position WGS84: 52 43.386 N, 5 1.232 W	**Water Depth:** 92m
Type of Vessel: SS Cargo	**Surveyed Length:** 62m
Date of Loss: 24 Mar 1917	**Circumstances:** Explosives from UC65

Phase 2: Method of wreck identification:

Positional Analysis:
This wreck plots 5nm west of the position given in *War Losses* (1990). The position in UC65's KTB cannot be accurately plotted.

Dimensional Analysis:
KORSNAES (ex-TAIFUN) was 61.30m x 8.60m (*Lloyd's Register* 1915-16).

Archival Analysis:
KORSNAES turned to the south to outrun the attacking U-boat, without success. Five bombs were placed in the ship, two in the after hold. The ship sank stern first in about one minute (TNA ADM 137/1308).

Phase 2 Conclusion:
KORSNAES (MBES/Archive). Positionally, dimensionally, and archivally this wreck is consistent with KORSNAES.

Phase 1: UKHO Analysis:

Original UKHO Name:
9972 UNKNOWN

Archival details:
N/A

MBES Analysis:
Upright SS or MV pointing south. Two forward holds and possibly one aft. 62m x 8.2m.

Additional Evidence:
Located by Risdon Beazley in 1976 but apparently not identified (UKHO).

Phase 1 Conclusion:
UNKNOWN

9975 RFA OSAGE (MBES/ARCHIVE)

Original UKHO: 9975 ROSE MARIE (PROBABLY)	**Phase 1 Analysis:** UNKNOWN
Position WGS84: 52 43.182 N, 5 25.696 W	**Water Depth:** 75m
Type of Vessel: RFA Motor Tanker	**Surveyed Length:** 62m
Date of Loss: 18 Dec 1940	**Circumstances:** Bombed

Phase 2: Method of wreck identification:

Positional Analysis:
The wreck lies 20nm from the position given in *War Losses* (1989) which is 4nm NE of the Arklow LV. According to the master, there was visual contact with the Arklow LV shortly before the attack on the ship. This occurred at around 0800. The ship did not sink until 1100, and no sinking position is reported. It is noted also that the ship lost steering as soon as the attacking plane opened fire (TNA ADM 199/2135).

Dimensional Analysis:
RFA OSAGE (ex-MV FEROL) was 61m x 10.4m (*Lloyd's Register* 1939-40). Twin shift, oil engines. The wreck is dimensionally similar to OSAGE and appears to be a MV with machinery aft, as per OSAGE.

Archival Analysis:
The damage reported by the master states that the 3rd bomb exploded under the starboard quarter. This is consistent with the most obviously damaged area on this wreck (TNA ADM 199/2135).

Phase 2 Conclusion:
RFA OSAGE (MBES/ARCHIVE). The damage to the vessel and its dimensions and design are consistent with OSAGE. There are no other MVs of similar dimensions in the archival list in this area. The positional discrepancy can only be explained by the timeline of the sinking. Under most circumstances the positional difference would rule out this identification. However, in this case the wreck's dimensions and rarity, coupled with the confused nature of the sinking description permit this attribution to be made, but with the caveat that the positional distance is unexplained.

Phase 1: UKHO Analysis:

Original UKHO Name:
9975 ROSE MARIE (PROBABLY)

Archival details:
ROSE MARIE was 85.34m x 12.31m (Larn & Larn 2002).

MBES Analysis:
Partially collapsed shipwreck pointing south. Appears to have bridge aft. 62m long.

Additional Evidence:
First surveyed in 1981.

Phase 1 Conclusion:
UNKNOWN. Wreck too small to be ROSE MARIE.

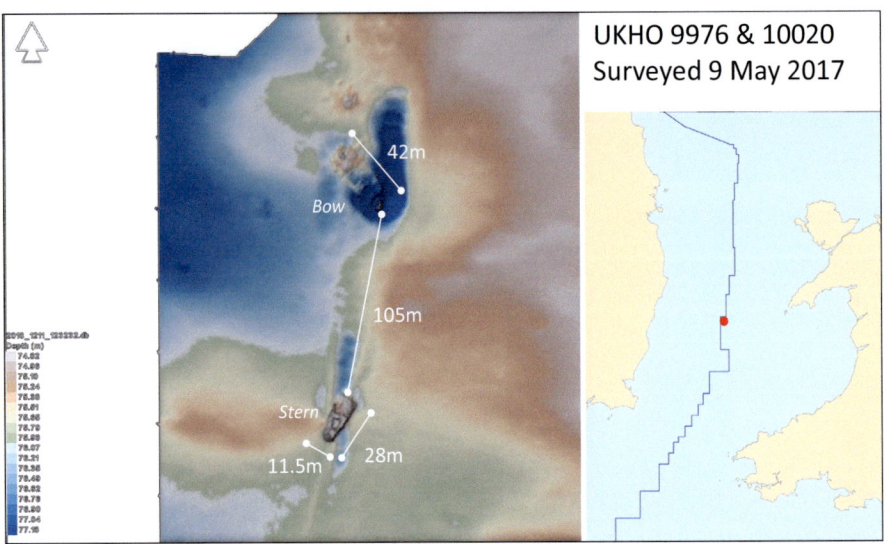

9976 & 10020 EMBLETON (MBES/Archive)

Original UKHO: 9976 BOSCOWAN (PROBABLY)	**Phase 1 Analysis:** UNKNOWN
Position WGS84: **9976**: 52 43.394 N,5 25.547 W **10020**: 52 43.307 N,5 25.547 W	**Water Depth:** 74m
Type of Vessel: SV Barque	**Surveyed Length:** 70m (approx)
Date of Loss: 21 Jul 1900	**Circumstances:** Collision with SS CAMPANIA

Phase 2: Method of wreck identification:

Positional Analysis:
This wreck plots 15.6nm NNW of the position given in *Shipping Casualties* (BOT 1900-1901 Appendix C Table1 p142), being 27nm NE of the Tuskar Rock. It should be noted this wreck lies on the course of SS CAMPANIA, the ship which collided with EMBLETON.

Dimensional Analysis:
EMBLETON was 69.00m x11.00m (*Lloyd's Register* 1883)

Archival Analysis:
The liner CAMPANIA collided with EMBLETON in fog, killing 11. EMBLETON was cut is half just aft of the mainmast, sinking immediately (*New York Times* 23 Jul 1900).

Phase 2 Conclusion:
EMBLETON (MBES/Archive). This wreck is archivally and dimensionally consistent with EMBLETON. It lies across SS CAMPANIA's course. Of note is the damage to the wreck, consistent with the report of the sinking and the seeming absence of boilers showing in the MBES scan.

Phase 1: UKHO Analysis:

Original UKHO Name:
9976 BOSCOWAN (sic) (PROBABLY)

Archival details:
BOSCAWEN was 85m long (Larn & Larn 2002). 2.7nm from *War Losses* (1990) position and 6.6nm from U-boat (KTB position). Ship was torpedoed between the forecastle and the fore hold (TNA ADM 199/2964).

MBES Analysis:
Shipwreck in two halves 105m apart with 10020 being to the south. Two halves of the wreck total approx. 70m in length, which due to the very broken nature of this site may well not be all of it.

Additional Evidence:
First surveyed in 1978 (UKHO).

Phase 1 Conclusion:
UNKNOWN. Damage to wreck does not match sinking report. Wreck looks to be too small for BOSCOWAN.

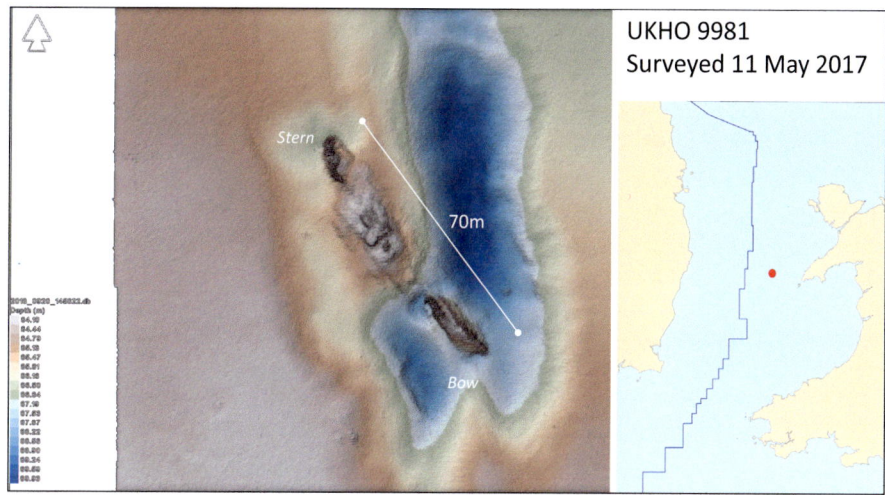

9981 KNUT (MBES/Archive)

Original UKHO: 9981 PERSEUS (PROBABLY)	**Phase 1 Analysis:** UNKNOWN
Position WGS84: 52 49.868 N, 5 5.065 W	**Water Depth:** 85m
Type of Vessel: SS Coaster	**Surveyed Length:** 70m (approx)
Date of Loss: 23 Dec 1942	**Circumstances:** Mine

Phase 2: Method of wreck identification:

Positional Analysis:
The wreck lies 3.5nm SW of the position given in the Official Danish list of shipping losses in WW2 (Ministeriet for Handel, Industri og Sofort 1950, p80-81). The position given in *War Losses* (1989) is obviously incorrect, being inside Cardigan Bay, where a vessel voyaging from Belfast to Mumbles (As KNUT was) would not be found.

Dimensional Analysis:
KNUT was 72.5m x 11.3m. This is consistent with the dimensions of this wreck.

Archival Analysis:
The ship was travelling south when it struck a mine which exploded on the starboard side next to No1 hold. The ship rapidly sank by the bow (Ministeriet for Handel, Industri og Sofort 1950, p80-81).

Phase 2 Conclusion:
KNUT MBES/Archive. Positionally, dimensionally, and archivally this wreck is consistent with KNUT.

Phase 1: UKHO Analysis:

Original UKHO Name:
9981 PERSEUS (PROBABLY)

Archival details:
Wreck is 5.3nm from *War Losses* position (1989). PERSEUS was 66m long (*Lloyd's Register* 1940-41). Master's report states ship was torpedoed by aircraft aft, blowing off the stern (TNA ADM 199/2136).

MBES Analysis:
Coaster of approx. 70m long. Bow and stern lie on port side. Central portion is upright. Two boilers and a probable engine present on DTM. Points to the SE. Possible that bow section has broken away from wreck due to scouring effect.

Additional Evidence:
First surveyed in 1982 (UKHO).

Phase 1 Conclusion:
UNKNOWN. The presence of the stern on this wreck site eliminates it as a candidate for PERSEUS.

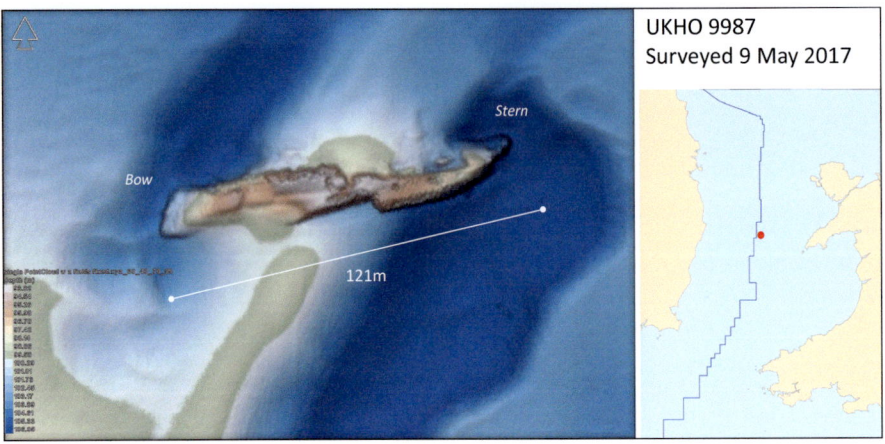

9987 SNOWDON RANGE (MBES/Archive)

Original UKHO: 9987 AGBERI (POSSIBLY)	**Phase 1 Analysis:** UNKNOWN
Position WGS84: 52 53.151 N, 5 18.746 W	**Water Depth:** 110m
Type of Vessel: SS Cargo	**Surveyed Length:** 121m (approx)
Date of Loss: 28 Mar 1917	**Circumstances:** Torpedo and Explosives by UC65

Phase 2: Method of wreck identification:

Positional Analysis:
The wreck plots 3.5nm SW of the position given on UC65's KTB chart and 10.2nm north of the vague position given by the master of 20-25nm west of Bardsey Island (TNA ADM 137/1308).

Dimensional Analysis:
SNOWDON RANGE was 118.87m x 15.90m (*Lloyd's Register* 1915-16).

Archival Analysis:
Philadelphia to Liverpool with a cargo of ammunition and grain. The ship was initially torpedoed on the port side of the engine room. The crew abandoned with four drowning. The U-boat used the survivors' boat to place bombs in three holds and ransack the ship for charts, the gun breech, projectiles etc. The Admiralty noted the codebooks had been left on board because the cabinet was damaged and could not be opened (TNA ADM 137/1308).

Phase 2 Conclusion:
SNOWDON RANGE (MBES/Archive). This wreck is positionally, archivally, and dimensionally consistent with SNOWDON RANGE.

Phase 1: UKHO Analysis:

Original UKHO Name:
9987 AGBERI (POSSIBLY)

Archival details:
AGBERI was 112.85m x 15.01m (Larn & Larn 2000). Enroute from Dakar for Liverpool (*War Losses* 1990, p191). U87 was sunk in the counterattack after it torpedoed AGBERI and has been identified 8.7nm away. The position reported for the loss of AGBERI was north of Bardsey Island, 14nm NE of this wreck and would lie within torpedo range of U87.

MBES Analysis:
MBES seems to show steamship broken in two amidships. Overall length is approx. 121m long and at least 14m wide.

Additional Evidence:
Located in 1980 (UKHO).

Phase 1 Conclusion:
UNKNOWN. This wreck cannot be AGBERI because it is too far from the recorded sinking position and the wreck of U87. The wreck also appears to be too long.

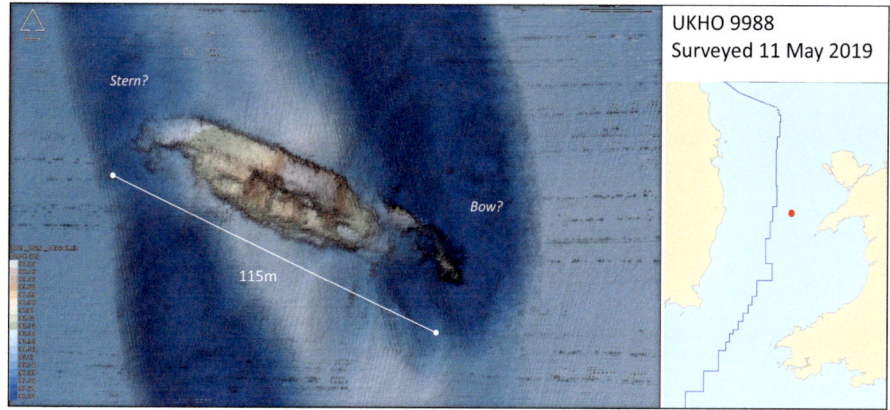

UKHO 9988
Surveyed 11 May 2019

9988 AGBERI (MBES/Archive)

Original UKHO: 9988 UNKNOWN		**Phase 1 Analysis:** UNKNOWN
Position WGS84: 52 55.768 N, 5 4.581 W		**Water Depth:** 94m
Type of Vessel: SS Liner		**Surveyed Length:** 115m (approx)
Date of Loss: 25 Dec 1917		**Circumstances:** Torpedoed by U87

Phase 2: Method of wreck identification:

Positional Analysis:
This wreck is 6.2nm from the position given in *War Losses* (1990). U87 was peremptorily destroyed after sinking this ship, so there is no KTB position. The wreck actually lies 1nm SE of the wreck of U87 and therefore seems to be most likely associated with the sinking of that U-boat.

Dimensional Analysis:
AGBERI was 112.85m x 15.01m (*Lloyd's Register* 1915-16).

Archival Analysis:
Dakar for Liverpool in convoy. Torpedoed amidships while in convoy. Struck port side at 1425, destroying the boats amidships. The crew and passengers evacuated in the other boats. Sank in 25 minutes at 1510. One injury. (TNA ADM 137/2963). U87 was sunk by HMS P56 which sighted U87's periscope at 1510 and dropped two depth charges, which brought the sub to the surface, which was then rammed, slicing off the stern. Both P56 and escort BUTTERCUP fired at the sub as it sank (TNA ADM 137/4147).

Phase 2 Conclusion:
AGBERI (MBES/Archive). Dimensionally, positionally, and archivally, this wreck is consistent with AGBERI.

Phase 1: UKHO Analysis:

Original UKHO Name:
9988 UNKNOWN

Archival details:
N/A

MBES Analysis:
Large wreck probably lying on its starboard side having slumped to the south and broken into the large scours around it. Length is approx. 115m.

Additional Evidence:
Located by Risdon Beazley in 1957, identity not given on wreck card (UKHO).

Phase 1 Conclusion:
UNKNOWN

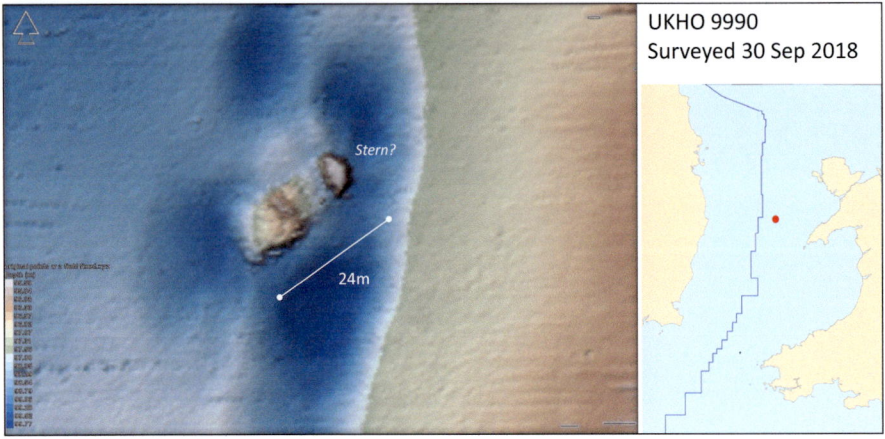

Stern?

24m

9990 DERELICT (Half of 9994) (MBES/Archive)

Original UKHO: 9990 UNKNOWN	**Phase 1 Analysis:** UNKNOWN
Position WGS84: 52 56.534 N, 5 9.714 W	**Water Depth:** 102
Type of Vessel: Unknown derelict	**Surveyed Length:** 24m
Date of Loss: 3 Dec 1951	**Circumstances:** Sunk by the Irish Corvette MAEV

Phase 2: Method of wreck identification:

Positional Analysis:
This wreck plots 3.9nm SE of 9994 and is possibly the other half of that wreck.

Dimensional Analysis:
The identity of the derelict is not given in the UKHO wreck card for 9994.

Archival Analysis:
UKHO records a derelict being sunk by the Irish Corvette MAEV at the position of 9994.

Phase 2 Conclusion:
9990 DERELICT (Half of 9994) (MBES/Archive). The wreck is considered likely to be the other half of 9994. Both wrecks share the same 9.4m width.

Phase 1: UKHO Analysis:

Original UKHO Name:
9990 UNKNOWN

Archival details:
N/A

MBES Analysis:
Small wreck, possibly the stern section of a ship 24m long and 9.4m wide.

Additional Evidence:
Located in 1996 and classified as a small wreck (UKHO).

Phase 1 Conclusion:
UNKNOWN

9995 PAROS (MBES/Archive)

Original UKHO: 9995 MEMPHIAN (POSSIBLY)	**Phase 1 Analysis:** UNKNOWN
Position WGS84: 52 49.731 N, 5 35.791 W	**Water Depth:** 66m
Type of Vessel: SS Cargo	**Surveyed Length:** 94m (approx)
Date of Loss: 17 Aug 1915	**Circumstances:** Torpedo and gunfire from U38

Phase 2: Method of wreck identification:

Positional Analysis:
This wreck plots 2.9nm south of the position given in the master's account of the loss (TNA ADM 137/2959). The U-boat position cannot be discerned from the KTB.

Dimensional Analysis:
PAROS was 106.68m x 14.07m (*Lloyd's Register* 1915-16).

Archival Analysis:
Steaming for Manchester (approx NE), the sub was spotted on the port quarter as it opened fire. The ship swung away to place the sub astern and was then torpedoed in the starboard side of the engine room. The survivors made for Bardsey and last saw the ship afloat, but down by the head (TNA ADM 137/1131).

Phase 2 Conclusion:
PAROS (MBES/Archive). The wreck is positionally and archivally consistent with PAROS. The difference in length can only be accounted for by the seemingly collapsed area of the stern to the north of the wreck, which might point to propeller salvage. This attribution is not certain. An alternative would be ADENWEN (initially discounted because that wreck is expected to be pointing approx. NE) or possibly GRELDON.

Phase 1: UKHO Analysis:

Original UKHO Name:
9995 MEMPHIAN (POSSIBLY)

Archival details:
MEMPHIAN was torpedoed in 1917 while bound for the USA (Larn & Larn 2001). The *War Losses* (1990) position for the sinking plots approx. 10nm north of the wreck. The U-boat position (KTB) is similar. The ship was 121.99m x 15.92m (*Lloyd's Register* 1916-17).

MBES Analysis:
Large wreck lying collapsed on starboard side and pointing south. Stern section appears truncated, so the overall length could be up to approx. 108m. Unlikely to be any longer.

Additional Evidence:
Located in 1982 (UKHO).

Phase 1 Conclusion:
UNKNOWN. This wreck appears to be too short to be MEMPHIAN.

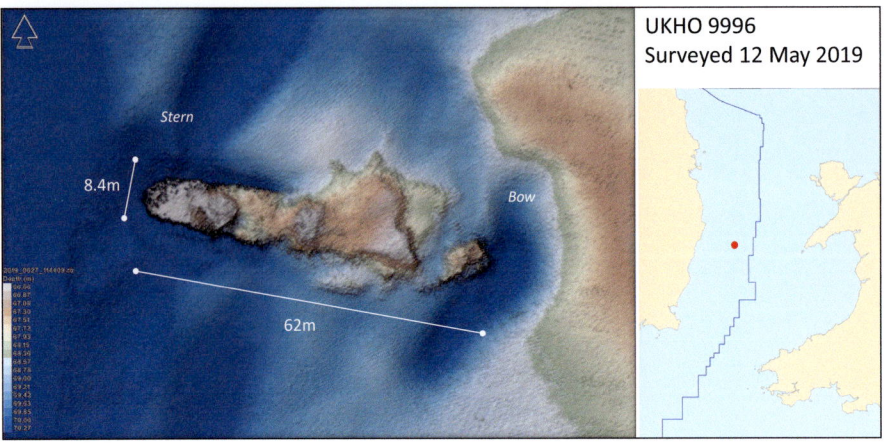

9996 DROMEDARY (MBES/Archive)

Original UKHO: 9996 GRELDON (POSSIBLY)	**Phase 1 Analysis:** UNKNOWN
Position WGS84: 52 50.784 N, 5 35.42 W	**Water Depth:** 67m
Type of Vessel: SS Cargo	**Surveyed Length:** 62m (approx)
Date of Loss: 8 Apr 1873	**Circumstances:** Foundered

Phase 2: Method of wreck identification:

Positional Analysis:
According to Larn & Larn (2002) SS DROMEDARY foundered 5nm off the Arklow lightship when the stern shaft burst. Problematically the reference in *Shipping Casualties* (1873) appears nonsensical and cannot be validated.

Dimensional Analysis:
DROMEDARY was 59.53m x 8.07m (*Lloyd's Register* 1872). The register is stamped with a note saying this ship foundered.

Archival Analysis:
As per positional analysis. The ship was carrying a cargo of coal and foundered without loss of life.

Phase 2 Conclusion:
DROMEDARY (MBES/Archive). By the slimmest of evidence this wreck might be DROMEDARY, based on dimensions and the unverified position in Larn & Larn (2002). An alternative wreck would be DAGALI sunk in 1909.

Phase 1: UKHO Analysis:

Original UKHO Name:
9996 GRELDON (POSSIBLY)

Archival details:
GRELDON was torpedoed in 1917 while bound for Italy (Larn & Larn 2001). The *War Losses* (1990) position for the sinking plots approx. 10nm north of the wreck. The U-boat position (KTB) is similar. The ship was 99.06m x 14.32m (*Lloyd's Register* 1916-17).

MBES Analysis:
Upright wreck with intact stern area and fore part broken up. Approx. 62m x 8.4m.

Additional Evidence:
Located in 1981 (UKHO).

Phase 1 Conclusion:
UNKNOWN. This wreck is too small to be GRELDON.

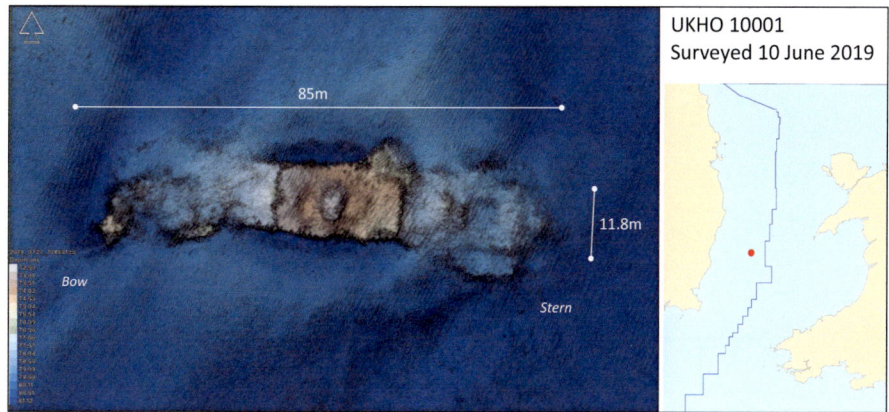

85m

11.8m

Bow

Stern

10001 LADOGA (MBES/Archive)

Original UKHO: 10001 UNKNOWN		**Phase 1 Analysis:** UNKNOWN	
Position WGS84: 52 38.178 N, 5 39.068 W		**Water Depth:** 77m	
Type of Vessel: SS Cargo		**Surveyed Length:** 85m (approx)	
Date of Loss: 16 Apr 1918		**Circumstances:** Torpedoed by UB73	

Phase 2: Method of wreck identification:

Positional Analysis:
This wreck lies 9.5nm SW of the position given in UB73's KTB. The ship was sunk with all hands shortly after its escort ship turned back to base. The Admiralty estimated a position of 40-50nm north of The Smalls (TNA ADM 137/1494).

Dimensional Analysis:
LADOGA was 83.82m x 12.19m (*Lloyd's Register* 1916-17).

Archival Analysis:
Bilbao for Maryport with a cargo of iron ore. (TNA ADM 137/1494). The U-boat KTB states the ship was hit in the centre and the boilers exploded, with the ship sinking immediately. The name was not determined. This attack followed on from the one on LODANER. That wreck (10022) lies 17nm north.

Phase 2 Conclusion:
LADOGA (MBES/Archive). Dimensionally and archivally this wreck is consistent with LADOGA. The positional data from archive suggest the ship must be in this area. This is the best match, but this case is not certain. The best alternative would be 9941, but this case is favoured due to proximity to LODANER (10022), with 9941 being unfeasibly far away.

Phase 1: UKHO Analysis:

Original UKHO Name:
10001 UNKNOWN

Archival details:
N/A

MBES Analysis:
Upright cargo vessel with collapsed holds fore and aft, machinery amidships. 85m x 11.8m.

Additional Evidence:
Located in 1980 (UKHO).

Phase 1 Conclusion:
UNKNOWN

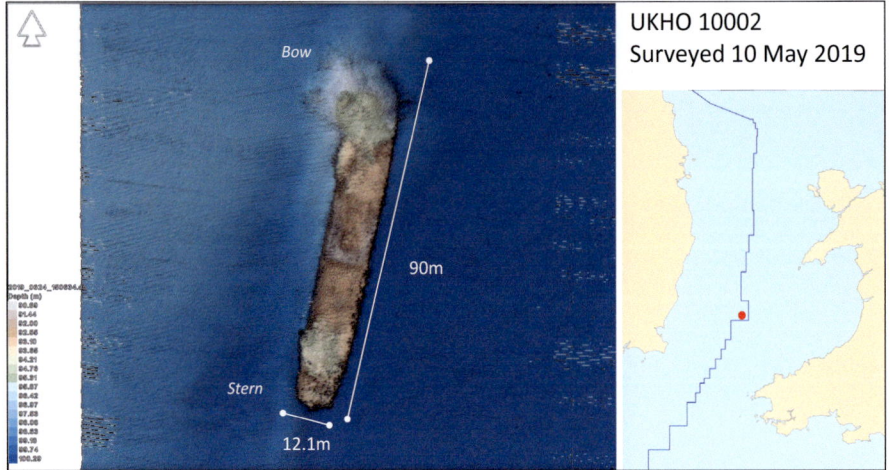

10002 POSEIDON (MBES/Archive)

Original UKHO: 10002 UNKNOWN		**Phase 1 Analysis:** UNKNOWN	
Position WGS84: 52 25.25 N, 5 27.117 W		**Water Depth:** 97m	
Type of Vessel: SS Cargo		**Surveyed Length:** 90m	
Date of Loss: 25 Mar 1917		**Circumstances:** Explosives from UC65	

Phase 2: Method of wreck identification:

Positional Analysis:
This wreck plots 9.4nm west of the position given in *War Losses* (1990) and 5.1nm south of the position given on UC65's KTB chart.

Dimensional Analysis:
POSEIDON was 90.52m x 12.21m (*Lloyd's Register* 1915-16).

Archival Analysis:
Les Falaise to Barrow carrying iron ore. Stopped by the U-boat and bombs placed in holds 1 and 3 and in the stokehold (TNA ADM 137/1308).

Phase 2 Conclusion:
POSEIDON (MBES/Archive). This wreck positionally, dimensionally and archivally consistent with POSEIDON.

Phase 1: UKHO Analysis:

Original UKHO Name:
10002 UNKNOWN

Archival details:
N/A

MBES Analysis:
Upright wreck of a cargo ship with bows being the most damaged area. 90m x 12.1m. Holds fore and aft with machinery amidships.

Additional Evidence:
Located in 1980 (UKHO).

Phase 1 Conclusion:
UNKNOWN

Bow

Stern

104m

10003 ANT CASSAR (MBES/Archive)

Original UKHO: 10003 HALBERDIER (POSSIBLY)	**Phase 1 Analysis:** UNKNOWN
Position WGS84: 52 30.783 N, 5 27.583 W	**Water Depth:** 82m
Type of Vessel: SS Cargo	**Surveyed Length:** 104m
Date of Loss: 27 Aug 1918	**Circumstances:** Torpedoed by UB118

Phase 2: Method of wreck identification:

Positional Analysis:
The wreck plots 3.1nm from the position given in the master's report (TNA ADM 137/2964). The position given in South West Approach (TNA ADM 137/1501) was known to be wrong, being 7nm too far to the west. The wreck plots 14.8nm NW of the position given in UB118's KTB.

Dimensional Analysis:
ANT CASSAR was 104.10m x 14.20m (*Lloyd's Register* 1918-19).

Archival Analysis:
Under government service carrying coal from Glasgow to Milford Haven thence onwards. The ship was hit by two torpedoes on the starboard side, under the bridge and then in the engine room. It rolled to starboard and then sank by the bow. All crew evacuated (TNA ADM 137/1501).

Phase 2 Conclusion:
ANT CASSAR (MBES/Archive). This wreck positionally, dimensionally, and archivally consistent with ANT CASSAR.

Phase 1: UKHO Analysis:

Original UKHO Name:
10003 HALBERDIER (POSSIBLY)

Archival details:
HALBERDIER was torpedoed in 1918 (Larn & Larn 2000). The *War Losses* (1990) position plots 19.3nm north of this wreck. However, the U-boat (KTB) position is much nearer at 5nm NW of this wreck. HALBERDIER was 70.10m x 10.66m (*Lloyd's Register* 1917-18).

MBES Analysis:
Very collapsed wreck of a cargo ship which lies over on its starboard side. 104m long. Width is difficult to measure due to slope but measurements in the points suggest it is between 12.5m and 14.5m.

Additional Evidence:
Located in 1980 (UKHO).

Phase 1 Conclusion:
UNKNOWN. This wreck is too long to be HALBERDIER.

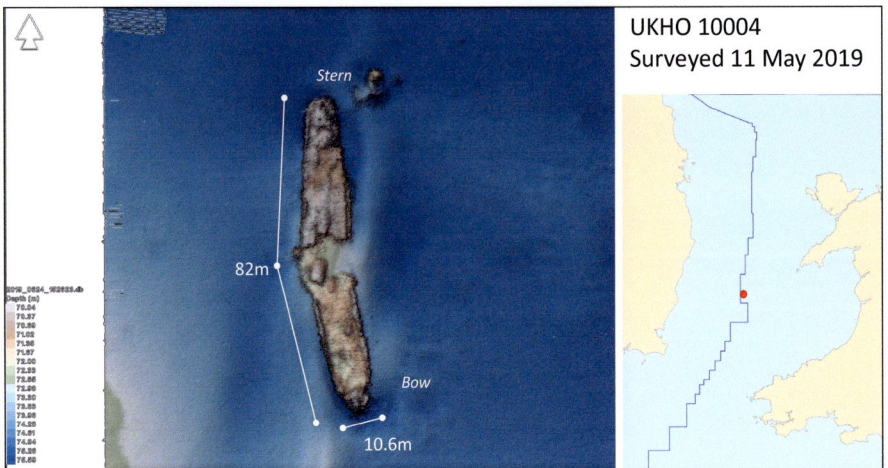

10004 CONINGBEG (MBES/Archive)

Original UKHO: 10004 UNKNOWN	**Phase 1 Analysis:** UNKNOWN
Position WGS84: 52 34.873 N, 5 26.082 W	**Water Depth:** 75m
Type of Vessel: SS Cargo Liner	**Surveyed Length:** 82m
Date of Loss: 17 Dec 1917	**Circumstances:** Torpedoed by U62

Phase 2: Method of wreck identification:

Positional Analysis:
The wreck plots 4.2nm NW of the position given in U62's KTB. There is no accurate Allied position, as the ship sank alone with all hands.

Dimensional Analysis:
CONINGBEG was 82.44m x 10.99m (*Lloyd's Register* 1917-18).

Archival Analysis:
Liverpool for Waterford. Disappeared without trace and considered a war loss (TNA ADM 137/1361). The ship was hit amidships and violently exploded, sinking by the head. It was gone within three minutes (KTB). This description closely matches that given in the U-boat commander's later autobiography (Hashagen 1931, p244-253).

Phase 2 Conclusion:
CONINGBEG (MBES/Archive). Positionally, dimensionally, and archivally, this wreck is consistent with CONINGBEG. The damage pattern on the ship is of particular note in this case.

Phase 1: UKHO Analysis:

Original UKHO Name:
10004 UNKNOWN

Archival details:
N/A

MBES Analysis:
Upright steamship wreck, broken amidships, prominent boiler and engine showing. Forward holds visible. Stern is the most intact area. 82m x 10.6m.

Additional Evidence:
Located in 19080 (UKHO).

Phase 1 Conclusion:
UNKNOWN

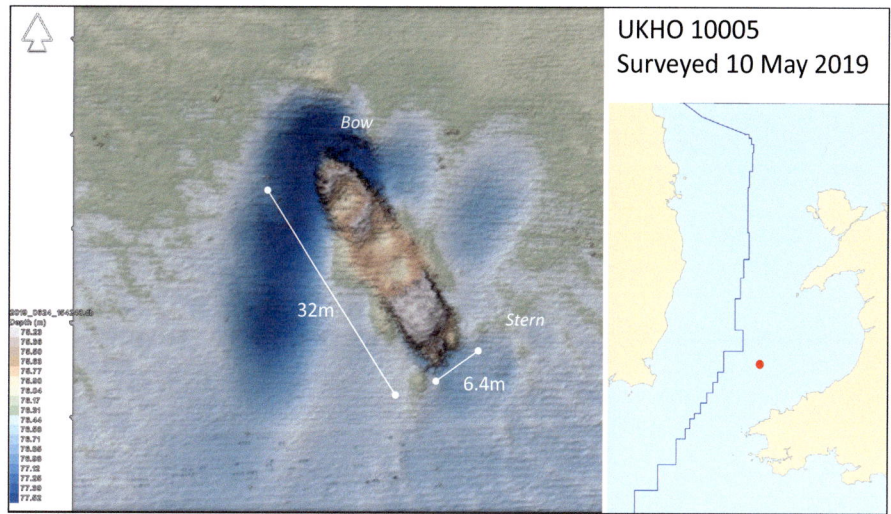

UKHO 10005
Surveyed 10 May 2019

10005 GARTHLOCH (MBES/Archive)

Original UKHO: 10005 UNKNOWN		**Phase 1 Analysis:** UNKNOWN	
Position WGS84: 52 19.755 N, 5 12.598 W		**Water Depth:** 75m	
Type of Vessel: MV Coaster		**Surveyed Length:** 32m	
Date of Loss: 14 Mar 1928		**Circumstances:** Foundered	

Phase 2: Method of wreck identification:

Positional Analysis:
This wreck lies 3.26nm from the reported position (*The Times* 16 Mar 1928).

Dimensional Analysis:
GARTHLOCH (ex-JOSEPH PEASE, ex-SAND DREDGER No1) was 30.93m x 6.14m, twin screw (*Lloyd's Register* 1889). Later converted to oil engine (*Lloyd's Register* 1918-19).

Archival Analysis:
Abandoned at sea with a cargo of flour (*Wreck Returns* 1928, p6). Five men including the master rescued. The vessel had developed engine trouble (*The Times* 16 Mar 1928).

Phase 2 Conclusion:
GARTHLOCH (MBES/Archive). Dimensionally and positionally this wreck is consistent with GARTHLOCH. The stern appears to have possibly been designed for twin propellers, as described in *Lloyd's Register* in the 1889 when the vessel was originally built, later modifications notwithstanding.

Phase 1: UKHO Analysis:

Original UKHO Name:
10005 UNKNOWN

Archival details:
N/A

MBES Analysis:
Upright wreck of a FV or small coaster with aft machinery and pointing NW. 32m x 6.4m.

Additional Evidence:
Located in 1980 (UKHO).

Phase 1 Conclusion:
UNKNOWN

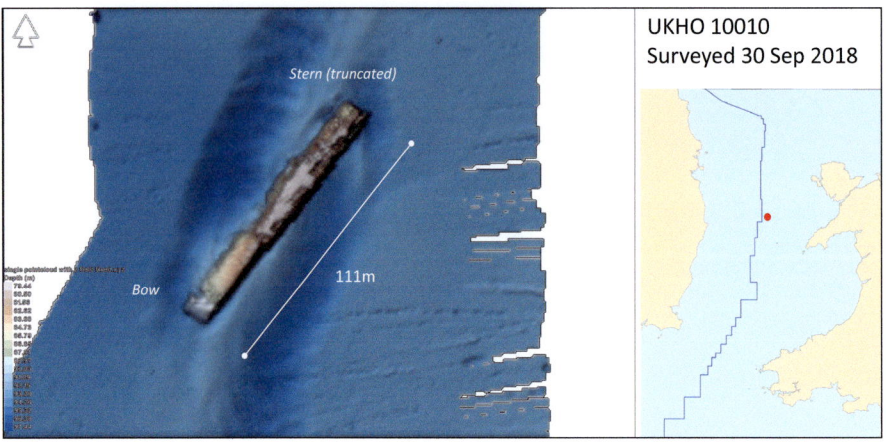

Stern (truncated)

111m

Bow

10010 MEMPHIAN (MBES/Archive)

Original UKHO: 10010 UNKNOWN	**Phase 1 Analysis:** UNKNOWN
Position WGS84: 52 59.801 N, 5 15.497 W	**Water Depth:** 96m
Type of Vessel: SS Cargo	**Surveyed Length:** 111m (not complete)
Date of Loss: 8 Oct 1917	**Circumstances:** Torpedoed by U96

Phase 2: Method of wreck identification:

Positional Analysis:
This wreck plots 17nm east of the position given in *War Losses* (1990) and 14nm east of the position given by the U-boat (KTB). It should be noted that this wreck drifted east for at least six hours before sinking.

Dimensional Analysis:
MEMPHIAN was 121.99m x 15.92m (*Lloyd's Register* 1916-17).

Archival Analysis:
The U-boat (KTB) states that it fired two torpedoes at MEMPHIAN, both of which hit the stern. The ship was still afloat as it went dark but looked certain to sink. Ship torpedoed at 1530 and 1534. The second torpedo blew 30ft of the stern clean away. Ship abandoned by 1530. 1.5nm to the north another ship was then seen to blow up and sink (This was SS GRELDON, torpedoed just prior to MEMPHIAN and sinking after it was torpedoed (KTB)). The survivors used a drogue to remain near MEMPHIAN until 2100, when the drogue snapped off. There was a gale blowing from the west and the survivors then ran with the wind before entering Bardsey Sound at 0400 (TNA ADM 137/1362).

Phase 2 Conclusion:
MEMPHIAN (MBES/Archive). Dimensionally, positionally, and archivally, this wreck is consistent with MEMPHIAN. The gale blew the ship to the east for at least six hours, but possibly much longer. No witnesses saw it sink. This wreck can be seen in the MBES data to be truncated at the stern, consistent with it having been blown off as described.

Phase 1: UKHO Analysis:

Original UKHO Name:
10010 UNKNOWN

Archival details:
N/A

MBES Analysis:
Large shipwreck lying possibly on starboard side, 111m long but looks truncated at the stern end. Not much superstructure showing in pointcloud. Split in hull 40m from bow.

Additional Evidence:
Located in 1981.

Phase 1 Conclusion:
UNKNOWN

10011 ROSE MARIE (MBES/Archive)

Original UKHO: 10011 UNKNOWN	**Phase 1 Analysis:** UNKNOWN
Position WGS84: 52 49.95 N, 5 19.583 W	**Water Depth:** 96m
Type of Vessel: SS Cargo	**Surveyed Length:** 85m (approx)
Date of Loss: 5 Jan 1918	**Circumstances:** Torpedoed by U61

Phase 2: Method of wreck identification:

Positional Analysis:
This wreck plots 4.1nm NE of the position given by U61 (KTB) and 6.7nm east of the master's reported position (TNA ADM 137/1488).

Dimensional Analysis:
ROSE MARIE was 85.34m x 12.20m (*Lloyd's Register* 1917-18).

Archival Analysis:
Stern disintegrated after the torpedo hit, the bow raised up and the ship quickly sank (KTB). Scapa to Barry (Admiralty Collier 1845). Ship was hit port side aft and sank within five minutes with one casualty (TNA ADM 137/1488).

Phase 2 Conclusion:
ROSE MARIE (MBES/Archive). Positionally, dimensionally, and archivally this wreck is consistent with ROSE MARIE. Of note is the heavily collapsed stern.

Phase 1: UKHO Analysis:

Original UKHO Name:
10011 UNKNOWN

Archival details:
N/A

MBES Analysis:
Upright steamship possibly pointing north. Bows broken down. High mid-section. Difficult to measure due to collapse but approx. 85-90m long.

Additional Evidence:
First surveyed in 1982 (UKHO).

Phase 1 Conclusion:
UNKNOWN

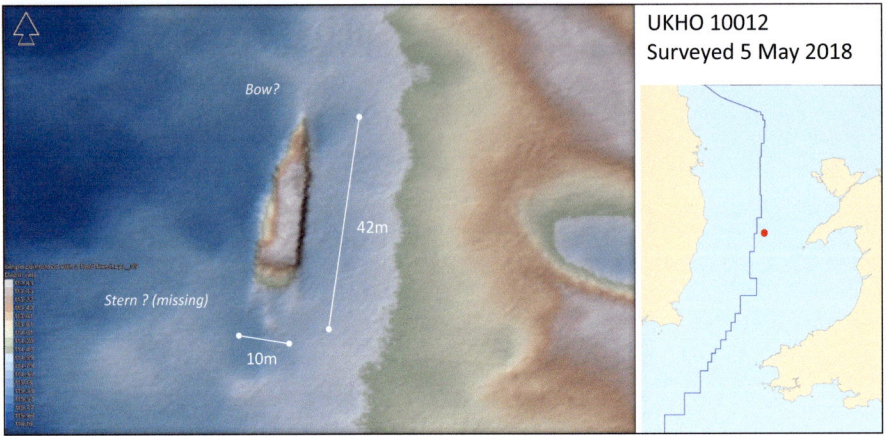

Bow?

42m

Stern ? (missing)

10m

10012 FORTUNA (MBES/Archive)

Original UKHO: 10012 UNKNOWN	**Phase 1 Analysis:** UNKNOWN
Position WGS84: 52 52.618 N, 5 16.847 W	**Water Depth:** 120m
Type of Vessel: SV Cargo	**Surveyed Length:** 42m (incomplete)
Date of Loss: 28 October 1927	**Circumstances:** Explosion & fire

Phase 2: Method of wreck identification:

Positional Analysis:
This wreck plots 10nm south of the position given in *Wreck Returns* (1927 Qtr. 4, p11).

Dimensional Analysis:
FORTUNA was an SV of 80.46m x 8.96m (*Lloyd's Register* 1915-16).

Archival Analysis:
Liverpool for South Georgia with coal & general cargo. Caught fire after an explosion (*Wreck Returns* 1927 Qtr. 4, p11). Ship had encountered gales since leaving Liverpool for South Georgia whaling station. The explosion took place in the cabin, blowing off the poop deck and setting the ship ablaze. The crew tried to fight the fire, but later abandoned the vessel prior to it beginning to capsize. Six killed, 19 survivors (*The Times* 1 Nov 1927).

Phase 2 Conclusion:
FORTUNA (MBES/Archive). This vessel is broadly consistent on dimensions, position, and archive. However, it should be noted that this case is not certain, relying on the after part of the ship being burned out lost or buried. However, the wreck resembles an SV with no discernible machinery present and this wreck is the best candidate found during the research.

Phase 1: UKHO Analysis:

Original UKHO Name:
10012 UNKNOWN

Archival details:
N/A

MBES Analysis:
Small wreck broken down at the southerly end. Difficult to discern bow/stern. Looks like it may be upside-down.

Additional Evidence:
Located in 1981. Described as upright and intact.

Phase 1 Conclusion:
UNKNOWN

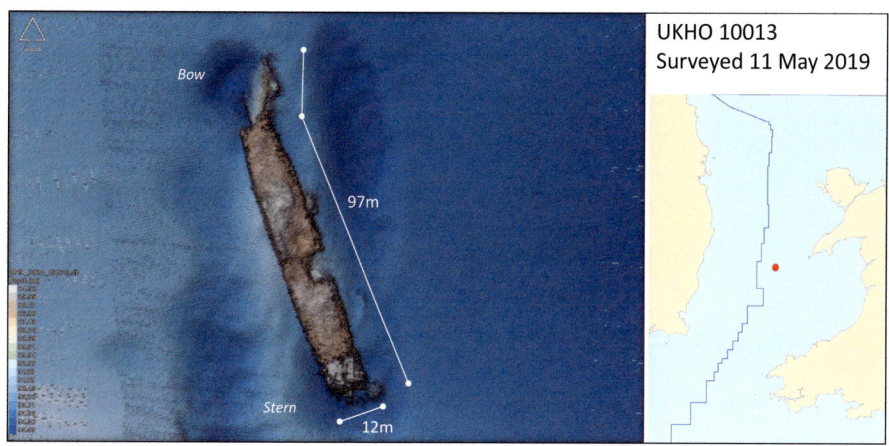

Bow

97m

Stern

12m

10013 RFA VITOL (MBES/Archive)

Original UKHO: 10013 UNKNOWN	**Phase 1 Analysis:** UNKNOWN
Position WGS84: 52 41.184 N, 5 16.063 W	**Water Depth:** 94m
Type of Vessel: RFA Tanker	**Surveyed Length:** 97m (approx)
Date of Loss: 7 Mar 1918	**Circumstances:** Torpedoed by U110

Phase 2: Method of wreck identification:

Positional Analysis:
The reported position (Hepper 2006, p123) plots 5.6nm SE of this wreck. U110 was destroyed on its patrol, so there is no evidence from the German side.

Dimensional Analysis:
RFA VITOL was 97.5m x 12.70m (National Maritime Museum).

Archival Analysis:
RFA VITOL was sunk by U110 on 7 Mar 1918. Published sources say the ship was 104.39m x 12.65m (Puddefoot 2010, p172). In fact, it was shorter. The plans retained at the National Maritime Museum reveal its length was actually 97.5m. This loss is absent from Larn & Larn (2000), *War Losses* (1990), and Admiralty (1947). Survivors stated the ship was initially torpedoed in the starboard side between engine room and boilers. After the vessel was abandoned a 2nd explosion was heard from the lifeboats, but not witnessed (TNA ADM 137/1491).

Phase 2 Conclusion:
RFA VITOL (MBES/Archive). The wreck is positionally, archivally, and dimensionally consistent with RFA VITOL.

Phase 1: UKHO Analysis:

Original UKHO Name:
UNKNOWN

Archival details:
N/A

MBES Analysis:
Wreck probably upright with collapsed bows to north. Noticeable hole amidships. Difficult to measure length accurately. 97m x 12m approx.

Additional Evidence:
Located in 1980 (UKHO).

Phase 1 Conclusion:
UNKNOWN

10014 ARDRI (MBES/Archive)

Original UKHO: 10014 UNKNOWN	**Phase 1 Analysis:** UNKNOWN
Position WGS84: 52 43.503 N, 5 7.181 W	**Water Depth:** 85m
Type of Vessel: SS Cargo	**Surveyed Length:** 52m
Date of Loss: 21 Jan 1936	**Circumstances:** Foundered

Phase 2: Method of wreck identification:

Positional Analysis:
This wreck lies 1.9nm SE of the position given in the clydeships.co.uk database (https://www.clydeships.co.uk/view.php?official_number=&imo=&builder=&builder_eng=&year_built=&launch_after=&launch_before=&role=&propulsion=&category=&owner=&port=&flag=&disposal=&lost=&ref=20967&vessel=CORAL).

Dimensional Analysis:
ARDRI was 50.34m x 7.95m (*Lloyd's Register* 1935-6).

Archival Analysis:
London to Glasgow with a cargo of cement (*Wreck Returns* 1936 Qtr. 1, p5). All crew rescued and landed at Cardiff on 23 Jan (*The Times* 24 Jan 1936).

Phase 2 Conclusion:
ARDRI (MBES/Archive). Positionally, dimensionally, and archivally, this wreck is consistent with ARDRI.

Phase 1: UKHO Analysis:

Original UKHO Name:
10014 UNKNOWN

Archival details:
N/A

MBES Analysis:
Upright wreck of a small coaster with the forward section being collapsed. 52m x 7.6m.

Additional Evidence:
Located in 1981 (UKHO).

Phase 1 Conclusion:
UNKNOWN

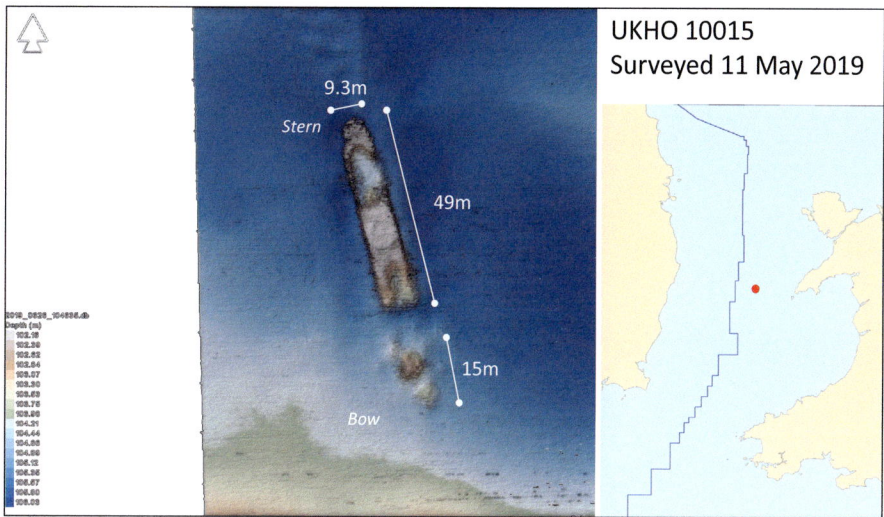

10015 GLYNWEN (MBES/Archive)

Original UKHO: 10015 UNKNOWN	**Phase 1 Analysis:** UNKNOWN
Position WGS84: 52 48.111 N, 5 12.519 W	**Water Depth:** 105m
Type of Vessel: SS Cargo	**Surveyed Length:** 64m (approx)
Date of Loss: 13 Oct 1940	**Circumstances:** Foundered

Phase 2: Method of wreck identification:

Positional Analysis:
This wreck is reported as lost in the Irish Sea enroute from Workington to Devonport, having struck a submerged object (*Wreck Returns* 1940 Qtr. 4, p9).

Dimensional Analysis:
GLYNWEN was 64.01m x 9.75m, well-decked (*Lloyd's Register* 1940-41).

Archival Analysis:
This wreck is reported as lost in the Irish Sea enroute from Workington to Devonport (*Wreck Returns* 1940 Qtr. 4, p9). Due to Irish neutrality its passage, as seen in other cases would have taken it through the eastern side of the study area. It is not listed in *War Losses*, but it is listed in *Wreck Returns* as a marine accident.
The drawing of the ship (LHEC LRF-PUN-W399-0009-P) shows it had one larger stern hold and two forward of the bridge, similar to the DTM. Also, the quarterdeck was 33m long (*Lloyd's Register* 1940-41) and is measured at exactly that length on the DTM.

Phase 2 Conclusion:
GLYWEN (MBES/Archive). Dimensionally, positionally, and archivally, but only the slimmest of acceptable evidence, derived from the ship plans, the wreck is consistent with GLYNWEN. It is in the right area (although vast) and is of the correct dimensions and heading.

Phase 1: UKHO Analysis:

Original UKHO Name:
10015 UNKNOWN

Archival details:
N/A

MBES Analysis:
Upright wreck of a cargo vessel with holds fore and aft of central bridge. Fore part of ship has broken off and collapsed, making accurate length measurement difficult. 64m x 8.3m.

Additional Evidence:
Located in 1980 (UKHO).

Phase 1 Conclusion:
UNKNOWN

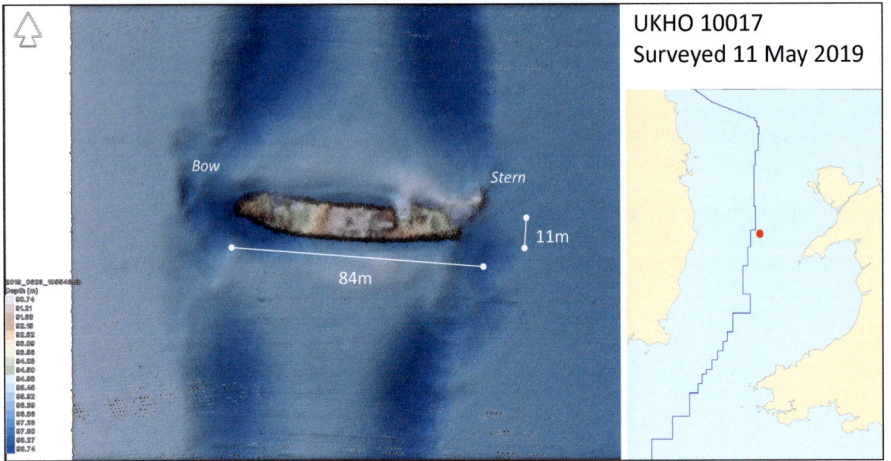

10017 FORMBY (MBES/Archive)

Original UKHO: 10017 UNKNOWN	**Phase 1 Analysis:** UNKNOWN
Position WGS84: 52 55.968 N, 5 16.447 W	**Water Depth:** 97m
Type of Vessel: SS Cargo Liner	**Surveyed Length:** 84m (approx)
Date of Loss: 15 Dec 1917	**Circumstances:** Torpedoed by U62

Phase 2: Method of wreck identification:

Positional Analysis:
This wreck lies 5.6nm NE of the position given by U62 (KTB). The ship was lost with all hands and its sinking position was unknown to the Admiralty.

Dimensional Analysis:
FORMBY was 82.34m x 11.02m (*Lloyd's Register* 1917-18).

Archival Analysis:
Waterford for Liverpool. Disappeared without trace and considered a war loss (TNA ADM 137/1361). The ship was hit in the engine room. It was gone within 3-4 minutes. Target course was estimated as 230deg (KTB).

Phase 2 Conclusion:
FORMBY (MBES/Archive). Positionally, dimensionally, and archivally this wreck is consistent with FORMBY.

Phase 1: UKHO Analysis:

Original UKHO Name:
10017 UNKNOWN

Archival details:
N/A

MBES Analysis:
Upright wreck of a cargo vessel with bows to west. Stern broken away giving an approx. length of 84m and width of 11m.

Additional Evidence:
Located in 1980 (UKHO).

Phase 1 Conclusion:
UNKNOWN

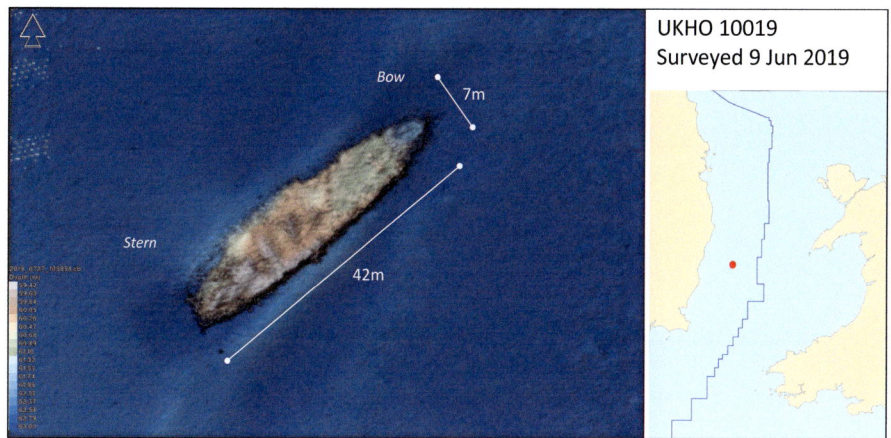

10019 ONYX (MBES/Archive)

Original UKHO: 10019 UNKNOWN	**Phase 1 Analysis:** UNKNOWN
Position WGS84: 52 40.86 N, 5 43.794 W	**Water Depth:** 62m
Type of Vessel: SS Cargo	**Surveyed Length:** 42m
Date of Loss: 4 Nov 1898	**Circumstances:** Foundered

Phase 2: Method of wreck identification:

Positional Analysis:
This wreck lies 4.3nm north of the position given in *Shipping Casualties* (BOT 1898-99, p110) and *Wreck Returns* (1898 Qtr. 4, p6).

Dimensional Analysis:
ONYX was 43.40m x 7.64m (*Lloyd's Register* 1898-99).

Archival Analysis:
Foundered Neath to Dublin with a cargo of Coal (*Wreck Returns* 1898 Qtr. 4, p6).

Phase 2 Conclusion:
ONYX (MBES/Archive). Positionally, dimensionally, and archivally this wreck is consistent with ONYX.

Phase 1: UKHO Analysis:

Original UKHO Name:
10019 UNKNOWN

Archival details:
N/A

MBES Analysis:
Small wreck of a FV or coaster, upright and in one piece, bows point NE. 42m x 7m.

Additional Evidence:
Located in 1980 (UKHO).

Phase 1 Conclusion:
UNKNOWN

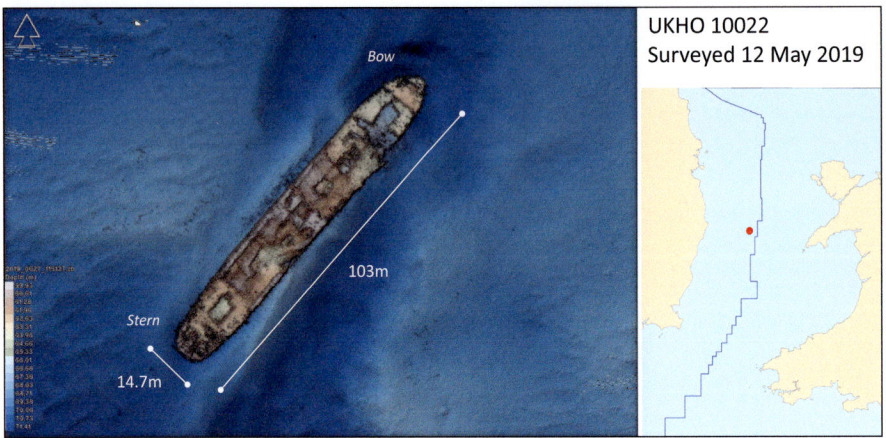

10022 LODANER (MBES/Archive)

Original UKHO: 10022 UNKNOWN	**Phase 1 Analysis:** UNKNOWN
Position WGS84: 52 54.3 N, 5 29.883 W	**Water Depth:** 67m
Type of Vessel: SS Cargo	**Surveyed Length:** 103m
Date of Loss: 16 Apr 1918	**Circumstances:** Torpedoed by UB73

Phase 2: Method of wreck identification:

Positional Analysis:
This wreck plots 6nm west of the position given by UB73 (KTB). There were no survivors, and the Admiralty did not know where this ship had been lost (TNA ADM 137/1493).

Dimensional Analysis:
LODANER was 100.90m x 14.90m (*Lloyd's Register* 1918-19).

Archival Analysis:
UB73 struck the ship with two torpedoes. One hit forward, the other aft. Ship immediately sank so that when UB73 surfaced the ship had already gone. It was noted the ship was painted entirely grey (KTB). Brest for Glasgow. Disappeared without trace (TNA ADM 137/1493). The attribution to this attack sinking LODANER was made by the German official historian (Spindler 1965, p57).

Phase 2 Conclusion:
LODANER (MBES/Archive) Positionally, dimensionally, and archivally this wreck is consistent with LODANER. It should be noted that the course of this vessel is correct for a northward bound zig-zagging ship (as described in the KTB). Also, the bridge of this ship was 45.1m long (*Lloyd's Register* 1918-19), matching the length recorded on the MBES scan.

Phase 1: UKHO Analysis:

Original UKHO Name:
10022 UNKNOWN

Archival details:
N/A

MBES Analysis:
Upright wreck of a cargo vessel with bows pointing to NW. Appears to have three holds forward of central bridge and two holds aft 103m x 14.7m.

Additional Evidence:
Located in 1980 (UKHO).

Phase 1 Conclusion:
UNKNOWN

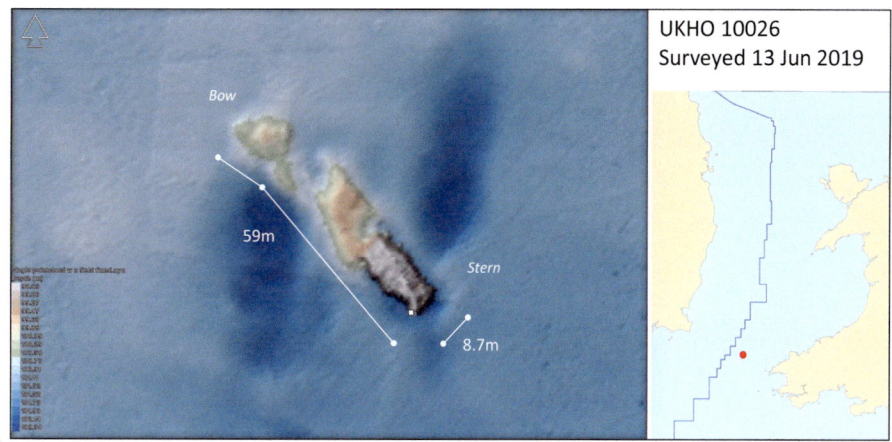

10026 GEORGE BEWLEY (MBES/Archive)

Original UKHO: 10026 UNKNOWN	**Phase 1 Analysis:** UNKNOWN
Position WGS84: 52 0.457 N, 5 38.845 W	**Water Depth:** 101m
Type of Vessel: Iron SV	**Surveyed Length:** 59m (approx)
Date of Loss: 11 May 1884	**Circumstances:** Collision with SS CORMORANT

Phase 2: Method of wreck identification:

Positional Analysis:
"*Off the Tuskar*" (*The Times* 13 May 1884) and (BOT 1883-84 Appendices, p134).

Dimensional Analysis:
GEORGE BEWLEY was 65.70m x 10.40m (*Lloyd's Register* 1883).

Archival Analysis:
Liverpool for Antofagasta. Three killed. (BOT 1883-84 Appendices, p134). Hit deeply into the fore hatch and sank in six minutes (*The Times* 13 May 1884).

Phase 2 Conclusion:
GEORGE BEWLEY (MBES/Archive). Dimensionally and archivally this wreck is consistent with GEORGE BEWLEY. Its actual position cannot be reconciled from the evidence found during the research. The evidence in this case therefore is not conclusive, but the damage to the wreck is consistent with the description given in *The Times* and the wreck is in the right area and dimensionally correct.

Phase 1: UKHO Analysis:

Original UKHO Name:
10026 UNKNOWN

Archival details:
N/A

MBES Analysis:
Wreckage of a broken up shipwreck, approx. 59m x 8.7m.

Additional Evidence:
First surveyed in 1983. Considered broken up at that time (UKHO).

Phase 1 Conclusion:
UNKNOWN

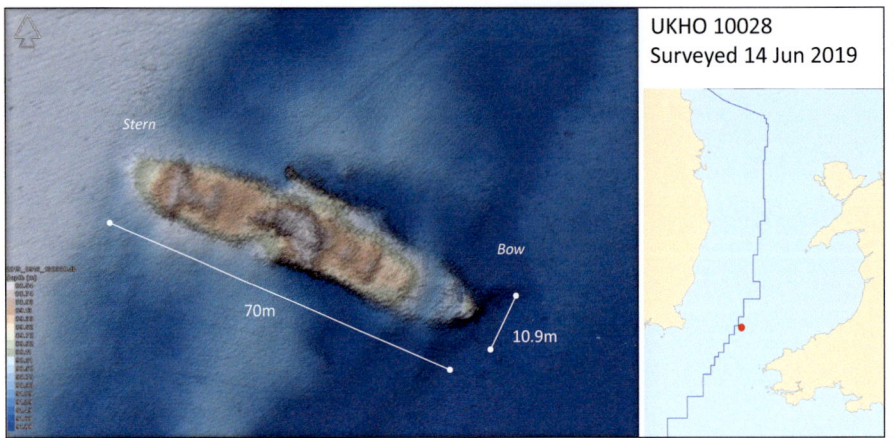

Stern

Bow

70m

10.9m

10028 HAVNA (MBES/Archive)

Original UKHO: 10028 UNKNOWN	**Phase 1 Analysis:** UNKNOWN
Position WGS84: 52 8.028 N, 5 41.741 W	**Water Depth:** 92m
Type of Vessel: SS Cargo	**Surveyed Length:** 70m
Date of Loss: 2 Mar 1918	**Circumstances:** Torpedoed by UB65

Phase 2: Method of wreck identification:

Positional Analysis:
This wreck lies 6.5nm NW of the position given by UB65 (KTB) and 3nm north of the position given in *War Losses* (1990).

Dimensional Analysis:
HAVNA was 69.60m x 11.00m (*Lloyd's Register* 1917-18).

Archival Analysis:
Seville for Maryport carrying iron ore. Ship sank in 30 seconds. Ten killed and seven survivors picked from rafts. Torpedo hit on the starboard side in the after part of the engine room (TNA ADM 137/1491).

Phase 2 Conclusion:
HAVNA (MBES/Archive). Positionally, dimensionally, and archivally this wreck is consistent with HAVNA. Of note is the obvious damage on the wreck, consistent with the account of the sinking.

Phase 1: UKHO Analysis:

Original UKHO Name:
10028 UNKNOWN

Archival details:
N/A

MBES Analysis:
Wreck of a SS with machinery amidships and holds fore and aft. Upright, points SE. Dimensions of 70m x 10.9m.

Additional Evidence:
First surveyed in 1983 (UKHO).

Phase 1 Conclusion:
UNKNOWN

10031 FERGA (MBES/Archive)

Original UKHO: 10031 UNKNOWN		**Phase 1 Analysis:** UNKNOWN	
Position WGS84: 52 26.443 N, 5 3.988 W		**Water Depth:** 69m	
Type of Vessel: SS Cargo		**Surveyed Length:** 61m	
Date of Loss: 14 Feb 1917		**Circumstances:** Gunfire from UC65	

Phase 2: Method of wreck identification:

Positional Analysis:
The wreck plots 9.7nm SE of the position given in *War Losses* (1990) and 10 nm west of the position given by the master (TNA ADM 137/1305). It is not clear why these two recorded positions have such a wide discrepancy . The U-boat position cannot be accurately discerned from the KTB.

Dimensional Analysis:
FERGA was 60.04m x 9.27m (*Lloyd's Register* 1916-17).

Archival Analysis:
Fired on by the submarine. Stopped and evacuated. UC65 then fired eight shots into the engine room and dived in sight of the oncoming HMT NORMAN, which then picked up the survivors (TNA ADM 137/1305).

Phase 2 Conclusion:
FERGA (MBES/Archive). Positionally, dimensionally, and archivally, this wreck is consistent with FERGA. Of note is the layout of the hatches which matches that seen in the ship plans in the project archive.

Phase 1: UKHO Analysis:

Original UKHO Name:
10031 UNKNOWN

Archival details:
N/A

MBES Analysis:
Upright cargo vessel with four holds. Dimensions of 61m x 9m.

Additional Evidence:
Located in 1985 (UKHO).

Phase 1 Conclusion:
UNKNOWN

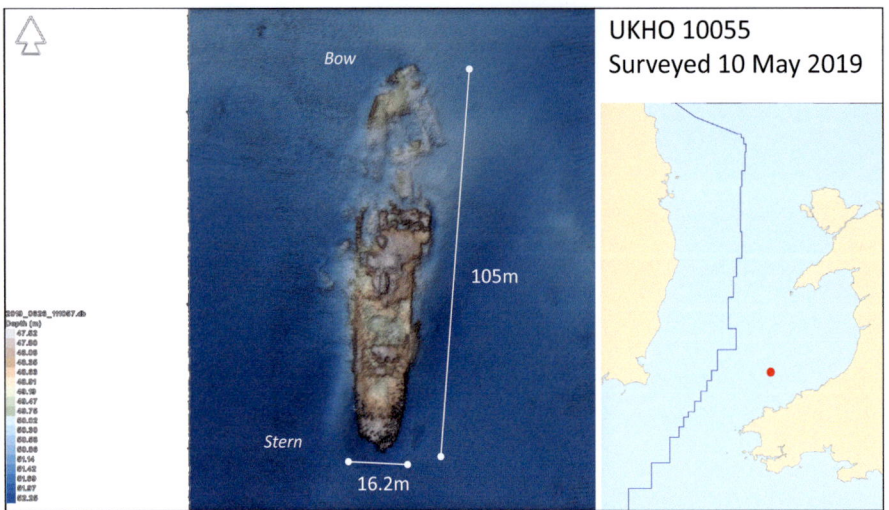

10055 EMPIRE PANTHER (MBES/Archive)

Original UKHO: 10055 UNKNOWN	**Phase 1 Analysis:** UNKNOWN
Position WGS84: 52 14.673 N, 4 58.566 W	**Water Depth:** 47m
Type of Vessel: SS Cargo	**Surveyed Length:** 105m (approx)
Date of Loss: 1 Jan 1943	**Circumstances:** Mined

Phase 2: Method of wreck identification:

Positional Analysis:
The reported positions for this incident vary widely. However, the Admiralty daily diary states a tug was vectored to a position 297deg off Strumble Head to take the ship in tow (TNA ADM 199/2254). UKHO also records the ship finally sinking in this area at a position 13nm SW of this wreck 8nm off Strumble Head.

Dimensional Analysis:
EMPIRE PANTHER (ex-WEST QUECHEE) was 125.00m x 16.60m (*Lloyd's Register* 1941-42).

Archival Analysis:
Struck a mine under No2 hold. The ship settled by the head, folding up between No2 and No4 hatches. The forward well deck was awash by the time the ship was evacuated. It then drifted for at least two hours in a SW force 6 wind before sinking (TNA ADM 199/2144).

Phase 2 Conclusion:
EMPIRE PANTHER (MBES/Archive). The evidence in this case is not conclusive due to positional discrepancies and the dimensions of the wreck being shorter than EMPIRE PANTHER. This is a large wreck, and the candidates are limited, this being the best fit and no more. It is possible for the ship to have drifted NE in the prevailing wind and to have foreshortened as it sank, which would be consistent with the way in which the ship is described as being damaged. The width is correct for EMPIRE PANTHER.

Phase 1: UKHO Analysis:

Original UKHO Name:
10055 UNKNOWN

Archival details:
N/A

MBES Analysis:
Upright cargo steamship wreck with scotch boilers and single engine. Wreck lies N/S with bows to north. Dimensions of 105m x 16.2m.

Additional Evidence:
Located in 1945 (UKHO).

Phase 1 Conclusion:
UNKNOWN

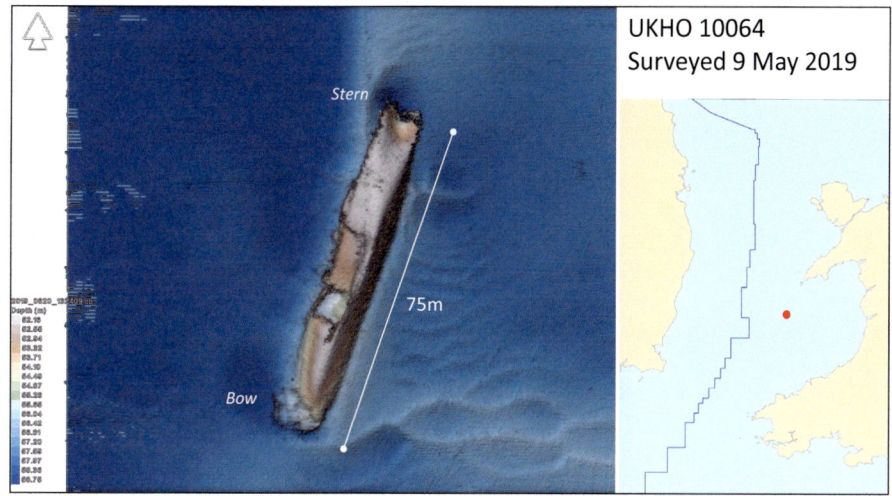

UKHO 10064
Surveyed 9 May 2019

10064 SOMERSET COAST (MBES/Archive)

Original UKHO: 10064 ROBERT EGGLETON (POSSIBLY)	**Phase 1 Analysis:** UNKNOWN
Position WGS84: 52 34.229 N, 4 52.952 W	**Water Depth:** 54m
Type of Vessel: SS Cargo	**Surveyed Length:** 75m
Date of Loss: 21 Apr 1917	**Circumstances:** Collision with SS SOUND FISHER

Phase 2: Method of wreck identification:

Positional Analysis:
This wreck lies 9.6nm south of the position given in *Wreck Returns* (1917 Qtr. 2, p7).

Dimensional Analysis:
SOMERSET COAST was 76.32m x 10.97m (*Lloyd's Register* 1915-16).

Archival Analysis:
Bristol to Liverpool with general cargo, sunk in collision (*Wreck Returns* 1917 Qtr. 2, p7). Hit by SS SOUND FISHER. No lives lost (BOT 1921, p35).

Phase 2 Conclusion:
SOMERSET COAST (MBES/Archive) Positionally, dimensionally, and archivally this wreck is consistent with SOMERSET COAST. Damage could be consistent with collision. However, the orientation of the wreck is problematic. An alternative candidate site could be 10070 UNKNOWN.

Phase 1: UKHO Analysis:

Original UKHO Name:
10064 ROBERT EGGLETON (POSSIBLY)

Archival details:
ROBERT EGGLETON was sunk in Dec 1917 (*War Losses* 1990) and was 88.45m x 11.92m (*Lloyd's Register* 1915-16).

MBES Analysis:
Wreck of a cargo vessel with large hole on port side. Wreck lies on N/S and is on its starboard beam ends, bows to south. Wreck is 75m long. Centre section appears to have collapsed.

Additional Evidence:
Located in 1980 (UKHO).

Phase 1 Conclusion:
UNKNOWN. The wreck is too short in length to be ROBERT EGGLETON.

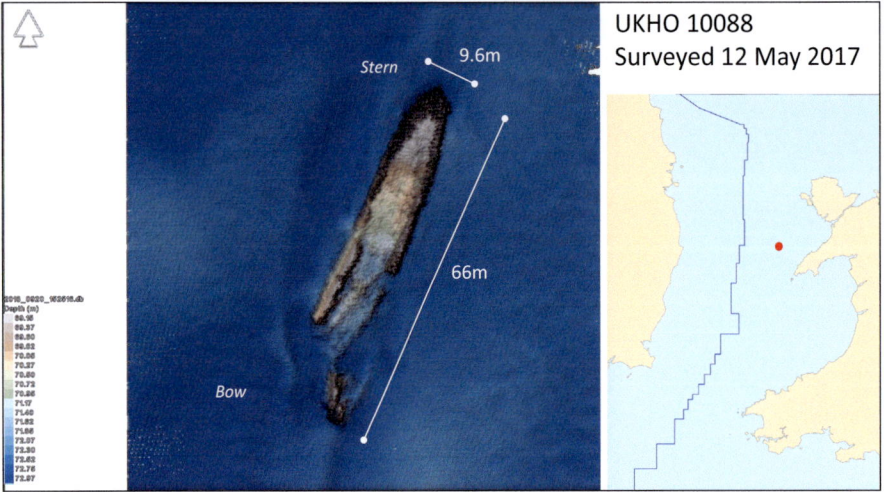

10088 DUKE OF CONNAUGHT (MBES/Archive)

Original UKHO: 10088 TARBETNESS	**Phase 1 Analysis:** UNKNOWN
Position WGS84: 52 57.918 N, 4 55.766 W	**Water Depth:** 71m
Type of Vessel: SV Cargo	**Surveyed Length:** 66m (approx)
Date of Loss: 2 Feb 1887	**Circumstances:** Collision with SS DRAGOMAN

Phase 2: Method of wreck identification:

Positional Analysis:
This wreck lies 5.8nm north of the position given in *Shipping Casualties* (BOT 1886-7 Appendix C Table 1, p147).

Dimensional Analysis:
DUKE OF CONNAUGHT (Ex-CITY OF MADRAS) was 63.70m x 9.80m (*Lloyd's Register* 1883).

Archival Analysis:
London to Maryport. 11 killed (BOT 1886-7 Appendix C Table 1, p147). SS DRAGOMAN hit the DUKE OF CONNAUGHT on the starboard side by the main hatch, cutting eight feet into the hull. The ship sank in three minutes (BOT wreck report No 3129).

Phase 2 Conclusion:
DUKE OF CONNAUGHT (MBES/Archive). Positionally, dimensionally, and archivally this wreck is consistent with DUKE OF CONNAUGHT. Of note is the diver report of an SV (seemingly confirmed by the MBES scan) and the damage consistent with a collision.

Phase 1: UKHO Analysis:

Original UKHO Name:
10088 TARBETNESS

Archival details:
TARBETNESS was 100.73m x 14.63m (Larn & Larn 2000).

MBES Analysis:
Wreck with bows broken off. Points south. Approx. 66m x 9.6m. The decks have collapsed down into the hull, and it appears to have hit the seabed hard, bow first. No evidence in pointcloud of engine or boilers.

Additional Evidence:
Located in 1982. CONFIRMED as TARBETNESS by diver in 1990. Disputed by another diver in 2000 on the basis of wooden decks and much rigging seen (UKHO).

Phase 1 Conclusion:
UNKNOWN. Attribution to TARBETNESS probably due to 1990 diver report, which can now be discounted. Wreck is too small for TARBETNESS. Possibly an SV.

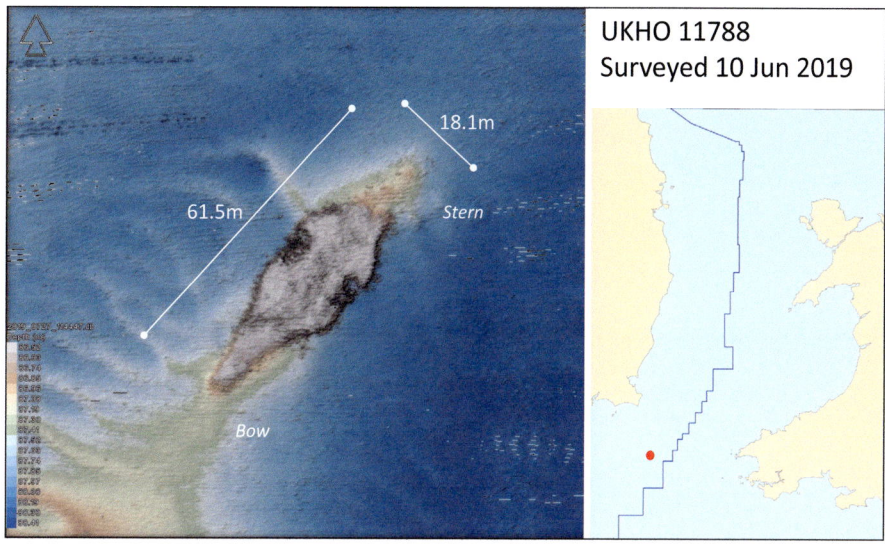

11788 HMS MERCURY (MBES/ARCHIVE)

Original UKHO: 11788 UC33	**Phase 1 Analysis:** UNKNOWN
Position WGS84: 51 53.019 N, 6 13.991 W	**Water Depth:** 84m
Type of Vessel: PSS Ferry converted to Minesweeper	**Surveyed Length:** 61.5m
Date of Loss: 24 Dec 1940	**Circumstances:** Foundered on tow after being mined

Phase 2: Method of wreck identification:

Positional Analysis:
HMS MERCURY foundered on tow after being mined off S Ireland. The position given in the subsequent Court Martial of the commanding officer plots 8nm from this wreck (TNA ADM 156/216).

Dimensional Analysis:
PSS MERCURY was 68m x 9.2m (*Lloyd's Register* 1939-40). The ship was a "boxed in" paddle steamer, so that the width recorded by *Lloyd's Register* was the width of the hull, not including the boxing around the paddle wheels.

Archival Analysis:
HMS MERCURY was the requisitioned PSS MERCURY built in 1934. This is one of only two paddle steamers in the entire MBES database for this research and the only one to be present south of Bardsey.

Phase 2 Conclusion:
HMS MERCURY (MBES/ARCHIVE). This wreck is positionally, dimensionally, and archivally consistent with being HMS MERCURY.

Phase 1: UKHO Analysis:

Original UKHO Name:
11788 UC33

Archival details:
The UCII class minelayer UC33 was rammed and sunk on 26 Sep 1917 in the St Georges Channel (TNA ADM 239/26).

MBES Analysis:
Wreck of what appears to be a paddle-wheeled vessel, probably upside-down. 62m x 17m.

Additional Evidence:
Located in 1978, reported as a submarine in 1982 (UKHO). No reason is given why this wreck is considered identified as UC33 on the UKHO wreck card.

Phase 1 Conclusion:
UNKNOWN. This wreck is not a submarine.

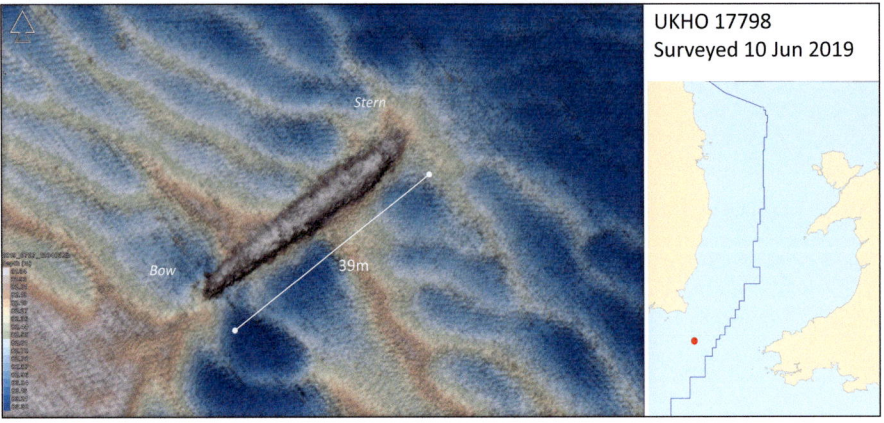

11798 UC33 (MBES/Archive)

Original UKHO: 11798 THORALF (POSSIBLY)		**Phase 1 Analysis:** UNKNOWN	
Position WGS84: 51 56.592 N, 6 14.47 W		**Water Depth:** 79m	
Type of Vessel: Submarine		**Surveyed Length:** 39m	
Date of Loss: 26 Sep 1917		**Circumstances:** Rammed by HMS P61	

Phase 2: Method of wreck identification:

Positional Analysis:
This wreck lies 1.6nm north of the position given by the Admiralty for the destruction of UC33 (TNA ADM 239/26).

Dimensional Analysis:
The dimensions of UC33's pressure hull was 39m x 3.7m. Although 49.35 x 5.22m as built (Gröner 1991), the outer skin appears to have rotted away, as is common with WW1-era U-boats of this type (McCartney 2014).

Archival Analysis:
UC33 was rammed on the port side abaft the conning tower and sank after an explosion. The commanding officer was picked up from the sea, identifying the submarine as UC33 (TNA ADM 239/26)

Phase 2 Conclusion:
UC33 (MBES/Archive). Dimensionally, positionally, and archivally there is little doubt that this is the wreck of UC33.

Phase 1: UKHO Analysis:

Original UKHO Name:
11798 THORALF (POSSIBLY)

Archival details:
THORALF was a timber built sailing vessel blown up by UB65 in 1918 (*War Losses* 1990). It was 75.30m x 10.40m (*Lloyd's Register* 1915-16).

MBES Analysis:
Wreck of submarine lying NE/SW. Pointcloud details show deck gun SW of conning tower and raised fore section.

Additional Evidence:
Located in 1978 (UKHO).

Phase 1 Conclusion:
UNKNOWN. This wreck is clearly not THORALF, because it is too small and too solid to be timber. It is the wreck of a small submarine.

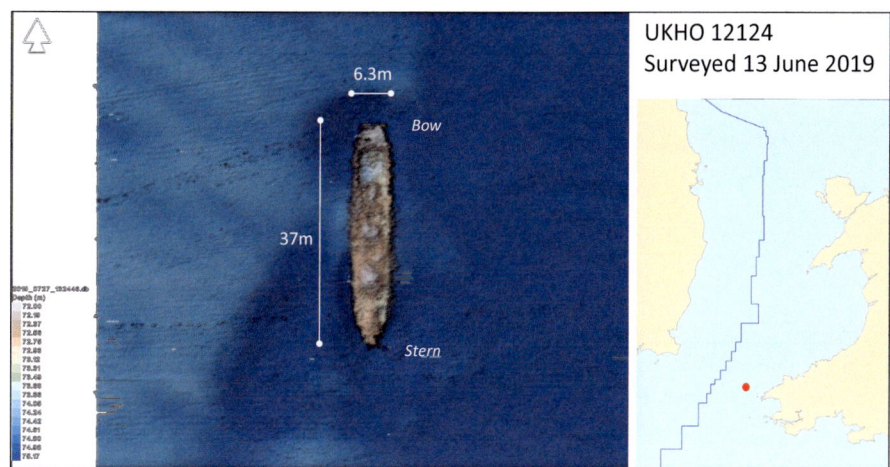

12124 CHATHAM (MBES/Archive)

Original UKHO: 12124 SAPPHIRE (POSSIBLY)	Phase 1 Analysis: UNKNOWN
Position WGS84: 51 57.491 N, 5 30.462 W	Water Depth: 76m
Type of Vessel: Dredger	Surveyed Length: 37m
Date of Loss: 18 Apr 1905	Circumstances: Foundered

Phase 2: Method of wreck identification:

Positional Analysis:
This wreck plots 3.5nm west of the position given in *Shipping Casualties* (BOT 1904-05 Appx C Table 1, p113).

Dimensional Analysis:
131 tons (BOT 1904-05 Appx C Table 1, p113), exact dimensions unknown.

Archival Analysis:
Lost on tow to Liverpool when cable parted. Abandoned by crew and left to sink (BOT 1904-05 Appx C Table 1, p113).

Phase 2 Conclusion:
CHATHAM (MBES/Archive) Positionally and archivally this wreck is consistent with CHATHAM. Of note is the single "hold" or bucket forward; a feature of vessels of this type.

Phase 1: UKHO Analysis:

Original UKHO Name:
12124 SAPPHIRE (POSSIBLY)

Archival details:
SAPPHIRE sank following a collision at a position which plots 5.7nm to the east of this wreck (UKHO). It was 61.01m x 9.90m (*Lloyd's Register* 1938-39).

MBES Analysis:
Upright small wreck with bows to south and an obvious hold in the stern. Wreck is 37m x 6.3m.

Additional Evidence:
Wreck located in 1945 (UKHO).

Phase 1 Conclusion:
UNKNOWN. This wreck is too small to be SAPPHIRE.

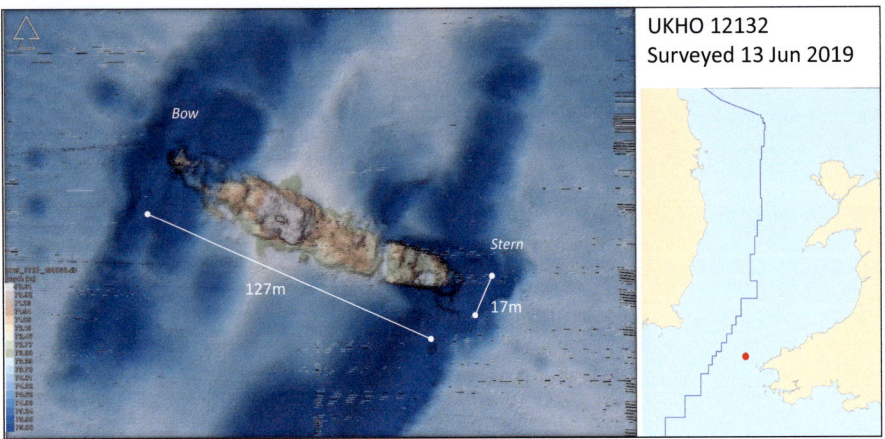

12132 BRYNYMOR (MBES/Archive)

Original UKHO: 12132 UNKNOWN	**Phase 1 Analysis:** UNKNOWN
Position WGS84: 51 57.674 N, 5 30.179 W	**Water Depth:** 76m
Type of Vessel: SS Cargo	**Surveyed Length:** 127m
Date of Loss: 14 Mar 1942	**Circumstances:** Collision with EMPIRE HAWKSBILL

Phase 2: Method of wreck identification:

Positional Analysis:
This wreck plots 2.1nm NW of the position recorded by UKHO (No 12114).

Dimensional Analysis:
BRYNYMOR was 124.96m x 17.06m (*Lloyd's Register* 1941-42).

Archival Analysis:
BRYNYMOR was sunk in collision "*off the Smalls*" after collision in convoy ONSM76 with EMPIRE HAWKSBILL (TNA ADM 199/2237). The position is also recorded as off the Bishops. The ship was in ballast (*Wreck Returns* Qtr. 1 1942, p8).

Phase 2 Conclusion:
BRYNYMORE (MBES/Archive). This wreck is positionally, dimensionally, and archivally consistent with BRYNYMOR.

Phase 1: UKHO Analysis:

Original UKHO Name:
12132 UNKNOWN

Archival details:
N/A

MBES Analysis:
Large shipwreck upright but heavily degraded. Central machinery with holds fore and aft. 127m x 17m.

Additional Evidence:
Wreck located in April 1945 (UKHO).

Phase 1 Conclusion:
UNKNOWN

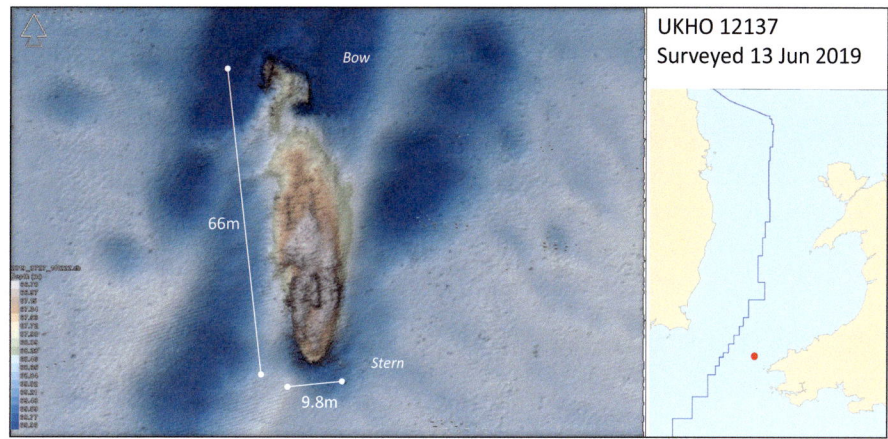

12137 NORFOLK COAST (MBES/Archive)

Original UKHO: 12137 UNKNOWN	**Phase 1 Analysis:** UNKNOWN
Position WGS84: 51 59.607 N, 5 26.963 W	**Water Depth:** 70m
Type of Vessel: MV Cargo	**Surveyed Length:** 66m (approx)
Date of Loss: 28 Feb 1945	**Circumstances:** Torpedoed by U1302

Phase 2: Method of wreck identification:

Positional Analysis:
This wreck plots 2nm NW of the position given in *War Losses* (1989) for the sinking of NORFOLK COAST.

Dimensional Analysis:
NORFOLK COAST was 61.10m x 10.10m (*Lloyd's Register* 1940-41).

Archival Analysis:
Sunk by U1302 (Rohwer 1999, p191). Cardiff for Liverpool. Six survivors picked up (TNA ADM 199/2312).

Phase 2 Conclusion:
NORFOLK COAST (MBES/Archive). This wreck is positionally, dimensionally, and archivally consistent with NORFOLK COAST.

Phase 1: UKHO Analysis:

Original UKHO Name:
12137 UNKNOWN

Archival details:
N/A

MBES Analysis:
Collapsed remains of a coaster/FV with engine to the stern, which points south. Holds collapsed. Approx. 66m x 9.8m.

Additional Evidence:
Located in in 1945 (UKHO).

Phase 1 Conclusion:
UNKNOWN

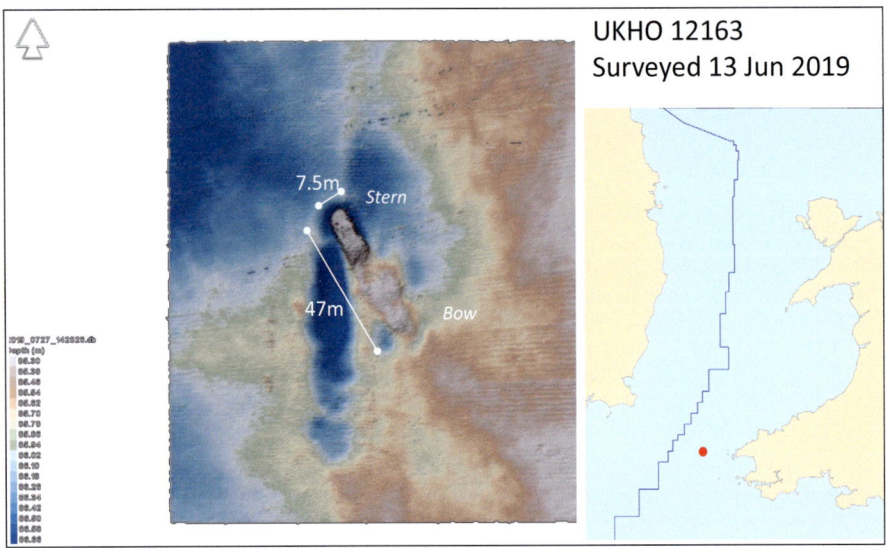

UKHO 12163
Surveyed 13 Jun 2019

7.5m
Stern
47m
Bow

12163 SAINT FINTAN (MBES/Archive)

Original UKHO: 12163 UNKNOWN		**Phase 1 Analysis:** UNKNOWN	
Position WGS84: 51 51.829 N, 5 37.376 W		**Water Depth:** 87m	
Type of Vessel: SS Cargo		**Surveyed Length:** 46m (approx)	
Date of Loss: 22 Mar 1941		**Circumstances:** Bombed	

Phase 2: Method of wreck identification:

Positional Analysis:
This wreck lies 4.5nm east of the position given in *War Losses* (1989).

Dimensional Analysis:
SAINT FINTAN was 43.6m x7.9m with machinery aft, well-decked (*Lloyd's Register* 1939-40).

Archival Analysis:
SAINT FINTAN was sunk by German bombing seven miles north, northwest of The Smalls. The crew of nine were all lost. The Admiralty daily diary records one body being recovered and identified (TNA ADM 199/2225).

Phase 2 Conclusion:
SAINT FINTAN (MBES/Archive). This wreck is positionally, dimensionally, and archivally consistent with SAINT FINTAN.

Phase 1: UKHO Analysis:

Original UKHO Name:
12163 UNKNOWN

Archival details:
N/A

MBES Analysis:
Wreckage of a coaster type vessel. Intact stern to NW. Flattened forward. 47m x 75m.

Additional Evidence:
Located in 1983 (UKHO).

Phase 1 Conclusion:
UNKNOWN

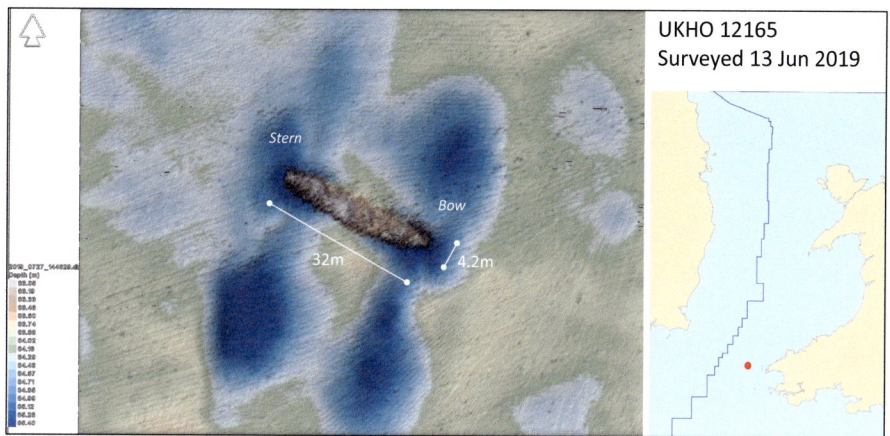

12165 ABERDEEN (MBES/Archive)

Original UKHO: 12165 DAN BEARD (AFTER PART) (POSSIBLY)	**Phase 1 Analysis:** UNKNOWN
Position WGS84: 51 54.958 N, 5 34.395 W	**Water Depth:** 85m
Type of Vessel: MFV	**Surveyed Length:** 32m
Date of Loss: 11 Mar 1941	**Circumstances:** Bombed

Phase 2: Method of wreck identification:

Positional Analysis:
Ship was sunk *"in Cardigan Bay"*.

Dimensional Analysis:
ABERDEEN was 31.75m x 6.22m (*Lloyd's Register* 1940-41).

Archival Analysis:
Two Belgian survivors picked up, eight died. Ship had been bombed and machine-gunned (TNA ADM 199/2225). Ship was not hit but made water rapidly after bombs fell close (*War Losses* 1989).

Phase 2 Conclusion:
ABERDEEN (MBES/Archive). Dimensionally and archivally this wreck is consistent with ABERDEEN. The positional information from archive is vague, so this attribution is made on the slimmest of acceptable evidence and is far from certain.

Phase 1: UKHO Analysis:

Original UKHO Name:
12165 DAN BEARD (AFTER PART) (POSSIBLY)

Archival details:
DAN BEARD was torpedoed in 1944, broke in half, with the fore part drifting ashore (UKHO). The after part sank in a position which plots 15nm NE of this wreck (*War Losses* 1989). It was 128.81m x 17.37m (*Lloyd's Register* 1943-44).

MBES Analysis:
Wreck of a small vessel, possibly a coaster or SV. Leans on its port side and is partially collapsed. 32m x 4.2m.

Additional Evidence:
Located in 1983 (UKHO).

Phase 1 Conclusion:
UNKNOWN. This wreck is clearly not half of DAN BEARD.

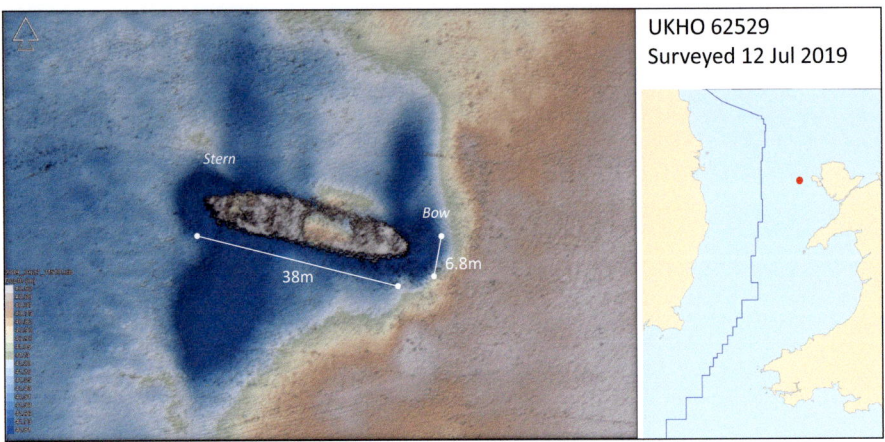

62529 CASTLEDOBBS (MBES/Archive)

Original UKHO: 62529 UNKNOWN	**Phase 1 Analysis:** UNKNOWN
Position WGS84: 53 16.443 N, 4 52.58 W	**Water Depth:** 43m
Type of Vessel: SS Cargo	**Surveyed Length:** 38m
Date of Loss: 7 Dec 1917	**Circumstances:** Collision with HMS P51

Phase 2: Method of wreck identification:

Positional Analysis:
The wreck lies 4.7nm north of the position given in *Shipping Casualties* (BOT 1921, p37).

Dimensional Analysis:
CASTLEDOBBS was 37.80m x 6.70m, well-decked, machinery aft (*Lloyd's Register* 1917-18).

Archival Analysis:
Cardiff to Belfast carrying coal. Collided with HMS P51, seven dead (BOT 1921, p37).

Phase 2 Conclusion:
CASTLEDOBBS (MBES/Archive). Dimensionally, positionally, and archivally this wreck is consistent with CASTLEDOBBS.

Phase 1: UKHO Analysis:

Original UKHO Name:
62529 UNKNOWN

Archival details:
N/A

MBES Analysis:
Upright wreck of a coaster or FV with single hold forward lying E/W with bows to east. 38m x 6.8m.

Additional Evidence:
Located by divers in 2003 and described as having boiler and engine (UKHO).

Phase 1 Conclusion:
UNKNOWN

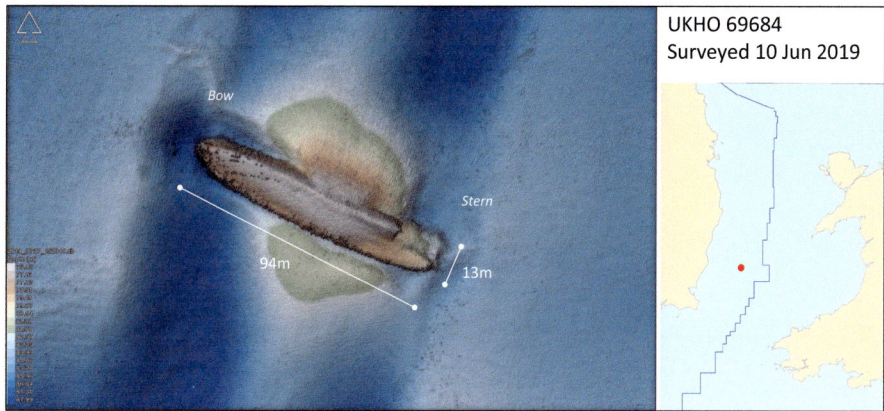

UKHO 69684
Surveyed 10 Jun 2019

69684 BIRCHWOOD (MBES/Archive)

Original UKHO: 69684 UNKNOWN	**Phase 1 Analysis:** UNKNOWN
Position WGS84: 52 29.071 N, 5 44.939 W	**Water Depth:** 87m
Type of Vessel: SS Cargo	**Surveyed Length:** 94m
Date of Loss: 3 Jan 1918	**Circumstances:** Torpedoed by U61

Phase 2: Method of wreck identification:

Positional Analysis:
This wreck lies 9.2nm north of the position given by the master (TNA ADM 137/1488) and in U61's KTB.

Dimensional Analysis:
BIRCHWOOD was 96.6m x 13.91m (*Lloyd's Register* 1917-18).

Archival Analysis:
On government service. Struck in engine room on the starboard side. Ship turned turtle and sank in four minutes (TNA ADM 137/1488).

Phase 2 Conclusion:
BIRCHWOOD (MBES/Archive). Positionally, dimensionally, and archivally this wreck is consistent with BIRCHWOOD. Of note is the fact the wreck is upside-down, possibly matching the description of the ship turning turtle when it sank.

Phase 1: UKHO Analysis:

Original UKHO Name:
69684 UNKNOWN

Archival details:
N/A

MBES Analysis:
Wreck of a SS or MV lying heavily on its starboard side beam ends, being almost upside-down. Stern to the west, which appears to have been damaged. 94m x 13m.

Additional Evidence:
Located in 2007 (UKHO).

Phase 1 Conclusion:
UNKNOWN

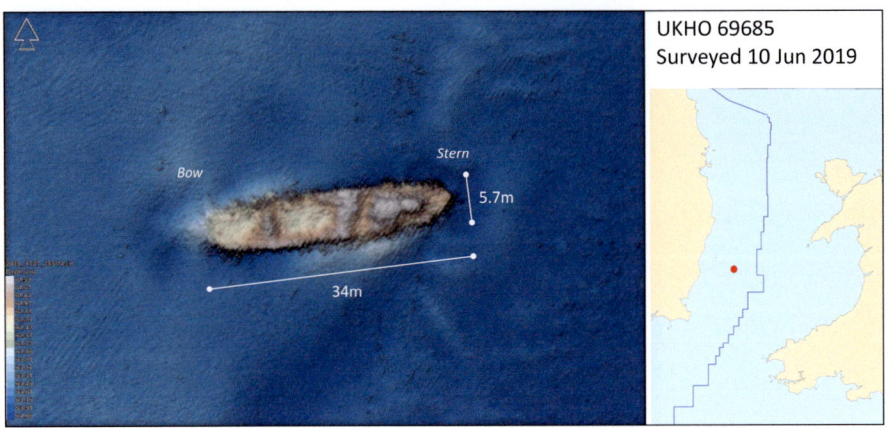

69685 MAGGIE (MBES/Archive)

Original UKHO: 69685 UNKNOWN	**Phase 1 Analysis:** UNKNOWN
Position WGS84: 52 38.962 N, 5 45.881 W	**Water Depth:** 62m
Type of Vessel: SS Cargo	**Surveyed Length:** 34m
Date of Loss: 17 Aug 1915	**Circumstances:** Gunfire from U38

Phase 2: Method of wreck identification:

Positional Analysis:
This wreck lies 1nm south of the position given in *War Losses* (1990). The U-boat position cannot be determined from U38's KTB.

Dimensional Analysis:
MAGGIE was 36.67m x 6.85 (*Lloyd's Register* 1915-16).

Archival Analysis:
Ship stopped, crew evacuated, U38 then pumped shells into both sides until the ship sank. Entire action took about 30 minutes (TNA ADM 137/1131).

Phase 2 Conclusion:
MAGGIE (MBES/Archive). Positionally, dimensionally, and archivally, the wreck is consistent with MAGGIE.

Phase 1: UKHO Analysis:

Original UKHO Name:
69685 UNKNOWN

Archival details:
N/A

MBES Analysis:
Upright wreck of a small coaster or FV pointing west. Holds forward with machinery aft. Dimensions of 34m x 5.7m.

Additional Evidence:
Located in 2007 (UKHO).

Phase 1 Conclusion:
UNKNOWN

69696 YTURRI BIDE (MBES/Archive)

Original UKHO: 69696 UNKNOWN	**Phase 1 Analysis:** UNKNOWN
Position WGS84: 52 35.376 N, 5 31.006 W	**Water Depth:** 76m
Type of Vessel: SS Cargo	**Surveyed Length:** 56m
Date of Loss: 16 May 1918	**Circumstances:** Gunfire by UB118

Phase 2: Method of wreck identification:

Positional Analysis:
This wreck lies 8.7nm north of the position given in *War Losses* (1990) and 10.3nm NW of the position given in UB118's KTB.

Dimensional Analysis:
YTURRI BIDE was 57.61m x 7.92m, well-decked (*Lloyd's Register* 1918-19).

Archival Analysis:
Maryport to Arvillas with a cargo of pitch. Sunk by gunfire. The report from the master is very cursory, due to language problems (TNA ADM 137/1516).

Phase 2 Conclusion:
YTURRI BIDE (MBES/Archive). Positionally, dimensionally, and archivally, this wreck is consistent with YTURRI BIDE. The DTM shows a similar layout to the painting on wrecksite.eu.

Phase 1: UKHO Analysis:

Original UKHO Name:
69696 UNKNOWN

Archival details:
N/A

MBES Analysis:
Upright wreck of a SS cargo with two holds forward. Bows are broken down to seabed and point west. 56m x 8.3m.

Additional Evidence:
Located in 2007 (UKHO).

Phase 1 Conclusion:
UNKNOWN

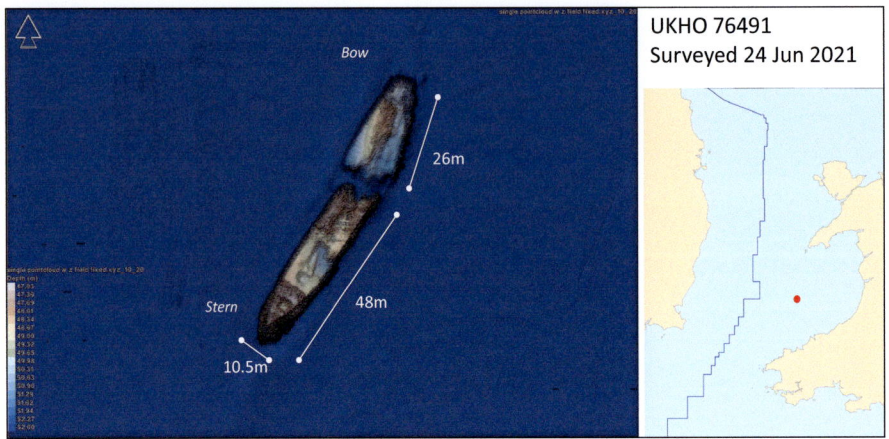

76491 AFTON (MBES/Archive)

Original UKHO: 76491 UNKNOWN	**Phase 1 Analysis:** UNKNOWN
Position WGS84: 52 22.619 N, 4 56.034 W	**Water Depth:** 55m
Type of Vessel: Cargo ship	**Surveyed Length:** 74m
Date of Loss: 15 Feb 1917	**Circumstances:** Explosives by UC65

Phase 2: Method of wreck identification:

Positional Analysis:
This wreck plots 2.6nm south of the position given in the master's report (TNA ADM 137/2961). The UC65 KTB is handwritten and gives few positional details. However, the accompanying chart places UC65 in close proximity to this wreck. This also matches with KYANITE (9914) which was sunk earlier in the day.

Dimensional Analysis:
AFTON was 73.86m x 10.41m (*Lloyd's Register* 1915-16).

Archival Analysis:
Torpedo missed the ship, overhauled by UC65 which fired into the ship, compelling it to stop. Bombs were then placed in the engine room and stokehold. The ship took over 30 minutes to sink after the bombs went off. UC65 then headed away to the SW. (TNA ADM 137/1305).

Phase 2 Conclusion:
AFTON (MBES/Archive). The wreck is positionally, dimensionally, and archivally consistent with AFTON.

Phase 1: UKHO Analysis:

Original UKHO Name:
76491 UNKNOWN

Archival details:
N/A

MBES Analysis:
Wreck of a cargo ship with holds fore and aft of central superstructure. Broken into two parts Points NE. Approx. 74m x 10.5m overall.

Additional Evidence:
Located in 2010 (UKHO).

Phase 1 Conclusion:
UNKNOWN

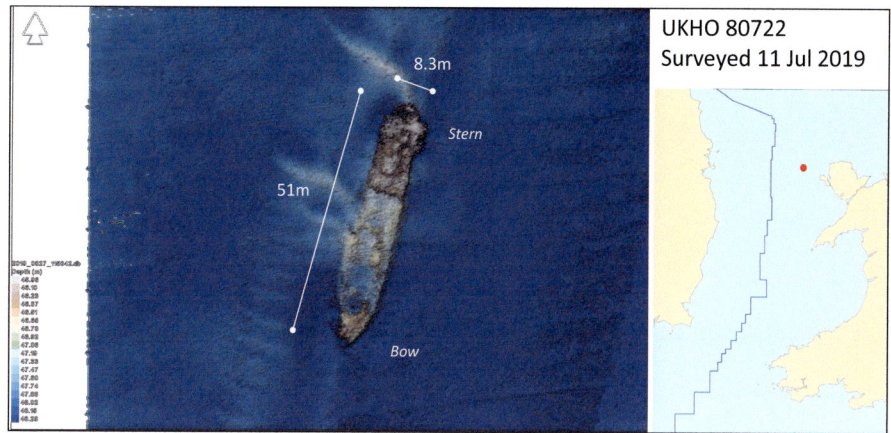

80722 SKERNAHAN (MBES/Archive)

Original UKHO: 80722 UNKNOWN		**Phase 1 Analysis:** UNKNOWN	
Position WGS84: 53 21.791 N, 4 54.379 W		**Water Depth:** 50m	
Type of Vessel: SS Cargo		**Surveyed Length:** 51m	
Date of Loss: 11 Aug 1916		**Circumstances:** Collision with SS YORKSHIRE	

Phase 2: Method of wreck identification:

Positional Analysis:
This wreck plots 3nm west of the position given in *Shipping Casualties* (BOT 1921, p 34).

Dimensional Analysis:
SKERNAHAN was 50.29m x 8.10m, well-decked, machinery aft (*Lloyd's Register* 1915-16).

Archival Analysis:
Limerick to Whitehaven in ballast. No lives lost (BOT 1921, p 34).

Phase 2 Conclusion:
SKERNAHAN (MBES/Archive). Positionally, dimensionally, and archivally this wreck is consistent with SKERNAHAN.

Phase 1: UKHO Analysis:

Original UKHO Name:
80722 UNKNOWN

Archival details:
N/A

MBES Analysis:
Upright wreck of a coaster, possibly SS. Bows point to the south, possibly two holds forward. Machinery aft. 51m x 8.3m.

Additional Evidence:
Located in 2013 (UKHO).

Phase 1 Conclusion:
UNKNOWN.

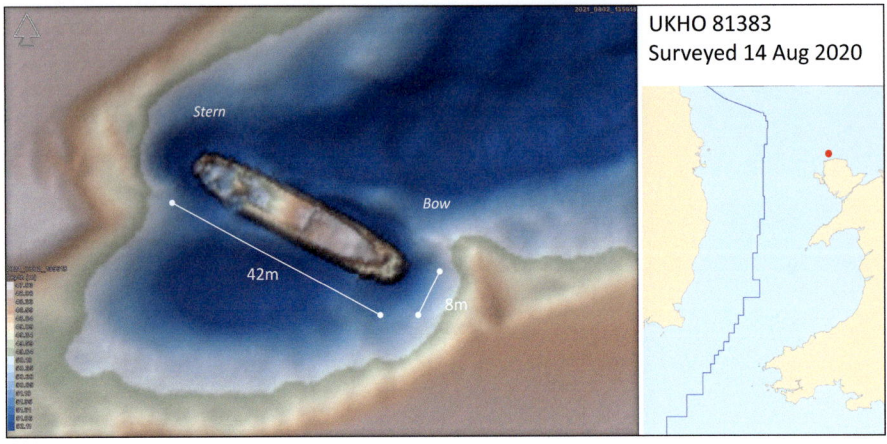

UKHO 81383
Surveyed 14 Aug 2020

81383 YEWS (MBES/Archive)

Original UKHO: 81383 UNKNOWN	**Phase 1 Analysis:** UNKNOWN
Position WGS84: 53 29.666 N,4 34.254 W	**Water Depth:** 53m
Type of Vessel: Coaster	**Surveyed Length:** 42m
Date of Loss: 19 Jan 1910	**Circumstances:** Collision with SS LA ROCHELLE

Phase 2: Method of wreck identification:

Positional Analysis:
This wreck lies 3.7nm east of the wreck (UKHO 7345) considered at present to be LA ROCHELLE. That wreck lies 3.4nm east of the reported collision point (BOT 1909-1910 Table 1 Appendix C, p120).

Dimensional Analysis:
YEWS was 43.43m x 7.11m (*Lloyd's Register* 1909-10). This is consistent with the dimensions of this wreck.

Archival Analysis:
A contemporary photo of this vessel in the project archive shows it had one hold forward, a central bridge and machinery aft. This could be considered consistent with the details seen on the scan of this wreck site.

Phase 2 Conclusion:
81383 YEWS (MBES/Archive). This wreck is positionally, dimensionally, and archivally consistent with YEWS.

Phase 1: UKHO Analysis:

Original UKHO Name:
81383 UNKNOWN

Archival details:
N/A

MBES Analysis:
Coaster wreck with bows to NW, 42m x 8m. Upright and intact. Hold forward.

Additional Evidence:
Located in 2013 (UKHO).

Phase 1 Conclusion:
UNKNOWN

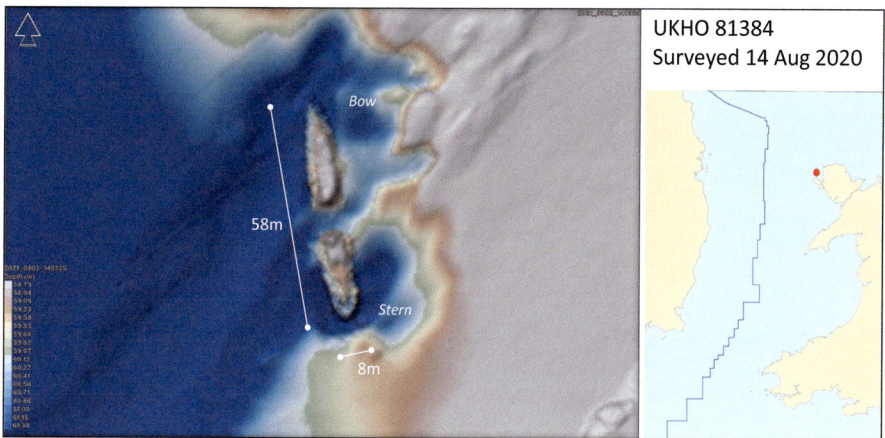

UKHO 81384
Surveyed 14 Aug 2020

81384 SAINT OLAF (MBES/Archive)

Original UKHO: 81384 UNKNOWN		**Phase 1 Analysis:** UNKNOWN	
Position WGS84: 53 25.951 N,4 43.724 W		**Water Depth:** 62m	
Type of Vessel: Coaster		**Surveyed Length:** 51m (approx)	
Date of Loss: 1 Dec 1900		**Circumstances:** Collision with SS VOLTAIC	

Phase 2: Method of wreck identification:

Positional Analysis:
This wreck plots 1.9nm from the position given as the collision point (BOT 1900-1901, p145).The ship was struck in the engine room and sank immediately. This wreck shows damage aft, where the engine room was situated.

Dimensional Analysis:
SAINT OLAF was 54.25m 8.25m (*Lloyd's Register* 1898-99). This is consistent with the wreck dimensions.

Archival Analysis:
The drawing of SAINT OLAF shows that the ship had its machinery placed aft of central raised bridge with a hold forward and a raised hold aft of the bridge. This is broadly consistent with the layout of this wreck.

Phase 2 Conclusion:
81384 SAINT OLAF (MBES/Archive). This wreck is positionally, archivally, and dimensionally consistent with SAINT OLAF.

Phase 1: UKHO Analysis:

Original UKHO Name:
81384 UNKNOWN

Archival details:
N/A

MBES Analysis:
Wreck of a coaster pointing north. Wreck is either in two portions of 28m (N section) and 23m (S section) or is 58m overall length. Width approx. 8m. Appears to have engine room in stern section.

Additional Evidence:
Located in 2013 (UKHO).

Phase 1 Conclusion:
UNKNOWN

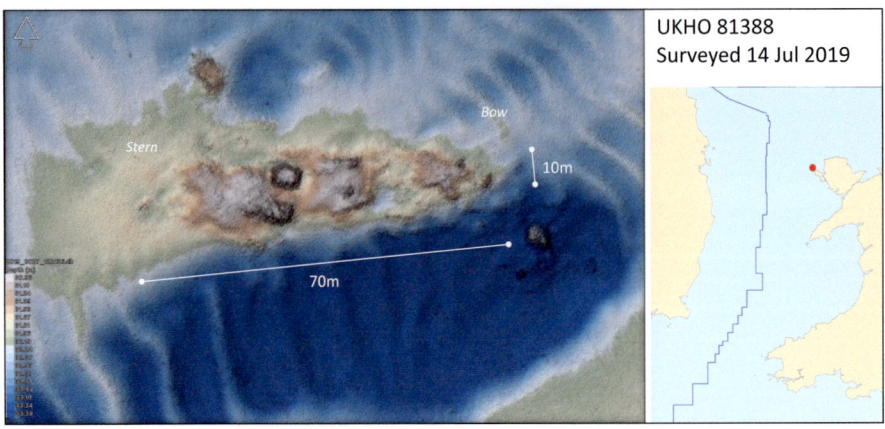

UKHO 81388
Surveyed 14 Jul 2019

81388 POCHARD (MBES/Archive)

Original UKHO: 81388 UNKNOWN	**Phase 1 Analysis:** UNKNOWN
Position WGS84: 53 19.744 N, 4 41.647 W	**Water Depth:** 36m
Type of Vessel: SS Cargo	**Surveyed Length:** 70m (approx)
Date of Loss: 7 Dec 1884	**Circumstances:** Foundered

Phase 2: Method of wreck identification:

Positional Analysis:
The wreck lies just off the North Stack (0.5nm) as described in *Shipping Casualties* (BOT 1884-85, p101).

Dimensional Analysis:
POCHARD was 79.29m x 10.08m (*Lloyd's Register* 1884-85).

Archival Analysis:
Foundered in very heavy weather, with 23 lives lost (BOT 1884-85, p101). Unnamed newspaper accounts on wrecksite.eu suggest the ship sank stern first in plain view of the Holyhead breakwater.

Phase 2 Conclusion:
POCHARD (MBES/Archive). Positionally the wreck is in the right area for POCHARD. The wreck cannot be measured accurately but is broadly the right dimensions. Diver reports of material of the right date add to the possibility (nothing more) that this wreck is POCHARD.

Phase 1: UKHO Analysis:

Original UKHO Name:
81388 UNKNOWN

Archival details:
N/A

MBES Analysis:
Wreck of steamship, heavily degraded lying E/W. 70m x 10m approx.

Additional Evidence:
Located in 2013 (UKHO). Divers report artefacts of the correct era (wrecksite.eu).

Phase 1 Conclusion:
UNKNOWN

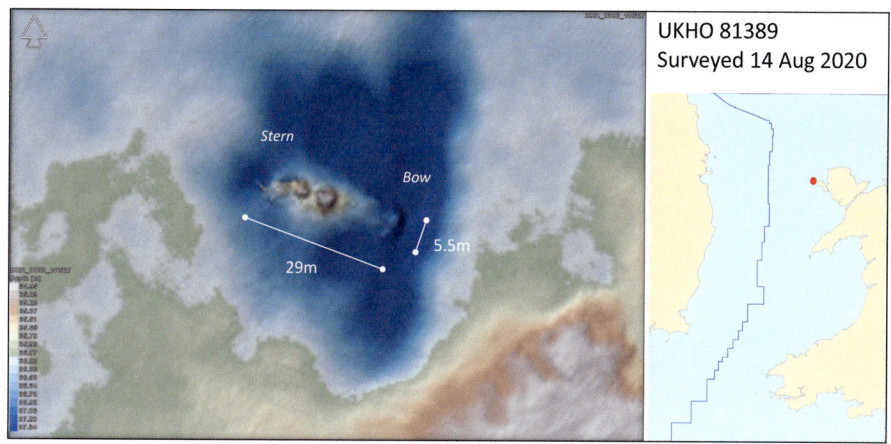

81389 ISLE OF ARRAN (MBES/Archive)

Original UKHO: 81389 UNKNOWN	**Phase 1 Analysis:** UNKNOWN
Position WGS84: 53 18.757 N,4 43.907 W	**Water Depth:** 38m
Type of Vessel: Small SS	**Surveyed Length:** 29m (approx)
Date of Loss: 25 Apr 1868	**Circumstances:** Collision with SS AUSTRALASIAN

Phase 2: Method of wreck identification:

Positional Analysis:
ISLE OF ARRAN was reported sunk in collision 7nm NW of Holyhead (BOT 1868, p33). Further details given in the *Liverpool Mercury* (27 Apr 1868) show this position to be unlikely. ISLE OF ARRAN had sheltered in Holyhead while on its way to Newport to collect coal. It had just left the harbour when it was run over by the AUSTRALASIAN which was taking an inshore route north, due to the conditions. This places the collision point close to Holyhead, where this wreck lies.

Dimensional Analysis:
ISLE OF ARRAN was listed as being of 110 tons displacement (estimated approx. 80ft/24m) and being a steamship. It does not appear in *Lloyd's Register* for 1867-8. This wreck is of comparable dimensions to the estimate and appears to have machinery within it.

Archival Analysis:
As given above.

Phase 2 Conclusion:
81389 ISLE OF ARRAN (MBES/Archive). Positionally, dimensionally, and archivally this wreck is consistent with ISLE OF ARRAN.

Phase 1: UKHO Analysis:

Original UKHO Name:
UNKNOWN

Archival details:
N/A

MBES Analysis:
Small steamship with bows to SE. Prominent boiler and engine. Wreck collapsed.

Additional Evidence:
Located in 2013 (UKHO).

Phase 1 Conclusion:
UNKNOWN

UKHO 82347
Surveyed 9 May 2019

82347 DUNSLEY (MBES/Archive)

Original UKHO: 82347 UNKNOWN	**Phase 1 Analysis:** UNKNOWN
Position WGS84: 52 31.603 N, 4 59.12 W	**Water Depth:** 73m
Type of Vessel: SS Cargo	**Surveyed Length:** 82m
Date of Loss: 30 Aug 1911	**Circumstances:** Foundered

Phase 2: Method of wreck identification:

Positional Analysis:
This wreck plots 6nm west of the position given in *Shipping Casualties* (BOT 1911-12, p85).

Dimensional Analysis:
DUNSLEY was 85.60m x 11.60m Well-decked (*Lloyd's Register* 1909-10).

Archival Analysis:
Appledore for Birkenhead in ballast. The Board of Trade inquiry into the loss of this ship dragged on into Jan 1912. It was established that the vessel had been deliberately over-insured, and the owners and officers had done nothing to save the vessel from sinking, raising obvious questions (*The Times* 15 Jan 1912). The ship is known to have flooded and sunk by the stern. Inspection of the after hatch was not carried out as stanchions had fallen into it (*The Times* 24 Nov 1911).

Phase 2 Conclusion:
DUNSLEY (MBES/Archive). Positionally, dimensionally, and archivally, this wreck is consistent with DUNSLEY.

Phase 1: UKHO Analysis:

Original UKHO Name:
82347 UNKNOWN

Archival details:
N/A

MBES Analysis:
Upright wreck of a cargo vessel with machinery aft and three holds forward. 82m x 11m.

Additional Evidence:
Located in 2015 (UKHO).

Phase 1 Conclusion:
UNKNOWN

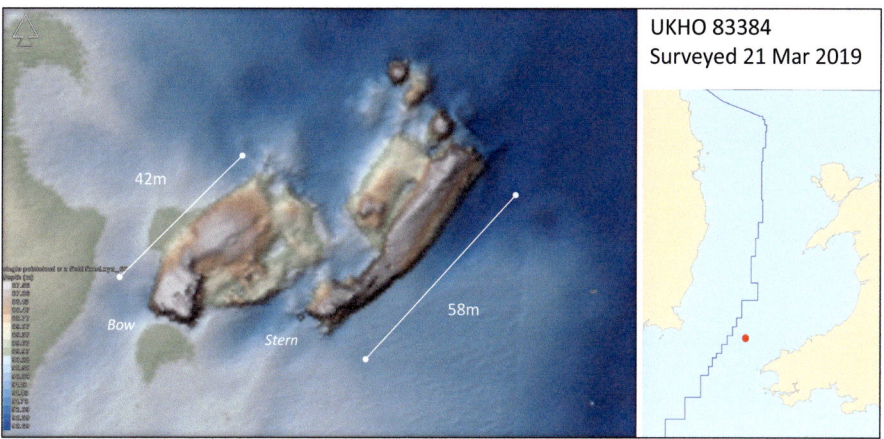

UKHO 83384
Surveyed 21 Mar 2019

83384 PENTWYN (MBES/Archive)

Original UKHO: 83384 UNKNOWN	Phase 1 Analysis: UNKNOWN
Position WGS84: 52 7.491 N, 5 34.446 W	**Water Depth**: 83m
Type of Vessel: SS Cargo	**Surveyed Length**: 100m (approx)
Date of Loss: 16 Oct 1918	**Circumstances**: Torpedoed by U90

Phase 2: Method of wreck identification:

Positional Analysis:
This wreck plots 4.5nm north of the position given in *War Losses* (1990), which is also the position given by the Admiralty (TNA ADM 137/1504).

Dimensional Analysis:
PENTWYN was 108.51m x 15.39m (*Lloyd's Register* 1916-17).

Archival Analysis:
Lagos for Liverpool in convoy. Torpedo opened a large hole in the starboard side and seems to have opened up the port side as well. The ship broke into two halves and sank by the centre (TNA ADM 137/1504).

Phase 2 Conclusion:
PENTWYN (MBES/Archive). Positionally, dimensionally, and archivally, this wreck is consistent with PENTWYN. Of note is the description of the ship breaking in two before sinking.

Phase 1: UKHO Analysis:

Original UKHO Name:
83384 UNKNOWN

Archival details:
N/A

MBES Analysis:
Wreck of SS which has been broken in half in the region of the machinery spaces, spilling the boilers onto the seabed. The wreck has folded back onto itself, so that the bow and stern both point to the SW. Overall length of wreck approx. 100m. Width is difficult to discern with accuracy.

Additional Evidence:
Located in 2015 (UKHO).

Phase 1 Conclusion:
UNKNOWN

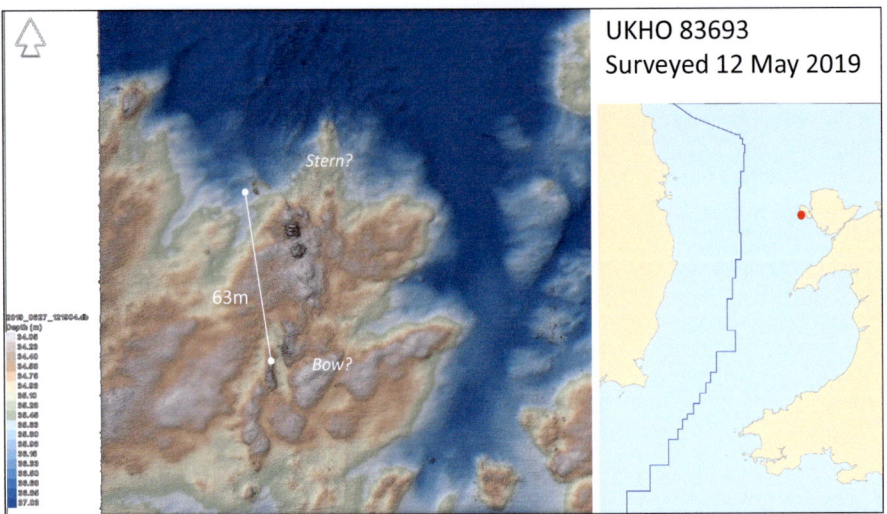

UKHO 83693
Surveyed 12 May 2019

83693 KANGAROO (MBES/Archive)

Original UKHO: 83693 UNKNOWN	**Phase 1 Analysis:** UNKNOWN
Position WGS84: 53 12.733 N, 4 40.202 W	**Water Depth:** 32m
Type of Vessel: SS	**Surveyed Length:** 63m (approx)
Date of Loss: 23 Jan 1862	**Circumstances:** Foundered

Phase 2: Method of wreck identification:

Positional Analysis:
This wreck plots 3nm north of the position given in *Shipping Casualties* (BOT 1862 Table 8, p14).

Dimensional Analysis:
According to wrecksite.eu Kangaroo was 51.10m x 7.10m. Dimensions are not given at this time in *Lloyd's Register*.

Archival Analysis:
KANGAROO foundered in a force 10 gale with 13 lives lost. Carrying general cargo (BOT 1862 Table 8, p14).

Phase 2 Conclusion:
KANGAROO (MBES/Archive). This wreck is positionally and archivally consistent with KANGAROO. This case is not certain, as only the outline of shipwreck can be seen in the MBES data. A single boiler and engine is consistent with KANGAROOS's most likely construction.

Phase 1: UKHO Analysis:

Original UKHO Name:
83693 UNKNOWN

Archival details:
N/A

MBES Analysis:
Scattered remains of an old steamship wreck lying in a rock outcrop. Evidence of a pair of features which could be a boiler and engine.

Additional Evidence:
Located in 2015 with high magnetometer reading (UKHO).

Phase 1 Conclusion:
UNKNOWN

Appendix 2 References

Admiralty, 1947. *Ships of the Royal Navy statement of losses during the Second World War*. London: HMSO.

Board of Trade, Marine Department (various dates). *Return of Shipping Casualties and Loss of Life*. London: HMSO.

Board of Trade, Marine Department (various dates). *Wreck Reports*. London: HMSO.

Gibson, Sir David, 1993. *Battle in the Irish Sea*. Liskeard: Maritime Books.

Grőner, E., 1991. *German Warships 1815-1945 Volume 2*. London: Conway.

Hashagen, E., 1931. *The Log of a U-boat Commander or U-boats Westward 1914-1918*. London: Putnam.

Haws, D., 1989. *Merchant Fleets 16: Ellerman Lines*. Hereford: TCL.

Hocking, C., 1969. *Dictionary of Disasters at Sea During the Age of Steam: Including sailing ships and ships of war lost in action 1824-1962. Vol I, A to L*. London: Lloyd's Register of Shipping.

Hocking, C., 1969. *Dictionary of Disasters at Sea During the Age of Steam: Including sailing ships and ships of war lost in action 1824-1962. Vol II, M to Z*. London: Lloyd's Register of Shipping.

Larn, B., & Larn, R., 2000. *Shipwreck Index of the British Isles, Volume 5: West Coast & Wales*. Redhill: Lloyd's Register.

Larn, B., & Larn, R., 2002. *Shipwreck Index of Ireland: Volume 6 – All Ireland. Part of the Shipwreck Index of the British Isles series*. Redhill: Lloyd's Register.

Lloyd's of London, 1989. *Lloyd's War Losses: The Second World War Volume 1: British, Allied and Neutral Merchant Vessels Destroyed by War Causes*. London: Lloyd's of London Press.

Lloyd's of London, 1990. *Lloyd's War Losses: The First World War: Casualties to Shipping Through Enemy Causes 1914-1918*. London: Lloyd's of London Press.

Lloyd's Register Foundation, Heritage & Education Centre (various dates). *Plan of SS Glynwen* LRF-PUN-W399-0009-P. London: Lloyd's Register.

Lloyd's Register of Shipping (various dates). *Lloyd's Register Wreck Returns*. London: Lloyd's Register of Shipping.

Lloyd's Register of Shipping (various dates). *Lloyd's Register of Shipping*. London: Lloyd's Register of Shipping.

McCartney, I., 2014. *The Maritime Archaeology of a Modern Conflict: Comparing the archaeology of German submarine wrecks to the historical text*. New York: Routledge.

Michael, C., 2008. *Wrecks of Liverpool Bay Volume 2*. Liverpool: Liverpool Marine Press.

Ministeriet for Handel, Industri og Sofort, 1950. *KRIGSFORLISTE OG KRIGSHAVAREREDE DANSKE SKIBE 3. September 1939 – 31. December 1949*. Kobenhaven: Det Kongelige Sokort-Arkiv.

Ministry of Defence, Director of Naval Warfare, 1973. *British Mining Operations 1939-1945, Volume 1. Naval Staff History BR1736(56)(1)*. Portsmouth.

Ministry of Defence, Director of Naval Warfare, 1977. *British Mining Operations 1939-1945, Volume 2. Naval Staff History BR1736(56)(2)*. Portsmouth.

Puddefoot, G., 2010. *Ready for Anything: The Royal Fleet Auxiliary, 1905-1950*. Barnsley: Seaforth.

National Archives and Records Administration (various dates), 1984. *Microfilmed Records of the German Navy. T1022*. Washington DC: NARA.

The National Archives (various dates). *South-West Approach: Submarines, June 1915*. ADM 137/1061. London.

The National Archives (various dates). *South West Approach: German Submarines August 1915*. ADM 137/1131. London.

The National Archives (various dates). *South West Approach: German Submarines; 6 – 13 February* 1917. *ADM 137/1304.* London.

The National Archives (various dates). *South West Approach: German Submarines, 14-28 February 1917. ADM 137/1305.* London.

The National Archives (various dates). *South West Approach: German Submarines; 22 – 31 March 1917. ADM 137/1308.* London.

The National Archives (various dates). *South West Approach: German Submarines; 24 – 30 April 1917. ADM 137/1311.* London.

The National Archives (various dates). *South West Approach: German Submarines, 18 November-7 December 1917.* ADM 137/1358. London.

The National Archives (various dates). *South West Approach: German Submarines, January 1918. ADM 137/1488.* London.

The National Archives (various dates). *South West Approach: German Submarines, 1-20 February 1918. ADM 137/1489.* London.

The National Archives (various dates). *South West Approach: German Submarines, 1-15 March 1918. ADM 139/1491.* London.

The National Archives (various dates). *South West Approach: German Submarines, 1-15 April 1918. ADM 137/1493.* London.

The National Archives (various dates). *South West Approach: German Submarines, 16-30 April 1918. ADM 137/1494.* London.

The National Archives (various dates). *South West Approach: German Submarines, August 1918. ADM 137/1501.* London.

The National Archives (various dates). *South West Approach: German Submarines, 1-15 September 1918. ADM 137/1502.* London.

The National Archives (various dates). *South West Approach: German Submarines, 16-30 September 1918. ADM 137/1503.* London.

The National Archives (various dates). *South West Approach: German Submarines, Octo-ber-November 1918. ADM 139/1504.* London.

The National Archives (various dates). *Irish Sea: Various, 1917. ADM 137/1361.* London.

The National Archives (various dates). *Irish Sea: German Submarines, 1917. ADM 137/1362.* London.

The National Archives (various dates). *Irish Sea: German Submarines, January – February 1918. ADM 137/1514* London.

The National Archives (various dates). *Irish Sea: German Submarines, March 1918. ADM 137/1515* London.

The National Archives (various dates). *Irish Sea: German Submarines, April-June 1918. ADM 137/1516.* London.

The National Archives (various dates). *Convoys: OL21 – OL41 ADM 137/2618.* London.

The National Archives (various dates). *Precis of reports of Homeward Convoys, May 1917 November 1918 ADM 137/2657.* London.

The National Archives (various dates). *British Merchant Vessels sunk and captured by the enemy, August 1914 to February 1918. ADM 137/2959 to ADM 137/2964.* London.

The National Archives (various dates). *Home Waters and Mediterranean. ADM 137/3975 to ADM 137/4019.* London.

The National Archives (various dates). *Assessments of results of attacks on German submarines. ADM 137/4147.* London.

The National Archives (various dates). *Court Martial of Temporary Lieutenant B Palmer RNVR for the hazard and loss of HMS MERCURY ADM 156/216.* London.

The National Archives (various dates). *Portland, Portsmouth, and Western Approaches: War Diaries ADM 199/1443.* London.

The National Archives (various dates). *Director of Torpedo, Anti-Submarine and Mine Warfare Division: proceedings of U-Boat Assessment Committee, reassessments, and identity of U-boats. ADM 199/1786.* London.

National Archives (various dates). *Director of Torpedo, Anti-Submarine and Mine Warfare Division: U-boats sunk or damaged and US Fleet Anti-Submarine Bulletins, ADM199/1789.* London.

The National Archives (various dates). *Survivors Reports Merchant Vessels,* Sept *1939 to May 1945. ADM 199/2130* to 199/2148. London.

The National Archives (various dates). *War diary summaries : situation reports. ADM 199/2195 to ADM 199/2325.* London.

The National Archives (various dates). *HMS COTILLION: report on passage of 7th LCT Flotilla and loss of LCT 326 . ADM 217/13.* London.

The National Archives (various dates). *Reported destruction of / damage to submarines Aug 1914 to Jan 1919. ADM 239/26.* London.

Rohwer, J., 1999. *Axis Submarine Successes of World War Two.* London: Greenhill.

Spindler, A., 1941. *Der Handelskrieg mit U-Booten*, Bd.4, *Januar bis Dezember 1917.* Berlin: Verlag von Mittler & Sohn.

Spindler, A., 1966. *Der Handelskrieg mit U-Booten,* Bd 5, *Januar bis November 1918.* Berlin: Verlag von Mittler & Sohn.

The Times (various dates). *The Times.* London: The Times.

Bibliography

Akermann. P., 2002. *The Encyclopaedia of British Submarines*. Penzance: Periscope Publishing.

Beesley, P., 1980. Special Intelligence and the Battle of the Atlantic: the British View, in Love Jr. R. W., (ed). *Changing Interpretations and New Sources in Navy History*. New York: Garland.

Bellamy, M., 2001. *The Shipbuilders: An Anthology of Scottish Shipyard Life*. Edinburgh: Birlinn.

Bennet, T., 1987. *Shipwrecks Around Wales Volume 1*. Newport: Happy Fish.

Bennet, T., 1992. *Shipwrecks Around Wales Volume 2*. Newport: Happy Fish.

Brady, K., McKeon, C., Lyttleton, J., & Lawler, I., 2012. *Warships, U-Boats & Liners: A Guide to Shipwrecks Mapped in Irish Waters by the Irish National Seabed Survey and INFOMAR Mapping Projects*. Dublin: The Stationery Office.

Blake, G., 1960. *Lloyd's Register of Shipping 1760-1960*. London: Lloyd's Register.

Carlisle, R., 2009. *Sovereignty at Sea: U.S. Merchant Ships and American Entry into World War 1*. Gainesville: University Press of Florida.

Chatterton, E. K., 1922. *Q-Ships and their Story*. London: Sidgwick & Jackson.

Chatterton, E. K., 1923. *The Auxiliary Patrol*. London: Sidgwick & Jackson.

Chatterton, E. K., 1934. *Danger Zone*. London: Rich & Cowan.

Conway, 1985. *Conway's All the World's Fighting Ships 1906-1921*. London: Conway.

Cock, R., and Rodger, N. A. M. 2006. *A Guide to the Naval Records in the National Archives of the UK*. London: University of London.

Corkill, A., 2013. *Hostile Sea: The U-boat Offensive around the Isle of Man during World War One*. Isle of Man: Hostile Sea.

Delgado, J. P., 1992. Recovering the past of USS Arizona: Symbolism, Myth and Reality. *Historical Archaeology*. Vol. 26 No.4, p69-80.

Evans, A. S., 1986. *Beneath the Waves: A History of HM Submarine Losses 1904-1971*. London: William Kimber.

Fenton, R. S., 1997. *Mersey Rovers: The coastal tramp ship owners of Liverpool and the Mersey*. Gravesend: World Ship Society.

Gibson, R. H., & Prendergast, M., 1931. *The German Submarine War 1914-1918*. London: John Constable Ltd.

Goddard, T., 1983. *Pembrokeshire Shipwrecks*. Swansea: Hughes & Sons.

Grove, E,. (ed), 1997. *The Defeat of the Enemy Attack on Shipping 1939-1945*. Farnham: Ashgate.

Halpern, G. P., 1994. *A Naval History of World War I*. Annapolis: Naval Institute Press.

Haws, D., & Middlemiss, N. L., (various dates). *Merchant Ships* (45 volumes). Cambridge: Patrick Stephens, & Newcastle Upon Tyne: Shield.

Holms, A. C., 1918. *Practical Shipbuilding; A treatise on the structural design and building of modern steel vessels. Volume I – Text.* London: Longmans.

Holms, A. C., 1918. *Practical Shipbuilding; A treatise on the structural design and building of modern steel vessels. Volume II – Diagrams and Illustrations.* London: Longmans.

Jordan, R. W., 1999. *The World's Merchant Fleets 1939.* London: Conway.

Keith, M. E., (ed), 2016. *Site Formation Processes of Submerged Shipwrecks.* Gainesville: UPF.

Kemp, P., (ed), 1994. *The Oxford Companion to Ships and the Sea.* Oxford: OUP.

Koerver, H. J., (ed), 2009. *Room 40: German Naval Warfare 1914-1918. Volume 1: The Fleet in Action.* Steinbach: LIS Reinisch.

Koerver, H. J., (ed), 2009. *Room 40: German Naval Warfare 1914-1918. Volume 2: The Fleet in Being.* Steinbach: LIS Reinisch.

Koerver, H. J., (ed), 2010. *German Submarine Warfare 1914-1918 in the eyes of British Intelligence: Selected Sources from the British National Archives.* Steinbach: LIS Reinisch.

Lindbaek, L., 1969. *Norway's New Saga of the Sea: The Story of her Merchant Marine in World War II.* New York: Exposition Press.

Llewellyn-Jones, M., 2006. *The Royal Navy and Anti-Submarine Warfare,* 1917-49. London: Routledge.

Lloyd's Register Foundation Heritage & Education Centre (various dates). *Infosheet No. 10: Lloyd's Register sources available to researchers.* London: Lloyd's Register.

Lloyd's List, 1984. *Lloyd's List: 250th Anniversary Special Supplement.* London: Lloyd's List.

MacRae, J. A. & Waine, C. V., 1990. *The Steam Collier Fleets.* Wolverhampton: Waine Research Publications.

McElwee, R., 1992. *The Last Voyages of the Waterford Steamers.* Waterford: The Book Centre.

Moreland, J., 2001. *Archaeology and Text.* London: Duckworth.

Paasch, H., 1977. *Illustrated Marine Encyclopedia 1890.* Watford: Argus.

Paasch, H., 1997. *Paasch's Illustrated Marine Dictionary.* London: Conway.

Parry, H., 1969. *Wreck and Rescue on the Coast of Wales.* Truro: Bradford Barton.

Pursey, H. J., 1957. *Merchant Ship Construction: Especially written for the Merchant Navy.* Glasgow: Brown & Ferguson.

Rankin, N., 2011. *Ian Fleming's Commandos. The story of 30 Assault Unit in WWII.* London: Faber & Faber.

Redknap, M., Rees, S., & Aberg, A., (eds) 2019. *Wales and the Sea: 10,000 years of Welsh maritime history.* Talybont: Y Lolfa.

Ritchie, L. A., (ed), 1992. *The Shipbuilding Industry: A Guide to Historical Records.* Manchester. MUP.

Rohwer, J., 1980. Ultra and the Battle of the Atlantic: the German View, in Love Jr. R. W., (ed), *Changing Interpretations and New Sources in Navy History.* New York: Garland.

Rothery, H. C., 1882. *A Digest of the Judgements of Board of Trade Inquiries.* London: Waterlow Bros.

Sims, Rear-Admiral, W. S., 1920. *The Victory at Sea.* London: John Murray.

Stichelbaut, B., 2005. The application of Great War aerial photography in battlefield archaeology: The example of Flanders. *Journal of Conflict Archaeology (Brill Academic Publishers)*, 1 (1), 235-243.

Still Jr, W. N., 2006. *Crisis at Sea: The United States Navy in European Waters in World War 1*. Gainesville: UPF.

Stokes, R., 2015. *Between the Tides: Shipwrecks of the Irish Coast*. Stroud; Amberley.

Technical History Section Admiralty, 1919a. *The Atlantic Convoy System 1917-1918*. C.B.1515(14). Portsmouth.

Technical History Section Admiralty, 1919a. *The Anti-Submarine Division of the Naval Staff December 1916 to November 1918*. C.B.1515(7). Portsmouth.

Tennent, A, J., 2006. *British and Commonwealth Merchant Ship Losses to Axis Submarines 1939-1945*. Stroud: Sutton.

The National Archives (various dates). *History of the First World War: various aspects of the unrestricted submarine warfare campaign. ADM 116/3421*. London.

The National Archives (various dates). *List of British merchant vessels attacked by enemy submarines; reports of actions and sinkings Mar 1915-Dec 1916. ADM 131/113*. London.

The National Archives (various dates). *Foreign merchant vessels attacked by enemy submarines; reports of actions and sinkings May 1915-Sep 1918. ADM 131/118*. London.

Thearle, S. J. P., 1910. *The Modern Practice of Shipbuilding in Iron and Steel Volume 1 – Text*. London: Collins' Clear Type.

Thearle, S. J. P., (undated). *The Modern Practice of Shipbuilding in Iron and Steel Volume II – Plates. 3rd Edition – Revised and Enlarged*. London: William Collins, Sons & Co.

Thomas, P. N., 1992. *British Ocean Tramps Volume 1. Builders & Cargoes*. Wolverhampton: Waine Research Publications.

Thomas, P. N., 1992. *British Ocean Tramps Volume 2. Owners & Their Ships*. Wolverhampton: Waine Research Publications.

Thomas, P. N., 1997. *British Steam Tugs*. Wolverhampton: Waine Research Publications.

Waine, C. V., 1977. *Steam Coasters and Short Sea Traders*. Wolverhampton: Waine Research Publications.

Walton, T., 1926. *Steel Ships: Their Construction and Maintenance*. London: Charles Griffin.

Wynne Jones, I., 1973. *Shipwrecks of North Wales*. Newton Abbot: David & Charles.

Index